Solving General Chemistry Problems

Solving
General Chemistry
Problems

FIFTH EDITION

R. Nelson Smith
Pomona College

Conway Pierce
Late of University of California, Riverside

W. H. FREEMAN AND COMPANY
San Francisco

Sponsoring Editor: Peter Renz
Project Editor: Nancy Flight
Manuscript Editor: Larry McCombs
Designer: Marie Carluccio
Production Coordinator: William Murdock
Illustration Coordinator: Cheryl Nufer
Artist: John Waller
Compositor: Bi-Comp, Inc.
Printer and Binder: The Maple-Vail Book Manufacturing Group

Library of Congress Cataloging in Publication Data
Smith, Robert Nelson, 1916–
    Solving general chemistry problems.

    First-4th ed. by C. Pierce and R. N. Smith published
under title: General chemistry workbook.
    Includes index.
    1. Chemistry—Problems, exercises, etc. I. Pierce,
Willis Conway, 1895–   joint author. II. Pierce,
Willis Conway, 1895–   General chemistry workbook.
III. Title.
QD42.S556 1980   540′.76   79-23677
ISBN 0-7167-1117-6

Printed in the United States of America

9 8 7 6 5 4 3 2

Contents

Preface

With the birth of the electronic hand calculator and the death of the slide rule, the teaching of general chemistry entered a new era. Unfortunately, the calculator did *not* bring automatic understanding of chemistry; analyzing and solving problems is just as difficult as ever. The need for detailed explanations, drill, and review problems is still with us.

The calculator came on the scene just as society was beginning to make major decisions based on statistical evidence. Which chemicals are "safe" and "noncarcinogenic"? At what level is a given pollutant "harmful"? More and more the chemist must decide what constitutes risk, and that decision is based on statistics. New analytical techniques can detect incredibly small amounts of materials, and the results of these analyses must be judged statistically. Students need to start as early as they can to think critically and with statistical understanding about their own work and that of others. The hand calculator makes it relatively easy to determine statistical significance, to plot data properly by the method of least squares, and to evaluate the reliability of quantities related to the slope and y intercept of such a plot. This book takes into account the impact of the calculator. It shows beginners how to use calculators effectively and, as they progress, to determine whether or not results are statistically significant.

The text will be useful to those beginning chemistry students who have difficulty analyzing problems and finding logical solutions, who have trouble with graphical representations and interpretation, or who have simply missed out on such items as logarithms and basic math operations. It will also be useful to the student who wants additional problems and explanations in order to gain a better understanding of concepts and to prepare for exams.

Instructors will find that the book does more than satisfy the self-help and tutorial needs of students who lack confidence or background preparation. It will supplement the weaker portions of the selected text and, in general, provide a much wider variety of problems. It will enable instructors to devote class time to a more complete discussion of general principles, because students will be able to obtain and study the details of problem solving from the book. In addition, it provides a basis for students to assess the reliability and quality of their quantitative lab work and explains how to treat data properly in graphical form and to assess the quality of the quantities derived from their graphs. Finally, the book presents a logical approach to the sometimes bewildering business of how to prepare compounds and how to predict whether a given reaction will occur.

Since *Solving General Chemistry Problems* is a supplement to the regular text and lab manual used in a beginning college chemistry course, it has been written so that chapters can be used in whatever order best suits the adopted text and the instructor's interests. Whatever interdependence exists between the chapters is the normal interdependence that would be found for similar material in any text.

Although the use of units is heavily emphasized throughout the text, it was decided *not* to make exclusive use of SI units; almost none of the current texts do so, and an informal poll of chemistry teachers showed little interest in making this change. Anyone strongly committed to another view may easily convert the answers to SI units or work the problems in whatever units are desired. The *methods* of calculation and the analytical approach will not be affected.

Some instructors may like to know the ways in which this fifth edition differs from the fourth. Two major changes are the addition of a chapter on chemical kinetics (Chapter 15) and the replacement of all the material on slide rules with a discussion of the efficient use and application of electronic hand calculators. Chapter 7 no longer considers specific gravity but instead discusses the application of bouyancy principles to accurate weighing. Graphical representation in Chapter 6 has been amplified to include the method of least squares and how to evaluate the reliability of the slopes and intercepts of best-fit lines. The material on thermochemistry (Chapter 14) has been expanded to include energy changes at constant volume as well as at constant pressure. Problems and concepts related to free energy are considered along with the energy obtained from electrochemical cells and are related to the entropies and enthalpies of reaction. Absolute entropies are also included. The application of Faraday's laws to electrolytic cells has been separated from the other electrochemical material and placed in a chapter of its own (Chapter 19). Chapter 9, on the sizes and shapes of molecules, now contains a refined set of rules for predicting molecular shape, and the rules are related to the hybrid orbitals that hold the atoms together. Even where the approach and general lines of reasoning have remained the same as in the fourth edition, the text has been substantially rewritten.

Conway Pierce, coauthor of the first four editions of the book and coauthor of the first four editions of *Quantitative Analysis,* published by John Wiley & Sons, died December 23, 1974. Professor Pierce was a valued friend, a stimulating teacher, and an original researcher whose contributions to chemistry and chemical education spanned more than fifty years. I hope that this edition of the book reflects the everchanging outlook of general chemistry while retaining the simple, direct, and clear expression for which Conway Pierce was noted.

R. Nelson Smith

October 1979

Solving General Chemistry Problems

1

Studying and
Thinking About Problems

For many of you, the first course in chemistry will be a new experience—perhaps a difficult one. To understand chemistry, you will have to work hundreds of problems. For many students, the mathematical side of the course may seem more difficult than it should, leading to unnecessary frustration. There appear to be two main sources of this difficulty and frustration; they center around (1) study habits, and (2) the way you analyze a problem and proceed to its solution. The following suggestions, taken seriously from the very beginning, may be of great help to you. For most people, improved study habits and problem-solving skills come only with practice and with a determined effort spread over a long time. It's worth it.

STUDY HABITS

1. Learn each assignment before going on to a new one. Chemistry has a vertical structure; that is, new concepts depend on previous material. The course is cumulative in nature. Don't pass over anything, expecting to learn it later. And don't postpone study until exam time. The message is this: keep up to date.

2. Know how to perform the *mathematical* operations you need in solving problems. The mathematics used in general chemistry is elementary, involving

only arithmetic and simple algebra. Nevertheless, if you don't understand it, you can expect troubles before long. So, before you can really get into chemistry, you need to master the mathematical operations in the first six chapters.

3. Don't think of your calculator as a security blanket that will bring you vision, light, and understanding about problems. Your calculator can minimize the tedium and time involved in the *mechanics* of a problem, thus leaving you more time to *think* about the problem. And, in principle, there is less likelihood of your making an arithmetical error with the calculator, but it won't help you at all in choosing the right method for solution. Many students make a substantial investment in a powerful calculator and then never learn to take advantage of its power and its time-saving capability. From the very beginning it will pay you to learn to use this incredible tool well and easily, so that you can devote your thinking time to understanding the principles and the problems. This book emphasizes the proper and efficient use of your calculator.

4. Minimize the amount of material you memorize. Limit memorization to the basic facts and principles from which you can reason the solutions of the problems. Know this smallish amount of factual material really well; then concentrate on how to use it in a logical, effective way. Too many students try to undertake chemistry with only a rote-memory approach; it can be fatal.

5. *Before* working homework problems, study pertinent class notes and text material until you think you fully understand the facts and principles involved. Try to work the problems without reference to text, class notes, or friendly assistance. If you can't, then work them with the help of your text or notes, or work with someone else in the class, or ask an upperclassman or the instructor. However, then be aware that you have worked the problems with a crutch, and that it's quite possible you still don't understand them. Try the same or similar problems again a few days later to see whether you can do them without any help, as you must do on an exam. Discussion of problems helps to fix principles in mind and to broaden understanding but, by itself, it doesn't guarantee the understanding you need to work them.

6. When homework assignments are returned and you find some problems marked wrong (in spite of your efforts), do something about it soon. Don't simply glance over the incorrect problem, kick yourself for what you believe to be a silly error, and assume you now know how to do it correctly. *Perhaps* it was just a silly mistake, but there's a good chance it wasn't. Rework the problem on paper (without help) and check it out. If you can't find the source of error by yourself, then seek help. There is often as much or more to be learned from making mistakes (in learning why you can't do things a certain way) as there is from knowing an acceptable way without full understanding. However, the time to learn from mistakes is *before* exams, on homework assignments.

7. In the few days before an examination, go through all the related homework problems. See if you can classify them into a relatively small number of *types* of problems. Learn how to recognize each type, and know a simple

straightforward way to solve that type. Recognition of the problem (not the mechanics of the solution) often is the biggest difficulty to be overcome. In most cases, there are only a few types of problems associated with a given topic.

8. Be sure that you understand material, rather than just being familiar with it (there's a huge difference!). See if you can write something about the topic in a clear, concise, and convincing manner, without any outside assistance. The act of writing is one of the best ways to fix an idea in your mind, and it is the same process that you use on an exam. Many students feel that repeated reading of an assignment is all that is needed to learn the material; unfortunately, that is true for only a few students. Most people will read the words the same way each time; if real understanding has not occurred by the second or third reading, further readings probably are a waste of time. Instead of going on to a fourth reading, search through the text and jot down on a piece of paper the words representing new concepts, principles, or ideas. Then, with the book closed, see if you can write a concise "three-sentence essay" on each of these topics. This is an oversimplified approach (and not nearly as easy as it sounds), but it does sharpen your view and understanding of a topic. It helps you to express yourself in an exam-like manner at a time when, without penalty, you can look up the things you don't know. If you need to look up material to write your essays, then try again a few days later to be sure you can now do it without help.

PROBLEM SOLVING

1. Understand a problem before you try to work it. Read it carefully, and don't jump to conclusions. Don't run the risk of misinterpretation. Learn to recognize the type of problem.

2. If you don't understand some words or terms in the problems, look up their meaning in the text or a dictionary. Don't just guess.

3. In the case of problems that involve many words or a descriptive situation, rewrite the problem using a minimum number of words to express the bare-bones essence of the problem.

4. Some problems give more information than is needed for the solution. Learn to pick out what is needed and ignore the rest.

5. When appropriate, draw a simple sketch or diagram (with labels) to show how the different parts are related.

6. Specifically pick out (a) what is given and (b) what is asked for.

7. Look for a relationship (a conceptual principle or a mathematical equation) between what is given and what is asked for.

8. Set up the problem in a concise, logical, stepwise manner, using units for all terms and factors.

9. Don't try to bend all problems into a mindless "proportion" approach that you may have mastered in elementary grades. There are many kinds of proportion, not just one. Problem solving based on proportions appeals to intuition, not logic. Its use is a hindrance to intellectual progress in science.

10. *Think* about your answer. See whether it is expressed in the units that were asked for, and whether it is reasonable in size for the information given. If not, check back and see if you can locate the trouble.

2

Number Notations, Arithmetical Operations, and Calculators

DECIMAL NOTATION

One common representation of numbers is *decimal notation*. Typical examples are such large numbers as 807,267,434.51 and 3,500,000, and such small numbers as 0.00055 and 0.0000000000000000248. Decimal notation is often awkward to use, and it is embarrassingly easy to make foolish mistakes when carrying out arithmetical operations in this form. Most hand calculators will not accept extremely large or extremely small numbers through the keyboard in decimal notation.

SCIENTIFIC NOTATION

Another common, but more sophisticated, representation of numbers is *scientific notation*. This notation minimizes the tendency to make errors in arithmetical operations; it is used extensively in chemistry. It is imperative that you be completely comfortable in using it. Hand calculators *will* accept extremely large or extremely small numbers through the keyboard in scientific notation. Ready and proper use of this notation requires a good understanding of the following paragraphs.

An *exponent* is a number that shows how many times a given number (called the *base*) appears as a factor; exponents are written as superscripts. For example, 10^2 means $10 \times 10 = 100$. The number 2 is the exponent; the number 10 is the base, which is said to be raised to the second power. Likewise, 2^5 means $2 \times 2 \times 2 \times 2 \times 2 = 32$. Here 5 is the exponent, and 2 is the base that is raised to the fifth power.

It is simple to express any number in exponential form as the product of some other number and a power of 10. For integral (whole) powers of 10, we have $1 = 1 \times 10^0$, $10 = 1 \times 10^1$, $100 = 1 \times 10^2$, $1000 = 1 \times 10^3$, and so on. If a number is not an integral power of 10, we can express it as the product of two numbers, with one of the two being an integral power of 10 that we can write in exponential form. For example, 2000 can be written as 2×1000, and then changed to the exponential form of 2×10^3. The form 2000 is an example of decimal notation; the equivalent form 2×10^3 is an example of scientific notation.

Notice that in the last example we transformed the expression

$$2000.0 \times 10^0$$

into the expression

$$2.0 \times 10^3$$

Notice that we shifted the decimal point *three* places to the *left,* and we also *increased* the exponent on the 10 by the same number, *three.* In changing the form of a number but not its value, we always follow this basic rule.

1. Decrease the lefthand factor by moving the decimal point to the *left* the same number of places as you *increase* the exponent of 10. An example is 2000 (i.e., 2000×10^0) converted to 2×10^3.

2. Increase the lefthand factor by moving the decimal point to the *right* the same number of places as you *decrease* the exponent of 10. An example is 0.005 (i.e., 0.005×10^0) converted to 5×10^{-3}.

PROBLEM:

Write 3,500,000 in scientific notation.

SOLUTION:

The decimal point can be set at any convenient place. Suppose we select the position shown below by the small x, between the digits 3 and 5. This gives as the first step

$$3 \, {}_x 500.000$$

Because we are moving the decimal point 6 places to the *left,* the exponent on 10^0 must be *increased* by 6, and so the answer is

$$3.5 \times 10^6$$

The number 3.5×10^6 might equally well be written as 35×10^5, 350×10^4, or 0.35×10^7, and so on. All of these forms are equivalent; for each calculation, we could arbitrarily set the decimal point at the most convenient place. However, the convention is to leave the lefthand number in the range between 1 and 10—that is, with a single digit before the decimal point. This form is known as *standard scientific notation*.

PROBLEM:
Write the number 0.00055 in standard scientific notation.

SOLUTION:
We want to set the decimal point after the first 5, as indicated by the small x:

$$0.0005_x 5$$

Because we are thus moving the decimal point four places to the *right,* we must *decrease* the exponent on the 10^0 by 4. Therefore the scientific notation is 5.5×10^{-4}.

You must be able to enter numbers easily and unerringly into your calculator using scientific notation. With most calculators, this would be accomplished as follows.

1. Write the number in standard scientific notation. *Standard* notation isn't required, but it is a good habit to acquire.

2. Enter the lefthand factor through the keyboard.

3. Press the exponent key (common key symbols are EEX and EE).

4. Enter the exponent of 10 through the keyboard. If the exponent is negative, also press the "change sign" key (common symbols are CHS and $+/-$). It is important that you *don't* press the $-$ key (i.e., the subtract key).

5. The lefthand factor of the desired scientific notation will occupy the lefthand side of the lighted display, while the three spaces at the righthand end of the display will show the exponent (a blank space followed by two digits for a positive exponent, or a minus sign followed by two digits for a negative exponent. If the exponent is less than 10, the first digit will be a 0—for example, 03 for 3).

Some calculators make it possible for you to choose in advance that all results be displayed in scientific notation (or decimal notation), regardless of which notation you use for entering numbers. You can also choose how many decimal places (usually up to a maximum of 8) will be displayed in decimal notation, or in the lefthand factor of scientific notation. If your calculator has this capability, you should learn to take advantage of it. [If you try to enter, in decimal notation, numbers that are too large (for many calculators, greater than 99,999,999) or too small (for many calculators, less than 0.00000001), you will find that not all of the digits are displayed for the large numbers, and that only 0 or only part of the digits is displayed for the small numbers. To avoid such errors, determine for *your* calculator the limits for entry by decimal notation. With most calculators, any number from $9.99999999 \times 10^{99}$ to 1×10^{-99} can be entered in scientific notation.]

MATH OPERATIONS IN SCIENTIFIC NOTATION

To use the scientific notation of numbers in mathematical operations, we must remember the *laws of exponents*.

1. *Multiplication*		$X^a \cdot X^b = X^{a+b}$
2. *Division*		$\dfrac{X^a}{X^b} = X^{a-b}$
3. *Powers*		$(X^a)^b = X^{ab}$
4. *Roots*		$\sqrt[b]{X^a} = X^{a/b}$

Because b (or a) can be a negative number, the first two laws are actually the same. Because b can be a fraction, the third and fourth also are actually the same. That is, $X^{-b} = 1/X^b$, and $X^{1/b} = \sqrt[b]{X}$.

In practice, we perform mathematical operations on numbers in scientific notation according to the simple rules that follow. It is not necessary to know these rules when calculations are done with a calculator (you need only know how to enter numbers), but many calculations are so simple that no calculator is needed, and you *should* be able to handle these operations when your calculator is broken down or not available. The simple rules are the following.

1. To multiply two numbers, put them both in standard scientific notation. Then multiply the two lefthand factors by ordinary multiplication, and multiply the two righthand factors (powers of 10) by the multiplication law for exponents—that is, by adding their exponents.

PROBLEM:
Multiply 3000 by 400,000.

SOLUTION:
Write each number in standard scientific notation. This gives

$$3000 = 3 \times 10^3$$
$$400,000 = 4 \times 10^5$$

Multiply:

$$3 \times 4 = 12$$
$$10^3 \times 10^5 = 10^{3+5} = 10^8$$

The answer is 12×10^8 (or 1.2×10^9).

___

If some of the exponents are negative, there is no difference in the procedure; the algebraic sum of the exponents still is the exponent of the answer.

___

PROBLEM:
Multiply 3000 by 0.00004.

SOLUTION:

$$3000 = 3 \times 10^3$$
$$0.00004 = 4 \times 10^{-5}$$

Multiply:

$$3 \times 4 = 12$$
$$10^3 \times 10^{-5} = 10^{3-5} = 10^{-2}$$

The answer is 12×10^{-2} (or 0.12).

___

2. To divide one number by another, put them both in standard scientific notation. Divide the first lefthand factor by the second, according to the rules of ordinary division. Divide the first righthand factor by the second, according to the division law for exponents—that is, by subtracting the exponent of the divisor from the exponent of the dividend to obtain the exponent of the quotient.

___

PROBLEM:
Divide 0.0008 by 0.016.

SOLUTION:
Write each number in standard scientific notation. This gives

$$0.0008 = 8 \times 10^{-4}$$
$$0.016 = 1.6 \times 10^{-2}$$

Divide:

$$\frac{8}{1.6} = 5$$

$$\frac{10^{-4}}{10^{-2}} = 10^{-4-(-2)} = 10^{-4+2} = 10^{-2}$$

The answer is 5×10^{-2} (or 0.05).

---

3. To add or subtract numbers in scientific notation, adjust the numbers to make all the exponents on the righthand factors the same. Then add or subtract the lefthand factors by the ordinary rules, making no further change in the righthand factors.

PROBLEM:
Add 2×10^3 to 3×10^2.

SOLUTION:
Change one of the numbers to give its exponent the same value as the exponent of the other; then add the lefthand factors.

$$2 \times 10^3 = 20 \times 10^2$$
$$3 \times 10^2 = \underline{\ \ 3 \times 10^2}$$
$$23 \times 10^2$$

The answer is 23×10^2 (or 2.3×10^3).

---

4. Because $10^0 = 1$ (more generally, $n^0 = 1$, for any number n), if the exponents in a problem reduce to zero, then the righthand factor drops out of the solution.

PROBLEM:
Multiply 0.003 by 3000.

SOLUTION:

$$0.003 = 3 \times 10^{-3}$$
$$3000 = 3 \times 10^3$$

Multiply:

$$3 \times 10^{-3} \times 3 \times 10^3 = 9 \times 10^{3-3} = 9 \times 10^0 = 9 \times 1 = 9$$

The use of these rules in problems requiring both multiplication and division is illustrated in the following example.

PROBLEM:
Use exponents to solve

$$\frac{2,000,000 \times 0.00004 \times 500}{0.008 \times 20} = ?$$

SOLUTION:
First, rewrite all numbers in standard scientific notation:

$$2,000,000 = 2 \times 10^6$$
$$0.00004 = 4 \times 10^{-5}$$
$$500 = 5 \times 10^2$$
$$0.008 = 8 \times 10^{-3}$$
$$20 = 2 \times 10^1$$

This gives

$$\frac{2 \times 10^6 \times 4 \times 10^{-5} \times 5 \times 10^2}{8 \times 10^{-3} \times 2 \times 10^1}$$

Dealing first with the lefthand factors, we find

$$\frac{2 \times 4 \times 5}{8 \times 2} = \frac{5}{2} = 2.5$$

The exponent of the answer is

$$\frac{10^6 \times 10^{-5} \times 10^2}{10^{-3} \times 10^1} = \frac{10^{6-5+2}}{10^{-3+1}} = \frac{10^3}{10^{-2}} = 10^{3-(-2)} = 10^{3+2} = 10^5$$

The complete answer is 2.5×10^5, or 250,000.

Approximate Calculations

Trained scientists often make mental estimates of numerical answers to quite complicated calculations, with an ease that appears to border on the miraculous. Actually, all they do is round off numbers and use exponents to reduce the calculation to a very simple form. It is quite useful for you to learn these methods. By using them, you can save a great deal of time in homework problems and on tests, and can tell whether an answer seems reasonable (i.e., whether you've made a math error).

PROBLEM:

We are told that the population of a city is 256,700 and that the assessed value of the property is $653,891,600. Find an approximate value of the assessed property per capita.

SOLUTION:

We need to evaluate the division

$$\frac{\$653,891,600}{256,700} = ?$$

First we write the numbers in standard scientific notation:

$$\frac{6.538916 \times 10^8}{2.56700 \times 10^5}$$

Round off to

$$\frac{6.5 \times 10^8}{2.6 \times 10^5}$$

Mental arithmetic gives

$$\frac{6.5}{2.6} = 2.5$$

$$\frac{10^8}{10^5} = 10^3$$

The approximate answer is 2.5×10^3, or $2500. This happens to be a very close estimate; the value obtained with a calculator is $2547.30.

PROBLEM:

Find an approximate value for

$$\frac{2783 \times 0.00894 \times 0.00532}{1238 \times 6342 \times 9.57}$$

SOLUTION:

First rewrite in scientific notation but, instead of using standard form, set the decimal points to make each lefthand factor as near to 1 as possible:

$$\frac{2.783 \times 10^3 \times 0.894 \times 10^{-2} \times 5.32 \times 10^{-3}}{1.238 \times 10^3 \times 6.342 \times 10^3 \times 0.957 \times 10^1}$$

Rewrite the lefthand factors rounding off to integers:

$$\frac{3 \times 10^3 \times 1 \times 10^{-2} \times 5 \times 10^{-3}}{1 \times 10^3 \times 6 \times 10^3 \times 1 \times 10^1}$$

Multiplication now gives

$$\frac{3 \times 1 \times 5 \times 10^{-2}}{1 \times 6 \times 1 \times 10^{7}} = \frac{15}{6} \times 10^{-9} = 2.5 \times 10^{-9} \text{ (approximate)}$$

After considerable practice, you will find that you can carry out such approximate calculations in your head. One useful way to get that practice is to make a regular habit of first estimating an approximate answer, and then checking your final exact answer against it to be sure that you are "in the right ballpark."

LOGARITHMS

A third way to represent a number is a condensed notation called a *logarithm*. The *common* logarithm of a number N (abbreviated $\log N$) is the power to which 10 (called the base) must be raised to give N. The logarithm therefore is an exponent.

When a number (N) is an integral power of 10, its logarithm is a simple integer, positive if N is greater than 1, and negative if N is less than 1. For example

$$N = \quad 1 \quad = 10^{0} \qquad \log 10^{0} = 0$$
$$N = \quad 10 \quad = 10^{1} \qquad \log 10^{1} = 1$$
$$N = \quad 1000 \quad = 10^{3} \qquad \log 10^{3} = 3$$
$$N = 0.0001 = 10^{-4} \qquad \log 10^{-4} = -4$$

When a number is not an integral power of 10, the logarithm is not a simple integer, and assistance is needed to find it. The most common forms of assistance are electronic hand calculators and log tables. With calculators, you simply enter into the keyboard the number (N) whose log you want, press the log key (or keys), and observe the log in the lighted display. For practice, and to make sure that you know how to use *your* calculator for this purpose, check that

for $N = 807,267,434.51$ $\qquad = 10^{8.90702}$, $\qquad \log N = 8.90702$

for $N = 3,500,000$ $\qquad = 10^{6.54407}$, $\qquad \log N = 6.54407$

for $N = 0.00055$ $\qquad = 10^{-3.25964}$, $\qquad \log N = -3.25964$

for $N = 0.0000000000000000248 = 10^{-16.60555}$, $\qquad \log N = -16.60555$

Remember that very large and very small numbers must be entered in scientific notation. In addition, if you have a TI-type calculator, you may need to know that you must press the INV and EE keys *after* entering the number in scientific notation and *before* pressing the log key if you wish to obtain the log to more than four decimal places.

Because logarithms are exponents, we have the following *logarithm laws* that are derived from the laws of exponents given on page 8. Let A and B be any two numbers.

Log of a product: $\log AB = \log A + \log B$

Log of a quotient: $\log \dfrac{A}{B} = \log A - \log B$

Log of a power (n): $\log A^n = n \log A$

Log of the nth root: $\log \sqrt[n]{A} = \log A^{1/n} = \dfrac{1}{n} \log A$

The logarithm of a number consists of two parts, called the characteristic and the mantissa. The *characteristic* is the portion of the log that lies before the decimal point, and the *mantissa* is the portion that lies after the decimal point. The significance of separating a logarithm into these two parts is evident when you apply the logarithm laws to the logs of numbers such as 2000, and 2, and 0.000002.

$$\log 2000 = \log (2 \times 10^3) = \log 2 + \log 10^3 = 0.30103 + 3 = 3.30103$$
$$\log 2 = \log (2 \times 10^0) = \log 2 + \log 10^0 = 0.30103 + 0 = 0.30103$$
$$\log 0.000002 = \log (2 \times 10^{-6}) = \log 2 + \log 10^{-6} = 0.30103 - 6 = -5.69897$$

Note that the characteristic is determined by the power to which 10 is raised (when the number is in standard scientific notation), and the mantissa is determined by the log of the lefthand factor (when the number is in scientific notation). It is these properties that make it so easy to find the logarithm of a number using a log table. Here is how you can do it.

1. Write the number (N) in standard scientific notation.

2. Look up the mantissa in the log table. It is the log of the lefthand factor in scientific notation, which is a number between 1 and 10. The mantissa will lie between 0 and 1.

3. The exponent of 10 (the righthand factor) is the characteristic of the log.

4. Add the mantissa and the characteristic to obtain log N.

PROBLEM:
Find the log of 203.

SOLUTION:

1. Write the number as 2.03×10^2.

2. In the log table, find 2.0 (sometimes written as 20) in the lefthand column. Read across to the column under 3. This gives log 2.03 = 0.3075.

3. Because the exponent of 10 is 2, the characteristic is 2.

4. Log 203 = log 2.03 + log 10^2 = 0.3075 + 2 = 2.3075.

PROBLEM:

Find the log of 0.000203.

SOLUTION:

1. Write the number as 2.03×10^{-4}.

2. As in the previous problem, find log 2.03 = 0.3075 (from the log table).

3. Because the exponent of 10 is −4, the characteristic is −4.

4. Log 0.000203 = log 2.03 + log 10^{-4} = 0.3075 − 4 = −3.6925.

Interpolation

The log tables of this book show only three digits for N. If you want the log of a four-digit number, you must estimate the mantissa from the two closest values in the table. This process is called *interpolation*. For example, to find the log of 2032, you would proceed as follows.

$$\text{Log } 2032 = \log (2.032 \times 10^3)$$

$$\text{Mantissa of } 2.04 = 0.3096$$

$$\text{Mantissa of } 2.03 = 0.3075$$

$$\text{Difference between mantissas} = 0.0021$$

The mantissa of 2.032 will be about 0.2 of the way between the mantissas of 2.03 and 2.04; therefore,

$$\text{Mantissa of } 2.032 = 0.3075 + (0.2 \times 0.0021) = 0.3075 + 0.0004 = 0.3079$$

$$\text{Log } 2032 = \log 2.032 + \log 10^3 = 3.3079$$

Most hand calculators will provide logs for nine-digit numbers (a number between 1 and 10 to eight decimal places), giving them to eight decimal places. It would require a huge book of log tables to give (with much effort) the equivalent information. Although you will normally use your calculator to deal with logs, you should be able to handle log problems with simple log tables when your calculator is broken down or not available.

ANTILOGARITHMS

The number that corresponds to a given logarithm is known as the *antilogarithm,* or antilog. Like logs, the antilogs are more easily obtained from a calculator than from a log table, but you should be able to use both methods.
  With a calculator you would find the antilog as follows.

1. Enter the given log through the keyboard. Use the "change sign" key after entry if the log is negative; *don't* use the − key (i.e., the subtract key).

2. Press the antilog key (or keys). On an HP-type calculator a common key symbol is 10^x; on a TI-type calculator you would usually press the INV and LOG keys, in that order.

3. The antilog appears in the lighted display.

Check your ability to find antilogs with your calculator, knowing that antilog 0.77815 = 6.00000, antilog 5.39756 = 2.49781 × 10^5, and antilog (−3.84615) = 1.42512 × 10^{-4}.

  With a log table you would find the antilog as illustrated by the following problems.

PROBLEM:
Find the antilog of 4.5502.

SOLUTION:
We want the number that corresponds to $10^{4.5502} = 10^{0.5502} \times 10^4$. Locate the mantissa, which is 0.5502, in a log table; then find the value of N that has this log. The mantissa 0.5502 lies in the row corresponding to 3.5 and in the column headed by 5. Therefore the number corresponding to $10^{0.5502}$ is 3.55, and the number we seek is 3.55 × 10^4.

PROBLEM:
Find the antilog of −6.7345.

SOLUTION:
We want the number that corresponds to $10^{-6.7345} = 10^{0.2655} \times 10^{-7}$. Note that we must have a *positive* exponent for the lefthand factor; the sum of the two exponents is still −6.7345. Locate the mantissa, which is 0.2655, in the log table; then

find the value of N that has this log. The mantissa 0.2655 lies in the row corresponding to 1.8, between the columns headed 4 and 5. In fact 0.2655 is 7/24, or approximately 0.3, of the way between 0.2648 and 0.2672. Therefore the number corresponding to $10^{0.2655}$ is 1.843, and the number we seek is 1.843×10^{-7}.

NATURAL LOGARITHMS

Numbers other than 10 could be used as the base for logarithms, but the only other base that is commonly used is e, an inexact number (like π) that has mathematical significance. Many laws of chemistry and physics are derived mathematically from physical models and principles and, as a result, involve logarithms with the base e. These logarithms are called *natural* logs. The natural log of a number N is abbreviated ln N. The value of e is 2.71828183. . . . You can always convert common logs to natural logs (or vice versa) if you know the conversion factor of 2.30258509 . . . (usually rounded to 2.303) and employ it in one of the following ways:

$$x = \ln N = 2.303 \log N$$
$$e^x = 10^{x/2.303}$$

Some calculators can provide natural logs directly, without any need to convert explicitly from one form to the other. If your calculator has this capability, you would simply enter through the keyboard the number whose natural log you desire, then press the natural log key (or keys), whose symbol is probably LN. On your own calculator you can check that ln 4762 = 8.46842, and that ln 0.0000765 = −9.47822.

Most calculators have a means of providing the antilns of natural logs, as follows.

1. Enter the given log through the keyboard. Use the "change sign" key after entry if the log is negative; *don't* use the − key (i.e., the subtract key).

2. Press the antiln key (or keys). On an HP-type calculator, a common key symbol is e^x; on a TI-type calculator you would usually press the INV and LN keys, in that order.

3. The antiln appears in the lighted display.

Using your own calculator, make sure that you can find the following antilns: antiln 1.09861 = 3.00000; antiln 13.47619 = 7.12254×10^5, and antiln $(−7.60354) = 4.98683 \times 10^{-4}$.

SOME OTHER BASIC MATH OPERATIONS

There are some additional math operations that you should be able to handle easily, either with calculators or with log tables. Three of these are discussed here.

Reciprocals

The quotient that results when any number is divided into one is said to be the *reciprocal* of that number. It is a common value that can be found on any calculator by making the required division, but almost every calculator has a "reciprocal" key (usually labeled $1/x$) that makes it even easier. All you do is enter through the keyboard the number whose reciprocal you want; then press the $1/x$ key, and the reciprocal appears in the lighted display. You should learn to take advantage of this key; it is very useful. With your own calculator verify that $1/83.6 = 0.01196$; $1/0.00000297 = 3.367 \times 10^5$; and $1/6.059 \times 10^7 = 1.650 \times 10^{-8}$.

Powers

We have discussed at length the usefulness of powers of 10 as part of scientific notation, but many practical problems involve the powers of other numbers. For example, the area of a circle involves the square of the radius, and the volume of a sphere involves the cube of the radius. Nearly every calculator yields the square of a number when you simply enter the number through the keyboard and then press the x^2 key; the square appears in the lighted display.

For powers other than 2 you will need to use the y^x key, as follows.

1. Enter through the keyboard the number you wish to raise to some power.

2. a. Press the ENTER key (on an HP-type calculator).
   b. Press the y^x key (on a TI-type calculator).

3. Enter through the keyboard the power to which you wish to raise the number, but *ignore* the minus sign if the exponent is a negative number. The number need not be an integer, and it may be less than one as well as larger than one.

4. If the power is negative, press the "change sign" key (common key symbols are CHS and $+/-$); *don't* press the $-$ key.

5. a. Press the y^x key (on an HP-type calculator).
   b. Press the $=$ key (on a TI-type calculator).

6. The answer will be found in the lighted display.

Ascertain with your calculator that $(7.452)^2 = 55.53230$; $(3.71 \times 10^{-5})^6 = 2.6076 \times 10^{-27}$; $(0.000429)^{3.59} = 8.138 \times 10^{-13}$; $(6.405)^{-3} = 3.086 \times 10^{-3}$. Only *positive* numbers normally can be raised to a power with the y^x key. If you try raising a negative number to a power, you will get an error message in the display.

To find the power of a number by means of a log table, you use the logarithm laws cited on page 14, as illustrated in the following problem.

PROBLEM:
Find the value of $(2530)^5$.

SOLUTION:
Use the logarithm law for powers ($\log A^n = n \log A$) to find the log of $(2530)^5$; then take the antilog to find the desired value.

$$\log 2530 = \log (2.53 \times 10^3) = \log 2.53 + \log 10^3$$
$$= 0.4031 + 3 = 3.4031$$
$$5 \log 2530 = (5)(3.4031) = 17.0155$$
$$(2530)^5 = \text{antilog } 17.0155 = 1.036 \times 10^{17}$$

Roots

The *root* of a number is the result of raising that number to a power of less than one. For simple cases we speak of the square root, cube root, fourth root, and so on of a number (N), corresponding to $N^{\frac{1}{2}}$, $N^{\frac{1}{3}}$, $N^{\frac{1}{4}}$, and so on. In the general case, the nth root of a number N is simply $N^{1/n}$, where n may be any number greater than one and is not limited to being an integer. Another common representation for these same roots is $\sqrt{N}$, $\sqrt[3]{N}$, $\sqrt[4]{N}$. Any calculator will give the square root directly. You just enter through the keyboard the number whose square root you want, then press the $\sqrt{x}$ key, and the square root appears in the lighted display.

For all other roots, it is simplest to use your calculator as follows, realizing that the y^x key can just as well be used for $y^{1/n}$ where $x = 1/n$.

1. Enter through the keyboard the number whose root you wish to take.

2. a. Press the ENTER key (on an HP-type calculator).
   b. Press the y^x key (on a TI-type calculator).

3. Enter through the keyboard the number (n) that corresponds to the root you wish to take. This need not be an integer.

4. Press the reciprocal ($1/x$) key.

5. a. Press the y^x key (on an HP-type calculator).
   b. Press the $=$ key (on a TI-type calculator).

6. The desired root will be found in the lighted display.

You should use your calculator to verify that $(726)^{\frac{1}{3}} = 26.944$; $(8.73 \times 10^{-5})^{\frac{1}{3}} = 4.436 \times 10^{-2}$; $(0.000000416)^{\frac{1}{5}} = 5.294 \times 10^{-2}$, $(6.591 \times 10^5)^{\frac{1}{4.27}} = 23.054$.

To find the root of a number by means of a log table, you use the logarithm laws cited on page 14, as illustrated in the following problem.

PROBLEM:
Find the value of $(2530)^{\frac{1}{5}} = \sqrt[5]{2530}$.

SOLUTION:
Use the logarithm law for roots ($\log A^{1/n} = 1/n \log A$) to find the log of $(2530)^{\frac{1}{5}}$. Then take the antilog to find the desired value.

$$\log 2530 = \log 2.53 + \log 10^3 = 0.4031 + 3 = 3.4031$$

$$\left(\frac{1}{5}\right) \log 2530 = \frac{3.4031}{5} = 0.6806$$

$$(2530)^{\frac{1}{5}} = \text{antilog } 0.6806 = 4.7931$$

Sequential Operations

Throughout this chapter, we have talked as though every number we wish to use with a calculator must be entered through the keyboard. Frequently, however, we wish to use the result of a just-performed calculation (still visible in the lighted display) as part of the next calculation step. Your calculator is able to handle a continuous series of calculations—including logs, powers, roots, and reciprocals as well as the basic operations of multiplication, division, addition, and subtraction. Each make of calculator differs somewhat in the procedure for sequential calculations, but it is extremely important that you learn to do them easily and efficiently. You should never have to copy down intermediate results and then reenter them later through the keyboard in order to complete the entire calculation. One way to minimize wasteful effort is to write down all of the operations in a single equation before starting to do any of the calculations. We shall not always do this in the illustrative problems of this book, because stepwise explanation of problems often is more important for our purposes than is the time saved by a maximally efficient mode of calculation. With practice, you will learn the best balance of these factors for you in solving problems.

EPILOGUE

For many students just starting in chemistry, there is unnecessary frustration and wasted time because of erroneous or inefficient use of new hand calculators. Learning chemistry is hard enough; lack of skill in using a relatively simple basic tool only compounds the difficulties. One of the main purposes of the following problems is to give you practice in correctly and efficiently doing the principal types of calculations you will deal with in general chemistry. Two other major types of numerical problems, dealing with reliability of measurements and graphing of data, are discussed in Chapters 5 and 6.

In each of the groups of problems given here you should be able to get the *correct* answers *rapidly* (without making a big intellectual production out of it) and *efficiently* (without performing a lot of unnecessary operations or writing down intermediate figures). Rapidity comes with practice, but correctness and efficiency may require reading your calculator's instruction manual, or consulting a friend with a calculator similar to yours. Making sure that you can do these things at the outset definitely will repay you in time saved later and in the quality of your performance in the course.

PROBLEMS A

1. Simple operations.
   (a) $37.237 + 170.04 + 6.404 + 0.395 =$
   (b) $406{,}732{,}465 - 9{,}357{,}530{,}622 =$
   (c) $76.42 - 37.91 - 53.85 + 28.56 - 94.09 + 22.34 + 6.06 - 40.77 =$
   (d) $372.375 \times 5{,}287{,}695.088 =$
   (e) $0.00000743275 \div 4{,}467{,}325.62 =$
   (f) $\dfrac{465.1 \times 372.7 \times 63.2 \times 7004.7}{518.75 \times 892.4} =$
   (g) $\dfrac{(0.000473)(-72.85) - (21.63)(-0.000625)}{(872.3)(-0.0345) - (643.62)(-0.759)} =$
   (h) $\dfrac{[(27 + 62) + (32 - 57)][(74 - 49) - (18 + 66)]}{[(376 + 422) - (857 - 62)]} =$

2. Reciprocals (use reciprocal key).
   (a) $\dfrac{1}{873.2} =$
   (b) $\dfrac{1}{0.0000362} =$
   (c) $1/4{,}267{,}625 =$
   (d) $\dfrac{1}{4.093 \times 10^{-16}} =$
   (e) $\dfrac{1}{6.023 \times 10^{23}} =$

3. Scientific notation and decimals (integral powers and roots of base 10).
   A. Express each of the following in standard scientific notation, and also give the answer in that notation.

(a) $0.000053 \times 0.00000000000087 =$

(b) $534{,}000{,}000{,}000{,}000 \times 8{,}700{,}000 =$

(c) $5{,}340{,}000{,}000{,}000{,}000 \div 0.000000000000087 =$

B. Perform the following computations.

(d) $\dfrac{2.783 \times 10^3 \times 0.894 \times 10^{-2} \times 5.32 \times 10^{-3}}{1.238 \times 10^8 \times 6.342 \times 10^2 \times 0.957 \times 10^{-6}} =$

(e) $\dfrac{8.52 \times 10^{-3} \times 7.394 \times 10^4 \times 23.16 \times 0.046}{1.637 \times 10^3 \times 4.5 \times 10^{-5} \times 1.25 \times 0.8954} =$

(f) $5.34 \times 10^{-5} + 8.7 \times 10^{-7} - 9.4 \times 10^{-6} =$

4. Powers and roots (general exponents for any base).

(a) $(5280)^2 =$

(b) $\sqrt{5280} =$

(c) $(0.00000000000528)^2 =$

(d) $(0.000000000528)^{\frac{1}{2}} =$

(e) $(1776)^{25} =$

(f) $(1977)^{\frac{1}{2}} =$

(g) $\sqrt[3]{4.7 \times 10^{13}} =$

(h) $(7.43 \times 10^5)^{-3} =$

(i) $(6.57 \times 10^{-9})^{-\frac{1}{2}} =$

(j) $\dfrac{1}{(5.5 \times 10^{-11})^{\frac{1}{2}}} =$

(k) $\dfrac{3721 \times 5.03 \times (210)^2}{47.6 \times 0.00326} =$

(l) $(6.72)^3(37.6)^{\frac{1}{2}} =$

(m) $\dfrac{0.093 \times (76)^3 \times 4.96}{(52)^2 \times \sqrt{0.0038}} =$

(n) $\left(\dfrac{5.73 \times 10^{-4} \times 3.8 \times 10^{10} \times 0.0067 \times 5.42 \times 10^6}{1.987 \times 0.082 \times 1.38 \times 10^{16}} \right)^{\frac{1}{2}} =$

(o) $\dfrac{(3.78)^2 \times 5.8 \times 10^{-2}}{6.6 \times 10^{-27} \times (4.2)^4} =$

(p) $\dfrac{8.035 \times 10^{-4} \times 0.000579 \times 45.45}{(3.51)^3 \times (4.2 \times 10^{-3})^2} =$

(q) $\dfrac{(45.1)^{-2}(3.21)^3}{(8.1)^{-3}(5.07)^2} =$

5. Logarithms and antilogarithms

A. Determine the logarithms of the following numbers.

(a) $\log 47.4 =$	(h) $\ln 3.82 =$
(b) $\log 367 =$	(i) $\ln 5385 =$
(c) $\log 0.0052 =$	(j) $\ln 0.000706 =$
(d) $\log (8.73 \times 10^{-7}) =$	(k) $\ln (4.64 \times 10^{-6}) =$
(e) $\log 1572 =$	(l) $\ln 239 =$
(f) $\log (3.627 \times 10^9) =$	(m) $\ln (6.375 \times 10^{27}) =$
(g) $\log (-365) =$	(n) $\ln (-4.37) =$

B. Determine the antilogarithms of the following numbers.

(o) antilog 7.74518 =   (u) antiln 3.87624 =

(p) antilog 0.30103 =   (v) antiln 0.69315 =

(q) antilog (−6.36315) =   (w) antiln (−2.04961) =

(r) antilog (−8.01106) =   (x) antiln (−96.35760) =

(s) antilog (1.64782) =   (y) antiln 0.00674 =

(t) antilog (−4.49807) =   (z) antiln (−2.30259) =

C. Express the following numbers in standard scientific notation.

(aa) $10^{7.74518}$ (ee) $10^{2.00724}$

(bb) $10^{0.30103}$ (ff) $10^{-4.49807}$

(cc) $10^{-6.36315}$ (gg) $10^{0.0000635}$

(dd) $10^{-8.01106}$

6. "Solving for X." Solve for whatever variable is used (X, V, R, etc.).

A. Simple manipulations.

(a) $\dfrac{7}{X} = \dfrac{4}{3}$

(b) $\dfrac{31 \times 12}{20V} = \dfrac{3}{5}$

(c) $\dfrac{8X}{3} = 24$

(d) $\dfrac{5}{9}(F - 32) = 45$

(e) $\dfrac{4\pi R^3}{3} = 720$

(f) $5P^{\frac{1}{2}} - 2 = 6$

(g) $(0.4)(550)(T_2 - 20) + (35)(T_2 - 20) = (0.1)(120)(100 - T_2)$

(h) $\dfrac{4X + 5}{32} = \dfrac{4 - 3X}{12}$

B. Problems involving logarithms. Solve for whatever variable is used.

(i) $\log X = 6.34797$

(j) $\log\left(\dfrac{1}{X}\right) = -1.72222$

(k) $\log\left(\dfrac{7}{Q}\right) = 2.53033$

(l) $\log\left(\dfrac{N^3}{3}\right) = 1.47654$

(m) $\log P = \dfrac{-2403.37}{T} + 9.183837$ (Find P for $T = 328$)

(n) $\Delta G = -(2.303)(1.987)(T)\log K$ (Find ΔG for $T = 298$ and $K = 4.65 \times 10^{-3}$)

C. Solving the quadratic equation. For the equation: $ax^2 + bx + c = 0$, the solution is

$$x = \frac{-b \pm \sqrt{b^2 - 4ac}}{2a}$$

Solve for the unknown variable in the following problems.
(o) $(M - 5)^2 = 49$
(p) $3x + 5 = 4x^2 + 3$
(q) $(x + 2)^2 = 3x + 6$
(r) $2x^2 + 3x = 5$
(s) $3n^2 - 7n = 6$
(t) $(m - 3)^2 = 5 - 2m$
(u) $(y + 4)^2 - 3 = 22$
(v) $2.35x^2 + 43.4x - 29.65 = 0$
(w) $4.75 \times 10^{-3}x^2 - 2.06 \times 10^{-2}x + 9.17 \times 10^{-3} = 0$

D. The repetitive use of stored constants. (These problems are ideally suited to a programmable calculator where, with a suitable program along with stored constants, one simply keys in the values of the independent variable and reads out the values of the calculated dependent variable. If your calculator is programmable, these would be good simple practice problems for programming.)

(x) Make a table of the factors needed to convert gas pressures expressed in R_t "mm of mercury" at $t°C$ to pressures expressed in R_o "torr" for each degree Celsius in the range 20°C to 30°C. Use the formula

$$R_o = R_t(1 - 6.30 \times 10^{-4}t)$$

(y) Make a table of the vapor pressures of water (P, in torr) at each 5°C interval in the range 0°C to 50°C, using the following formula: T (Kelvin) $= 273 + t°C$.

$$\text{Log } P = - \frac{2301.5382}{T} + 9.0961$$

(z) Make a table of the densities of water for each degree Celsius in the range 0°C to 15°C, using the formula

$$D_{H_2O} = \frac{0.99983960 + 1.8224944 \times 10^{-2}t - 7.92221 \times 10^{-6}t^2 - 5.544846 \times 10^{-8}t^3}{1 + 1.8159725 \times 10^{-2}t}$$

PROBLEMS B

7. Simple operations.
(a) $15.64 + 0.925 + 475.06 + 2.197 =$
(b) $732,379,454.22 - 1,625,486,915.64 =$
(c) $18.735 + 270.16 - 89.98 - 428.2 + 39.34 + 300.09 - 165.322 =$
(d) $536.297 \times 8,477,062.35 =$
(e) $6,323,576.819 \times 0.00000000844611 =$
(f) $\dfrac{475.3 \times 730.4 \times 273.8}{760.0 \times 298.2} =$
(g) $\dfrac{(-0.00704)(46.46) - (395.7)(-0.0001156)}{(220.6)(-0.0435) - (-62.413)(0.7807)} =$

(h) $\dfrac{[(18 - 46) + (93 - 62)][(60 - 27) - (43 + 19)]}{[(276 + 321) - (761 - 93)]} =$

8. Reciprocals (use reciprocal key).

(a) $\dfrac{1}{386.4} =$ (d) $\dfrac{1}{8.023 \times 10^{26}} =$

(b) $\dfrac{1}{3,621,422} =$ (e) $\dfrac{1}{2.056 \times 10^{-19}} =$

(c) $\dfrac{1}{0.000000007916} =$

9. Scientific notation and decimals (integral powers and roots of base 10).

A. Express each of the following in standard scientific notation, and also give the answer in that notation.

(a) $0.000000473 \times 0.00000926 =$

(b) $85,900,000,000,000,000 \times 27,300,000,000 =$

(c) $0.00000000000000000643 \div 90,600,000,000,000,000,000 =$

B. Perform the following computations.

(d) $\dfrac{4.691 \times 10^2 \times 8.035 \times 10^{-5} \times 7.3 \times 10^6}{6.09 \times 10^{16} \times 1.07 \times 10^{-9} \times 0.539 \times 10^{-3}} =$

(e) $\dfrac{7.307 \times 10^{-6} \times 432.1 \times 6.75 \times 10^2 \times 0.0572}{3.65 \times 2.63 \times 10^4 \times 0.0062 \times 5.15 \times 10^{-3}} =$

(f) $4.75 \times 10^{-3} + 8.06 \times 10^{-2} - 5.52 \times 10^{-4} - 9.97 \times 10^{-5} =$

10. Powers and roots (general exponents for any base).

(a) $(365.1)^2 =$

(b) $\sqrt{365.1} =$

(c) $(0.0000000000000365)^2 =$

(d) $(0.0000000000003651)^{\frac{1}{2}} =$

(e) $(1978)^{27} =$

(f) $(2001)^{\frac{1}{2}} =$

(g) $\sqrt[3]{5.4 \times 10^{11}} =$

(h) $(8.25 \times 10^7)^{\frac{1}{3}} =$

(i) $(4.62 \times 10^{-8})^{\frac{1}{2}} =$

(j) $\dfrac{1}{(3.97 \times 10^{-9})^{\frac{1}{4}}} =$

(k) $\dfrac{0.000537 \times (62.4)^3 \times 2134}{3.19 \times 10^{-5} \times 82.7} =$

(l) $(3.65)^4 (165.2)^{\frac{1}{4}} =$

(m) $\dfrac{(18.7)^3 (48.9)^2 (0.000326)}{\sqrt{426.5} \times (3.17)^2} =$

(n) $\left(\dfrac{4.03 \times 10^{-5} \times 0.000737 \times 6.02 \times 10^{14} \times 53.7}{3.07 \times 10^4 \times 0.0329 \times 21.63} \right)^{\frac{1}{2}} =$

(o) $\dfrac{(7.18 \times 10^{-3})^2 \times 4.26 \times 10^7}{5.55 \times 10^{16} \times (3.76)^5} =$

(p) $\dfrac{6.47 \times 10^{-7} \times 39.8 \times 0.0000427}{(5.33 \times 10^{-2})^3 \times (2.36)^4} =$

(q) $\dfrac{(26.5)^4 (3.77)^{-2}}{(48.2)^2 (5.61)^{-4}} =$

11. Logarithms and antilogarithms.
    A. Determine the logarithms of the following numbers.

    (a) log 63.2 =　　　　　　　　　　(h) ln 16.05 =
    (b) log 906 =　　　　　　　　　　　(i) ln 1.536 =
    (c) log 0.0000451 =　　　　　　　　(j) ln 0.000219 =
    (d) log (7.32 × 10⁻⁵) =　　　　　　(k) ln (5.77 × 10⁻⁹) =
    (e) log 43,620 =　　　　　　　　　　(l) ln 170,300 =
    (f) log (6.344 × 10¹¹) =　　　　　　(m) ln (4.696 × 10¹⁵) =
    (g) log (−27.6) =　　　　　　　　　(n) ln (−13.3) =

    B. Determine the antilogarithms of the following numbers.

    (o) antilog 4.37891 =　　　　　　　(u) antiln 2.19722 =
    (p) antilog 0.90309 =　　　　　　　(v) antiln 0.12345 =
    (q) antilog (−5.65432) =　　　　　　(w) antiln (−27.27546) =
    (r) antilog (−9.00576) =　　　　　　(x) antiln (−1.98765) =
    (s) antilog (2.55555) =　　　　　　　(y) antiln (0.00026) =
    (t) antilog (−0.17562) =　　　　　　(z) antiln (−7.77777) =

    C. Express the following numbers in standard scientific notation.

    (aa) $10^{4.37891}$　　　　　　　　　(ee) $10^{3.00525}$
    (bb) $10^{0.90309}$　　　　　　　　　(ff) $10^{0.49715}$
    (cc) $10^{-5.65432}$　　　　　　　　　(gg) $10^{-0.0000777}$
    (dd) $10^{-9.00576}$

12. "Solving for X." Solve for whatever variable is used (x, Y, Q, etc.).
    A. Simple manipulations.

    (a) $\dfrac{8.3}{2.5} = \dfrac{12}{x}$

    (b) $\dfrac{14}{6} = \dfrac{9 \times 27}{13Y}$

    (c) $\dfrac{12Q}{5} = 60$

    (d) $\dfrac{5}{9}(F - 32) = 70$

    (e) $\dfrac{\pi D^3}{6} = 235$

    (f) $15 = 4 + 7V^{\frac{1}{2}}$

    (g) $(0.17)(110)(100 - T_2) = (0.53)(480)(T_2 - 22) + (40)(T_2 - 22)$

    (h) $\dfrac{5x - 4}{19} = \dfrac{5 - 4x}{3}$

    B. Problems involving logarithms. Solve for whatever variable is used.

    (i) $\log Y = 4.83651$

    (j) $\log \left(\dfrac{1}{H}\right) = 5.17624$

    (k) $\log \dfrac{1}{Z^2} = -5.63257$

    (l) $\log \left(\dfrac{5}{M}\right) = 2.19607$

    (m) $\log P = \dfrac{-5342.19}{T} + 20.22784$　(Find P for $T = 298$)

(n) $\Delta G = -(2.303)(1.987)(T) \log K$ (Find ΔG for $T = 333$
and $K = 6.39 \times 10^{-5}$)

C. Solving the quadratic equation. For the equation: $ax^2 + bx + c = 0$, the solution is

$$x = \frac{-b \pm \sqrt{b^2 - 4ac}}{2a}$$

Solve for the unknown variable in each of the following problems.

(o) $(y + 7)^2 = 64$

(p) $5x^2 - 8 = 6x - 7$

(q) $(z - 3)^2 = 5z - 11$

(r) $4q^2 + 5q = 3$

(s) $7p^2 - 3p = 6$

(t) $(2N + 4)^2 = 7 - 4N$

(u) $(0.4V - 0.3)^2 = 0.16 - 0.24V$

(v) $8.59x^2 - 29.34x + 18.25 = 0$

(w) $6.47 \times 10^{-5}x^2 - 4.09 \times 10^{-3}x + 2.18 \times 10^{-2} = 0$

3

Use of Dimensions

When numbers are used to express results of measurements, the units of measurement should always be given. Too frequently these units, or dimensions, are assumed but not shown.

In using dimensions in calculations, we follow a few simple rules.

1. Every number that represents a measurement is given with its dimension—for example, 12 men, 16 feet, 5 miles.

2. Numbers that do not involve a measurement are written without a dimension. Examples are π (the ratio of the circumference of a circle to its diameter) and logarithms.

3. In addition and subtraction, all numbers must have the same dimensions. We can add 2 apples to 3 apples, but we cannot add 2 apples to 3 miles.

4. In multiplication and division, the dimensions of the numbers are multiplied and divided just as the numbers are, and the product or quotient of the dimension appears in the final result. Thus the product (6 men)(2 days) = 12 man-days, and the product (5 gal)(4 lbs/gal) = 20 lbs. Note that units common to both numerator and denominator cancel each other, just as do factors common to both.

A term frequently used with numbers is *per,* which shows how many units of one measurement correspond to *one* unit of another. A common method of calculation is to divide the total number of units of one property by the total number of units of another property to which it corresponds. Thus, if we are told that 4.2 gallons weigh 10.5 pounds, we express the relation by the ratio

$$\frac{10.5 \text{ lb}}{4.2 \text{ gal}} = 2.5 \frac{\text{lb}}{\text{gal}}$$

which is read as "2.5 pounds per gallon." A few examples will help in understanding the use of units.

PROBLEM:
If apples cost 30 cents per dozen, how many can be bought for 50 cents?

SOLUTION:
Set up an equation that will eliminate "cents" and give "apples" in its place, as follows.

(a) number of apples $= \left(\dfrac{50 \text{ cents}}{30 \frac{\text{cents}}{\text{doz}}} \right) \left(12 \frac{\text{apples}}{\text{doz}} \right)$

$= \dfrac{50 \times 12 \text{ apples}}{30} = 20 \text{ apples}$

or (b) number of apples $= (50 \text{ cents}) \left(\dfrac{1 \text{ doz}}{30 \text{ cents}} \right) \left(12 \frac{\text{apples}}{\text{doz}} \right)$

$= \dfrac{50 \times 1 \times 12 \text{ apples}}{30} = 20 \text{ apples}$

In (a) we divided 50 cents by 30 cents/doz, while in (b) we multiplied 50 cents by 1 doz/30 cents, the reciprocal of 30 cents/doz. These are equivalent operations and you should feel comfortable with either method. In each case, "cents" and "doz" cancel out, leaving only "apples," as we had hoped.

A somewhat more sophisticated method for setting up the problem is to use the negative exponent, -1, for units that appear in the denominator of a set of units. Thus the term "per dozen" may be written as doz^{-1}. The preceding problem could have been set up in the following form:

number of apples $= \left(\dfrac{50 \text{ cents}}{30 \text{ cents doz}^{-1}} \right) (12 \text{ apples doz}^{-1}) = 20 \text{ apples}$

PROBLEM:
Find the number of feet in 1.5 mi (one mile is 5280 ft).

SOLUTION:

To find the number of feet, we need the product shown in the equation

$$\text{feet} = 1.5 \cancel{\text{mi}} \times 5280 \, \frac{\text{ft}}{\cancel{\text{mi}}}$$

$$= 7920 \text{ ft}$$

Alternatively, we may use the solution

$$\text{feet} = 1.5 \cancel{\text{mi}} \times 5280 \text{ ft } \cancel{\text{mi}}^{-1}$$
$$= 7920 \text{ ft}$$

PROBLEM:

Find the number of gallons in 5 cu yd (also written as yd³), using the conversion factors

$$231 \text{ in}^3 = 1 \text{ gal}$$
$$3 \text{ ft} = 1 \text{ yd}$$
$$12 \text{ in} = 1 \text{ ft}$$

SOLUTION:

$$\text{gallons} = 5 \text{ yd}^3 \times \left(3 \, \frac{\text{ft}}{\text{yd}}\right)^3 \times \left(12 \, \frac{\text{in}}{\text{ft}}\right)^3 \times \frac{1 \text{ gal}}{231 \text{ in}^3}$$

$$= 5 \cancel{\text{yd}^3} \times 27 \, \frac{\cancel{\text{ft}^3}}{\cancel{\text{yd}^3}} \times 1728 \, \frac{\cancel{\text{in}^3}}{\cancel{\text{ft}^3}} \times \frac{\text{gal}}{231 \cancel{\text{in}^3}}$$

$$= 1010 \text{ gal}$$

PROBLEM:

Find a conversion factor F by which you can convert yd³ to gallons.

SOLUTION:

A conversion factor can be developed from other known factors by calculating a value of F that will satisfy the equation

$$\text{gal} = (\text{yd})^3 [F]$$

F must have such units that, when they are substituted in this equation, they will cancel yd³ and yield only gal as the net result, as follows:

$$\text{gal} = (\text{yd})^3 \left[\left(\frac{3 \cancel{\text{ft}}}{\text{yd}}\right)^3 \left(\frac{12 \cancel{\text{in}}}{\cancel{\text{ft}}}\right)^3 \left(\frac{1 \text{ gal}}{231 \cancel{\text{in}^3}}\right)\right]$$

$$F = 202 \, \frac{\text{gal}}{\text{yd}^3}$$

Thus, if you want to convert 20 yd³ to gallons, you merely multiply 20 by 202, as follows:

$$\text{gal} = (20 \ \cancel{\text{yd}^3}) \left(202 \ \frac{\text{gal}}{\cancel{\text{yd}^3}}\right) = 4040 \ \text{gal}$$

PROBLEM:

If a runner does the 100 yd dash in 10 seconds, what is his speed in miles per hour?

SOLUTION:

The first thing you should observe is that the units of the desired answer must be mi/hr, so the original information must be used in the ratio of length/time, or as 100 yd/10 sec. Once you have made the proper decision about how to use the units of the original data, follow the same procedure as in finding a conversion factor F:

$$\frac{\text{mi}}{\text{hr}} = \left(\frac{100 \ \cancel{\text{yd}}}{10 \ \cancel{\text{sec}}}\right) \left(\frac{1 \ \text{mi}}{1760 \ \cancel{\text{yd}}}\right) \left(\frac{60 \ \cancel{\text{sec}}}{\cancel{\text{min}}}\right) \left(\frac{60 \ \cancel{\text{min}}}{\text{hr}}\right) = 20.5 \ \frac{\text{mi}}{\text{hr}}$$

PROBLEMS A

1. Compute the number of seconds in the month of July.

2. Develop a factor to convert days to seconds.

3. A satellite is orbiting at a speed of 18,000 miles per hour. How many seconds does it take to travel 100 miles?

4. A traveler on a jet plane notes that in 30 seconds the plane passes 6 section-line roads (1 mile apart). What is the ground speed, in miles per hour?

5. A cubic foot of water weighs 62.4 lb. What is the weight of a gallon of water (231 cu in)?

6. For each of the following pairs of units, work out a conversion factor F that will convert a measurement given in one unit to a measurement given in the other, and show the simple steps used in your work.
   (a) ounces to tons
   (b) cubic inches to cubic yards
   (c) feet per second to miles per hour
   (d) tons per square yard to pounds per square inch
   (e) cents per pound to dollars per ton
   (f) seconds to weeks
   (g) cubic feet per second to quarts per minute
   (h) miles to fathoms (1 fathom = 6 ft)
   (i) yards to mils (1 mil = 1/1000 in)

PROBLEMS B

7. For each of the following pairs of units, follow the same procedure as that for Problem 6.
   (a) cubic feet to gallons
   (b) ounces per square foot to pounds per square yards
   (c) gallons per second to cubic yards per minute
   (d) tons per cubic foot to pounds per cubic inch
   (e) yards per second to inches per hour
   (f) dollars per pound to nickels per ounce
   (g) miles to mils (1 mil = 1/1000 in)
   (h) knots to miles per hour (1 knot = 101.5 feet per minute)
   (i) degrees of arc per second to revolutions per minute

8. An acre-foot of water will cover an acre of land with a layer of water one foot deep. How many gallons are in an acre-foot? Use the following factors: 1 acre = 4840 yd²; and 1 gal = 231 in³.

9. Municipal water is sold at 21 cents per 100 cu ft. What is the price per acre-foot?

10. In the Bohr model of the hydrogen atom, an electron travels in a circular orbit about the nucleus at approximately 5×10^6 miles per hour. How many revolutions per second does the electron make if the radius of the orbit is 2×10^{-9} inches?

11. A light-year is the distance that light travels in one year at a velocity of 186,000 miles per second. How many miles is it to the galaxy in Andromeda, which is said to be 650,000 light-years away?

12. A parsec is a unit of measure for interstellar space; it is equal to 3.26 light-years. How many miles are in one parsec?

Units of Scientific Measurements

PREFIXES

Scientific measurements range from fantastically large to incredibly small numbers, and units that are appropriate for one measurement may be entirely inappropriate for another. To avoid the creation of many different sets of units, it is common practice to vary the size of a fundamental unit by attaching a suitable prefix to it. Table 4-1 shows common metric prefixes and the multiples they indicate for any given unit of measurement. Thus a kilometer is 1000 meters, a microgram is 10^{-6} gram, and a nanosecond is 10^{-9} second.

SI UNITS

Except for temperature and time, nearly all scientific measurements are based on the metric system. In recent years, there has been a concerted international effort to persuade scientists to express all metric measurements in terms of just seven basic units, called SI units (for Système International). In addition to the seven basic SI units, there are seventeen other common units derived from them that have special names. However, despite the logical arguments that have been put forth for undeviating adherence to SI units, there has not been a strong popular move in this direction. For one thing, each scientist must cope

TABLE 4-1
Metric Prefixes

Prefix	Factor	Symbol	Prefix	Factor	Symbol
exa	10^{18}	E	centi	10^{-2}	c
peta	10^{15}	P	milli	10^{-3}	m
tera	10^{12}	T	micro	10^{-6}	μ
giga	10^{9}	G	nano	10^{-9}	n
mega	10^{6}	M	pico	10^{-12}	p
kilo	10^{3}	k	femto	10^{-15}	f
			atto	10^{-18}	a

with the vast accumulated literature and the history of common usage. Furthermore, two of the most common basic SI units (kilogram for mass, and meter for length) are inappropriately large for many scientific measurements. In this book, we follow the common usage that appears in almost all chemistry textbooks and the accumulated literature (such as handbooks). Even so, we do use many of the basic or derived SI units because common usage includes a number of them, though not all. (A complete list of the basic and derived SI units, and their symbols, is given on the next to last page of this text.)

METRIC AND ENGLISH UNITS

Table 4-2 shows how some of the basic metric units are related to units commonly used in English-speaking countries for nonscientific measurements. Although the United States, Great Britain, and Canada have officially resolved to convert to the metric system, it will be many years before the conversion is complete. In the meantime, you must learn to convert from one system to the other. The three conversion factors given in Table 4-2 (rounded off to 2.54

TABLE 4-2
Metric and English Units

Dimension measured	Metric unit	English unit	Conversion factor, F
Length	centimeter (cm)	inch (in)	$2.540 \frac{cm}{in}$
Volume	liter (ℓ)	quart (qt)	$0.9463 \frac{liter}{qt}$
Mass	gram (g)	pound (lb)	$453.6 \frac{g}{lb}$
Temperature	degree Celsius(°C)	degree Fahrenheit (°F)	$°C = \frac{5}{9}(°F - 32)$
Time	second (sec)	second (sec)	——

cm/in., 0.946 liter/qt, and 454 g/lb) should be memorized, for they will take care of essentially *all* the interconversions you will have to make. In addition, you should know the formulas for converting from Fahrenheit to Celsius temperatures and vice versa. The customary units of days, hours, minutes, and seconds are used in both systems for the measurement of time.

There is one metric unit with a special name that scientists frequently use because it permits the use of simple numbers when talking about the sizes of atoms and molecules. It is called an angstrom (Å); $1 \text{ Å} = 10^{-8}$ cm.

CONVERSION OF UNITS

The following problems illustrate conversions of units. Note that in all such computations it is important to include the dimensions of numbers, just as stressed in Chapter 3, and that the use of these dimensions helps to avoid errors.

PROBLEM:
Convert $\frac{5}{16}$ inch to millimeters.

SOLUTION:

$$\left(\frac{5}{16} \text{ in}\right)\left(2.54\frac{\text{cm}}{\text{in}}\right)\left(10 \ \frac{\text{mm}}{\text{cm}}\right) = 7.94 \text{ mm}$$

PROBLEM:
A trip takes 3 hrs. Express this in picoseconds.

SOLUTION:

$$(3 \text{ hrs})\left(60 \ \frac{\text{min}}{\text{hr}}\right)\left(60 \ \frac{\text{sec}}{\text{min}}\right)\left(\frac{1 \text{ psec}}{10^{-12} \text{ sec}}\right) = 1.08 \times 10^{16} \text{ psec}$$

PROBLEM:
Find the number of cubic nanometers in a gallon (231 in³).

SOLUTION:

$$(1 \text{ gal})\left(231 \ \frac{\text{in}^3}{\text{gal}}\right)\left(2.54 \ \frac{\text{cm}}{\text{in}}\right)^3\left(\frac{1 \text{ m}}{100 \text{ cm}}\right)^3\left(\frac{1 \text{ nm}}{10^{-9} \text{ m}}\right)^3 = 3.79 \times 10^{24} \text{ nm}^3$$

PROBLEM:
Express the velocity of 20 mi/hr in terms of cm/sec.

SOLUTION:

$$\text{Velocity} = \left(20 \ \frac{\text{mi}}{\text{hr}}\right)\left(5280 \ \frac{\text{ft}}{\text{mi}}\right)\left(12 \ \frac{\text{in}}{\text{ft}}\right)\left(2.54\frac{\text{cm}}{\text{in}}\right)\left(\frac{1 \ \text{hr}}{60 \ \text{min}}\right)\left(\frac{1 \ \text{min}}{60 \ \text{sec}}\right)$$

$$= 894 \ \frac{\text{cm}}{\text{sec}}$$

PROBLEM:
Find the conversion factor F by which you would convert lbs/ft^3 to kg/m^3.

SOLUTION:

$$\frac{\text{kg}}{\text{m}^3} = \left(\frac{\text{lbs}}{\text{ft}^3}\right)[F]$$

$$= \left(\frac{\text{lbs}}{\text{ft}^3}\right)\left[\left(454 \ \frac{\text{g}}{\text{lb}}\right)\left(\frac{1 \ \text{kg}}{1000 \ \text{g}}\right)\left(\frac{1 \ \text{ft}}{12 \ \text{in}}\right)^3\left(\frac{1 \ \text{in.}}{2.54 \ \text{cm}}\right)^3\left(100 \ \frac{\text{cm}}{\text{m}}\right)^3\right]$$

$$F = 16.03 \ \frac{\text{kg ft}^3}{\text{m}^3 \ \text{lb}}$$

TEMPERATURE

Three different scales are in common use for measurements of temperature: the Celsius scale (expressed in degrees Celsius, or °C), the Fahrenheit scale (expressed in degrees Fahrenheit, or °F), and the Kelvin scale (expressed in kelvins, or K). The Fahrenheit scale is commonly used in daily life and in engineering work. The Celsius scale is used in scientific work and is coming into common usage in daily life in English-speaking countries. The Kelvin scale (also called the absolute scale) is the SI choice for temperature measurements, and it is widely used in scientific work.

Table 4-3 compares the three scales. Each scale has a different zero point. The size of the unit is the same for the Celsius and Kelvin scales, but it is

TABLE 4-3
The Three Temperature Scales

	Scale		
Reference point	F	C	K
Boiling point of water	212°	100°	373.2
Freezing point of water	32°	0°	273.2
Difference (FP to BP)	180°	100°	100

different for the Fahrenheit scale. The *range* (temperature difference) from the freezing point to the boiling point of water is 180°F or 100°C or 100 K. Therefore, in measuring temperature changes or differences, 1°C = 1 K = 1.8°F, or 1°F = 100/180°C = 5/9°C = 5/9 K. Note that these formulas apply *only* to temperature differences or changes. They do not take into account the differing zero points of the three temperature scales. In order to convert a temperature measurement from one scale to another, we must have formulas that allow for *both* the differing unit sizes *and* the differing zero points.

The following is one logical approach that can be taken to derive a formula for converting Fahrenheit measurements to Celsius measurements.

1. From the Fahrenheit scale create a new Imaginary temperature scale (call it °I) that will give a reading of 0°I at freezing point of water (just like the Celsius scale). This can be accomplished by simply subtracting 32 from all of the readings on the Fahrenheit scale to give the conversion formula

$$°I = °F - 32$$

Note that the *size* of the units is the same on both scales.

2. The Imaginary scale and the Celsius scale both have the same zero point (the freezing point of water), but they differ in the size of their units. The readings on the Celsius scale will always be just 100/180 = 5/9 of the readings on the Imaginary scale, so the conversion formula from one to the other is

$$°C = 5/9°I$$

3. If the readings on the Imaginary scale are now expressed in terms of the readings on the Fahrenheit scale, we shall have the formula to convert from Fahrenheit to Celsius given in Table 4-2. That is,

$$°C = 5/9(°F - 32)$$

The units of the Kelvin scale are the same size as those of the Celsius scale, but the zero point of the Kelvin scale is 273.2 units lower than that of the Celsius scale. Therefore, K = °C + 273.2 = 5/9(°F - 32) + 273.2.

PROBLEM:
Convert 115°F to equivalent Celsius and Kelvin temperatures.

SOLUTION:
Substituting 115 in the Fahrenheit-to-Celsius conversion formula, we have

$$°C = \frac{5}{9}(°F - 32) = \frac{5 \times 83}{9} = 46.1$$

Using the Celsius-to-Kelvin formula, we have

$$K = °C + 273.2 = 46.1 + 273.2 = 319.3$$

Thus,

$$115°F = 46.1°C = 319.3 \text{ K}$$

PROBLEM:
Convert 30°C to the Fahrenheit scale.

SOLUTION:
Rearrange the conversion formula to give °F on the left:

$$\frac{9}{5}°C = °F - 32$$

$$°F = \frac{9}{5}°C + 32$$

Now we can substitute 30 for °C and solve for °F:

$$°F = \frac{9 \times 30}{5} + 32 = 86.0$$

Thus, 30°C = 86°F.

PROBLEMS A

When you work these problems, show the units in each step of the calculation, and show the units of the answers.

1. A brass bar is 2 × 3 × 6 cm. Find its area and volume.

2. A cylindrical rod is 2 cm in diameter and 12 inches long. Find its area and volume.

3. First-class postage is A cents for each ounce or fraction thereof. How much postage is required for a letter weighing 98 g? Give your answer in terms of A.

4. What is the weight, in pounds, of 20 kg of iron?

5. The distance from Paris to Rouen is 123 km. How many miles is this?

6. The regulation basketball may have a maximum circumference of $29\frac{1}{4}$ inches. What is its diameter in centimeters?

7. The longest and shortest visible waves of the spectrum have wavelengths of 0.000067 cm and 0.000037 cm. Convert these values to (a) angstroms, and (b) nanometers.

8. The wavelengths of X rays characteristic of certain metallic targets are (a) copper, 1.537395 Å; (b) chromium, 2.28503 Å; (c) molybdenum, 0.70783 Å; (d) tungsten, 0.20862 Å. Express these wavelengths in centimeters.

9. If the laboratory temperature is 21°C, what is the Fahrenheit temperature?

10. When the temperature gets to −50°F in Siberia, what would be the temperature on the Celsius and Kelvin scales?

11. Mercury freezes at −38.87°C. What are the freezing points on the Fahrenheit and Kelvin scales?

12. What is the Fahrenheit temperature at absolute zero (0 K)?

13. At what temperature do the Fahrenheit and Celsius scales have the same reading?

14. Light travels at a speed of 3.0×10^{10} cm/sec. A light-year is the *distance* that light can travel in a year's time. If the sun is 93,000,000 miles away, how many light-years is it from the earth?

15. If 1 ml of water is spread out as a film 3 Å thick, what area in square meters will it cover?

16. The area of a powdered material is 100 m²/g. What volume of water is required to form a film 10 Å thick over the surface?

17. An agate marble is placed in a graduated cylinder containing 35.0 ml of water. After the marble is added, the surface of the water stands at 37.5 ml. Find the diameter and surface area of the marble.

18. (a) If there are 6.02×10^{23} molecules in 18 ml of water, what is the volume occupied by one molecule? (b) If the molecules were little spheres, what would be the radius of a water molecule? (Give the answer in angstroms.)

19. If you should decide to establish a new temperature scale based on the assumptions that the melting point of mercury (−38.9°C) is 0°M and the boiling point of mercury (356.9°C) is 100°M, what would be (a) the boiling point of water in degrees M, and (b) the temperature of absolute zero in degrees M?

20. It has been found that the percentage of gold in sea water is 2.5×10^{-10}. How many tons of sea water would have to be processed in order to obtain 1.0 g of gold?

21. A solution contains 5.0 g of sodium hydroxide per liter. How many grams will be contained in 50 ml?

22. A solution contains 40 g of potassium nitrate per liter. How many milliliters of this solution will be needed in order to get 8.0 g of potassium nitrate?

23. The neck of a volumetric flask has an internal diameter of 12 mm. The usual practice is to fill a volumetric flask until the liquid level (meniscus) comes just to the mark on the neck. If by error one drop (0.050 ml) too much is added, at how many millimeters above the mark on the neck will the meniscus stand? Assume a plane (not curved) meniscus.

24. A soap bubble is 3.0 inches in diameter and made of a film that is 0.010 mm thick. How thick will the film be if the bubble is expanded to 15 inches in diameter?

25. (a) A cube measures 1.00 cm on an edge. What is the surface area? (b) The cube is crushed into smaller cubes measuring 1.00 mm on an edge. What is the surface area after crushing? (c) Further crushing gives cubes measuring 100 Å on an edge. What is the surface area now? How many football fields (160 ft × 100 yd) would this make?

26. The unit of viscosity (η) is called a poise. Viscosity is determined experimentally by measuring the length of time (t) for a certain volume (V) of liquid to run through a capillary tube of radius R and length L under a pressure P, according to the Poiseuille equation

$$\eta = \frac{P\pi t R^4}{8LV}$$

What are the units of a poise in the centimeter–gram–second (cgs) system?

27. The energy of a quantum of light is proportional to the frequency (ν) of the light. What must be the units of the proportionality constant (h) if $E = h\nu$? E is expressed in ergs, and ν has the units of sec^{-1}.

28. Show that the product of the volume of a gas and its pressure has the units of energy.

PROBLEMS B

29. What are the area and the volume of a bar measuring 2 × 4 × 20 cm?

30. What are the area and the volume of a cylindrical rod with hemispherical ends if the rod is 1.00 inch in diameter and has an overall length of 55.54 cm?

31. The driving distance between Los Angeles and San Francisco by one route is 420 miles. Express this in kilometers.

32. A common type of ultraviolet lamp uses excited mercury vapor, which emits radiation at 2537Å. Express this wavelength (a) in centimeters, and (b) in nanometers.

33. A certain spectral line of cadmium often is used as a standard in wavelength measurements. The wavelength is 0.000064384696 cm. Express this wavelength (a) in angstroms, and (b) in nanometers.

34. What must be the velocity in miles per hour of a jet plane if it goes at twice the speed of sound? (The speed of sound is 1000 ft/sec under the prevailing conditions.)

35. The domestic airmail rate is B cents per ounce or fraction thereof. How much postage will be required for a letter weighing 76 g? Give your answer in terms of B.

36. If the price of platinum is $851/oz (ounce avoirdupois), what is the cost of a crucible and cover weighing 12.356 g?

37. Liquid nitrogen boils at $-195.82°C$. What are the boiling points on the Fahrenheit and Kelvin scales?

38. How many kilograms are there in 5.00 megatons of limestone?

39. Gallium is unusual in that it boils at 1700°C and melts at 29.8°C. What are these temperatures on the Fahrenheit and Kelvin scales?

40. If 50.0 g of a substance S contains 6.02×10^{23} molecules, and if the cross-sectional area of each molecule is 20.0 Å², what is the surface area of a solid that needs 1.50 g of S to cover it with a layer one molecule thick?

41. (a) If there are 6.02×10^{23} molecules in 58.3 ml of ethyl alcohol, what is the volume occupied by one molecule? (b) If the molecules were little spheres, what would be the radius of an ethyl alcohol molecule? (Give the answer in angstroms.)

42. A regulation baseball is 9.00 inches in circumference. What is its diameter in centimeters?

43. A manufacturer of a glass-fiber insulation material impresses his potential customers with the "fineness" of his product (and presumably with its insulating qualities too) by handing them glass marbles ½ inch in diameter and stating that there is enough glass in a single marble to make 96 miles of glass fiber of the type used in his product. If this is true, what is the diameter of the insulating glass fibers?

44. The volume of a red blood cell is about 90 μm^3. What is its diameter in millimeters? (Assume that the cell is spherical.)

45. A ½-inch-diameter marble is placed in a graduated cylinder containing 10.0 ml of water. To what level will the liquid rise in the cylinder?

46. A solution contains 0.0500 g of salt per milliliter. How many milliliters of this solution will be needed if we are to get 100 g of salt?

47. A solution contains 1.00 g of sulfuric acid per 100 ml. How many grams of acid will be contained in 350 ml?

48. How many grams of sulfuric acid must be added to 500 g of water in order that the resulting solution be (by weight) 20 percent sulfuric acid?

49. The average velocity of a hydrogen molecule at 0°C is 1.84×10^5 cm/sec. (a) How many miles per hour (mph) is this? Molecular gas velocities are proportional to the square root of the absolute temperature. (b) At what temperature will the velocity of a hydrogen molecule be 100,000 mph?

50. A 10 ml graduated pipet has an 18-inch scale graduated in tenths of milliliters. What is the internal diameter of the pipet?

51. An error of 1.0 mm was made in adjusting the meniscus (liquid level) to the mark in a 10 ml volumetric pipet. The internal diameter of the pipet stem was

4.0 mm. Calculate the percentage error in the volume delivered, assuming a plane (not curved) meniscus.

52. To what temperature must a bath be heated so that a Fahrenheit thermometer will have a reading that is three times as large as that on a Celsius thermometer?

53. According to Newton's law, the force exerted by an object is equal to its mass times its acceleration. The unit of force needed to accelerate a mass of 1 g by 1 cm/sec^2 is called a dyne. What are the units of a dyne in the cgs system?

54. A budding young chemist decided to throw tradition overboard and include time in the metric system. To do this she kept the unit ''day'' to refer to the usual 24 hr time interval we know. She then subdivided the day into centidays, millidays, and microdays. Solve the following problems. (a) A 100 yd dash done in 9.7 sec took how many microdays? (b) A 50 min class period lasts for how many centidays? (c) A car going 60 mph goes how many miles per centiday? (d) What is the velocity of light in miles per milliday if it is 186,000 mi/sec? (e) What is the acceleration of gravity in centimeters per microday^2 if it is 980 cm/sec^2?

5

Reliability of Measurements

In pure mathematics or in counting, every number has an exact meaning. The figure 2, for example, means *precisely* two units (not *approximately* two units). In using numbers to express the results of measurements, however, we use numbers with inexact meanings, because no measurement is perfectly accurate. If we say that an object has a length of 2 m, we mean that it is approximately 2 m long. We would not be surprised to find that the actual length differs from 2.000000 m.

In expressing the results of measurements, we should use numbers in a way that indicates the reliability of the result. Suppose two persons measure the diameter of a dime with a centimeter scale. One person reports a result of 1.79 cm; the other finds the diameter to be 1.80 cm. Both would agree that the desired reading is near the 1.8 cm mark and just slightly toward the 1.7 cm side of that mark. However, one person has estimated the value as 1.79 cm, whereas the other feels that the edge of the dime is close enough to the mark to report a value of 1.80 cm. How should the uncertainty be expressed in reporting this result? It would be correct to report the result either as 1.79 cm or as 1.80 cm, but it would not be correct to give the average value of 1.795 cm. This last figure implies that the true value is known to lie near the middle of the range between 1.79 cm and 1.80 cm, which is not the case.

Suppose the measurements are repeated with a caliper whose vernier scale permits careful measurement to the nearest 0.01 cm and estimation to the nearest 0.001 cm. Now the two results might be 1.792 cm and 1.796 cm. It would now be proper to report the average value of 1.794 cm, because the third decimal place is reasonably well known. It would be even more informative to report the result as 1.794 ± 0.002 cm. The symbol ± is read as "plus or minus." It shows that the actual results vary by 0.002 cm in either direction from the reported average value.

In talking about the results of measurements, we distinguish between the accuracy and the precision of the results. The *accuracy* of a series of measurements tells how closely the average of the results agrees with the true value of the quantity that is measured. The *precision* of a series of measurements tells how nearly the repeated measurements yield the same result. For example, suppose that the markings on a centimeter scale are placed too far apart (as if the scale has been stretched). In this case, the results obtained in a series of measurements of the same object might be quite precise (different measurements would yield nearly the same answer), but they would be inaccurate (the average result would be far from the true value).

Measurements commonly involve *systematic errors*. These are errors that are reproducibly introduced in each measurement because of the construction, use, or calibration of the equipment (as in the case of the stretched scale). The precision of the results may give the illusion of accuracy in such cases. For this reason, it is desirable to make a measurement by various entirely different methods. If the results still show high precision (close agreement with one another), then it is unlikely that systematic errors exist. The accuracy of the measurement can also be tested by using the same measurement methods on a "standard sample" whose value has been certified by some reliable institution, such as the National Bureau of Standards.

Measurements also commonly involve *random errors*. These are errors whose size and direction differ from measurement to measurement; that is, they are unpredictable and unreproducible. They are commonly associated with the limited sensitivity of instruments, the quality of the scales being read, the degree of control over the environment (temperature, vibration, humidity, and so on), or human frailties (limitations of eyesight, hearing, judgment, and so on). We shall say much more about random error later in this chapter.

SIGNIFICANT FIGURES

All digits of a number that are reasonably reliable are known as *significant figures*. The number 1.79 has three significant figures: 1, 7, and 9. The number 1.794 has four significant figures.

The position of the decimal point in a measured value has nothing to do with the number of significant figures. The diameter of a dime may be given as 1.794 cm or as 17.94 mm. In either case, four significant figures are used.

PROBLEM:
A student weighs a beaker on a triple-beam balance, finding values of 50.32 g, 50.31 g, and 50.31 g in successive weighings. Express the average weight to the proper number of significant figures.

SOLUTION:
To obtain the average, add the weights and divide by 3.

$$
\begin{array}{rl}
 & 50.32 \text{ g} \\
 & 50.31 \text{ g} \\
 & \underline{50.31 \text{ g}} \\
\text{Total} = & 150.94 \text{ g} \\
\text{Average} = & 50.3133 \text{ g}
\end{array}
$$

Because the weighings disagree in the second decimal place, it is not proper to give the weight to more than the second decimal place. The average weight, to the proper number of significant figures, is 50.31 g.

Final Zero as Significant Figure

Final zeros after the decimal point are significant figures and are used to indicate the decimal place to which the measurements are reliable. Thus 1.0 cm indicates a length reliably known to tenths of a centimeter but not to hundredths of a centimeter, whereas 1.000 cm indicates a length reliably known to thousandths of a centimeter. A very common mistake is leaving out these zeros when the measured quantity has an integral value.

PROBLEM:
Give the value of a 10 g weight to the proper number of significant figures. The balance on which the weight is used will respond to weight differences of 0.0001 g.

SOLUTION:
The weight is given as 10.0000 g, to show that it is reliable to 0.0001 g.

When a number has no zeros after the decimal point, *final zeros before the decimal point* may or may not be significant, depending upon the usage. If we say that there are 1000 students enrolled in a school, all the zeros probably are significant. But if the population of a city is given as 360,000, the last two or three zeros are not significant, because daily changes make the population uncertain by perhaps several hundred persons. The final zeros in this case are used only to indicate the position of the decimal point. A convenient way to indicate reliability of a number that has final zeros before the decimal point is to express the number in standard scientific notation, using the correct number of significant figures in the nonexponential factor—for example, 3.60×10^5 for the city whose population is known only to the nearest 1000 persons.

Zeros Before a Number

If a number is less than one, the zeros following the decimal point and preceding other digits are *not* significant. The number 0.0032 has only two significant figures, 3 and 2. If 0.0032 m is written as 3.2 mm, we have the same two significant figures. Remember, however, that if a zero is added *after* the last digit, it *is* significant.

PROBLEM:
State the number of significant figures in each of the following measurements: 275; 2.75; 0.0275; 0.027500; 27,500.

SOLUTION:
275 has 3 significant figures: 2, 7, 5.
2.75 has 3 significant figures: 2, 7, 5.
0.0275 has 3 significant figures: 2, 7, 5.
0.027500 has 5 significant figures: 2, 7, 5, 0, 0.
27,500 (if we mean *exactly* 27,500) has 5 significant figures: 2, 7, 5, 0, 0.

Rounding Off Numbers

In the beaker-weighing problem on p. 45, we dropped the last two digits of the number 50.3133 because they were not significant. When we drop figures that are not significant, we say we have "rounded off the number." The following conventions commonly are used in rounding off numbers. Most hand calculators with advanced features follow these conventions, although many simpler calculators simply drop digits without any changes in the remaining digits.

1. If the leftmost digit being dropped is greater than 5, the last remaining digit is increased by 1. Thus, 5.263 would be rounded off to 5.3.

2. If the leftmost digit being dropped is less than 5, the last remaining digit is left unchanged. Thus, 5.236 would be rounded off to 5.2.

3. If the leftmost digit being dropped is 5, the last remaining digit is increased by 1. Thus, 5.250 would be rounded off to 5.3.

These rules avoid the systematic errors that would be introduced if measurements were always rounded off to the next largest or to the next smallest digit.

Significant Figures in Results of Calculations

The results of measurements often are used to calculate some other result. In such a case, the result of the calculation should be expressed with an appropriate number of significant figures to reflect the reliability of the original measurements. There are two rules for this procedure.

1. *Addition or subtraction.* The values to be added or subtracted should all be expressed with the same units. If they are expressed in scientific notation, they should all be expressed with the same power of 10. The result should be rounded off so that it has only as many digits after the decimal point as the number with the *fewest* digits after the decimal. For example, consider the sum of the following weights:

$$
\begin{array}{r}
13.8426 \text{ g} \\
764.5 \quad \text{ g} \\
\underline{7.08 \quad \text{ g}} \\
\text{Sum} = 785.4226 \text{ g}
\end{array}
$$

This mathematical sum is a very misleading statement because it contains seven significant figures. It implies that the total weight is known to the nearest 0.0001 g when, in fact, one of the weights being added is known only to the nearest 0.1 g and another only to the nearest 0.01 g. The weight known least reliably (to the fewest decimal places) limits the reliability of the sum. Therefore, the sum is properly expressed as 785.4 g. (Note that the number of significant figures in the weights is irrelevant in applying this rule for addition or subtraction. The proper result has four significant figures, although one of the weights being added has only three significant figures. It is the number with the fewest digits *after* the decimal point that determines the position of the least reliable digit in the answer.)

2. *Multiplication or division.* The product or quotient should be rounded off to the same number of significant figures as the least accurate number involved in the calculation. Thus, $0.00296 \times 5845 = 17.3$, but $0.002960 \times 5845 = 17.30$. However, this rule should be applied with some discretion. For example, consider the following multiplication:

$$0.00296 \times 5845 \times 93$$

The rule indicates that the result should be rounded off to two significant figures, so that the product would be 1600, or 1.6×10^3. However, an error of ± 1 in the value of 93 is not much more significant than an error of ± 1 in 102; we can say that 93 *almost* has three significant figures. Because the other numbers involved all have at least three significant figures, it would be reasonable to report the result of this multiplication as 1.61×10^3 (using three significant figures). Obviously, such decisions must be made by common sense rather than a hard and fast rule. The best situation would be to have a range of error for each number; then the calculation could be performed with all the largest or all the smallest values to see which digits in the result are reliable.

A case can be made for rounding off all numbers involved in a calculation before the calculation is actually made. However, in this age of hand calculators this procedure adds little except extra effort. The simplest approach probably is to enter all the numbers of a calculation through the keyboard without regard to significant figures. Then, by inspection at the end of the calculation, determine how many decimal places should be used (for addition and subtraction) or how many significant figures should be used (for multiplication and division) in the final result. *The calculator cannot make the decision about the proper number of decimal places or significant figures to use; only the operator can do this.* One of the most common errors in the use of calculators is to write down all of the digits that appear in the display as a result of a calculation, regardless of their significance. You must learn to *think* about your answers.

A pure number such as 3 or 4 has an unlimited number of significant figures (4.000000 . . . .), as does a defined quantity such as π (3.14159 . . .) or e (2.7182818 . . . .). Do not fall into the trap of excessively rounding off results that come from equations using pure or derived numbers. For example, if you want to find the volume of a sphere whose radius has been measured as 15.13 cm, you shouldn't round off the answer to 1×10^4 cm^3 just because you are going to use the formula $V = \frac{4}{3}\pi r^3$, in which 4, 3, and π each appears to have only one significant figure. It is the *measured* values that determine the number of significant figures. In this case, the volume should be expressed to four significant figures as 1.451×10^4 cm^3. Your calculator probably has a π key that will give with one stroke the value of π to 8 or 10 decimals.

DISTRIBUTION OF ERRORS

We have talked about significant figures and the general unreliability of the "last figure" of a measurement. Now we shall talk about just *how* unreliable these last figures are. Your experience has shown that really gross errors rarely occur in a series of measurements. Suppose you were able to make an infinite number of measurements on the same quantity (call it x). You would not be surprised if, on plotting each observed value of x against the frequency with which it occurred, you obtained a symmetrical curve similar to that shown in Figure 5-1. One of the advantages of making an infinite number of measurements is that the *average* ($\bar{x}$) of the values will be equal to the "true value" (μ), represented by the dotted vertical line drawn from the peak of the curve. As expected, the more a value of x deviates from μ, the less frequently it occurs. This curve is symmetrical because there is equal probability for + and − errors; it is called a *normal distribution curve*. If you made your measurements in a more careless manner or with a less sensitive measuring device, you would obtain a distribution curve more like that in Figure 5-2, shorter and broader but with the

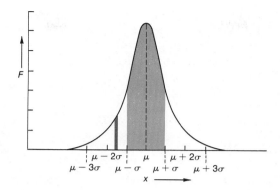

FIGURE 5-1
Normal distribution curve with a small standard deviation.

same general shape. The breadth of a distribution curve (the spread of results) is a measure of the reliability of the results. Two major ways of describing this spread are *average deviation* and *standard deviation*.

Average Deviation

The deviation of each individual measurement (x) from the average ($\bar{x}$) of all the measurements is found by simple subtraction; the deviation of the ith measurement is $x_i - \bar{x}$. We are interested in the *sizes* of the deviations, without regard to whether they are $+$ or $-$. That is, we are interested only in the absolute values of the deviation, $|x_i - \bar{x}|$. The average deviation is simply the average of these absolute values:

$$\text{Average deviation} = \frac{|x_1 - \bar{x}| + |x_2 - \bar{x}| + |x_3 - \bar{x}| + \cdots + |x_n - \bar{x}|}{n}$$

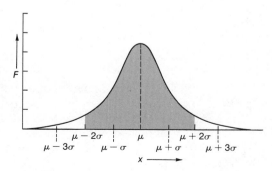

FIGURE 5-2
Normal distribution curve with a large standard deviation.

Simple expressions for $\bar{x}$ and average deviation can be made by using the symbol Σ, which means "the sum of" whatever follows it:

$$\bar{x} = \frac{\Sigma x_i}{n}$$

and

$$\text{Average deviation} = \frac{\Sigma |x_i - \bar{x}|}{n} \qquad (5\text{-}1)$$

In each case, the sum of n values is understood. One important point to remember: $\bar{x}$ may be calculated for any number of measurements, but only for an infinite number of measurements will $\bar{x} = \mu$, the "true value."

PROBLEM:
Five persons measure the length of a room, getting values of 10.325 m, 10.320 m, 10.315 m, 10.313 m, and 10.327 m. Find the average value and the average deviation.

SOLUTION:
Add the separate values and divide by 5 to get the arithmetical mean. Set opposite each value its deviation from the average, without regard to sign. Take the average of these deviations.

Measurement x_i	Deviation $|x_i - \bar{x}|$
10.325 m	0.005 m
10.320	0.000
10.315	0.005
10.313	0.007
10.327	0.007
$\Sigma x_i = 51.600$ m	0.024 m $= \Sigma|x_i - \bar{x}|$

$$\bar{x} = \frac{\Sigma x_i}{5} = 10.320 \text{ m} \qquad \frac{\Sigma|x_i - \bar{x}|}{5} = 0.0048 \text{ m} = \text{Average deviation}$$

The average is 10.320 m, with an average deviation of 0.0048 m. It is proper to write the average as 10.320 m, because the deviation affects digits in only the third decimal place. It is not correct to give the length as 10.32 m; this implies that the measurement is uncertain in the second decimal place. Average deviation is one of the simplest measures of reliability of measurements (the spread of experimental values), but a better estimate of reliability can be made with standard deviation.

Standard Deviation

The standard deviation is the square root of the *variance* (s^2). The variance is *almost* the same as the average of the squares of the deviations of the measurements from the average ($\bar{x}$); it is defined as

$$\text{variance} = s^2 = \frac{(x_1 - \bar{x})^2 + (x_2 - \bar{x})^2 + (x_3 - \bar{x})^2 + \cdots + (x_n - \bar{x})^2}{n - 1}$$

$$= \frac{\Sigma(x_i - \bar{x})^2}{n - 1} \tag{5-2}$$

For reasons that we need not discuss here, $n - 1$ is used as the denominator instead of n. Of course, for very large values of n (say, 1000), there is no appreciable difference between n and $n - 1$. Thus, for very large numbers of measurements, you really can say that the variance is the average of the squares of the deviations. Because the standard deviation is the square root of the variance, we have

$$\text{standard deviation} = s = \sqrt{\frac{\Sigma(x_i - \bar{x})^2}{n - 1}} \tag{5-3}$$

The standard deviation will have the same units as the original measurements, and the same units (but not the same value) as the average deviation.

PROBLEM:
Calculate the standard deviation (s) of the measurements made in the preceding problem.

SOLUTION:
As in the last problem, first find the average of the measurements, and then subtract it from each of the individual measurements to get the deviation. The sum of the squares of these deviations divided by $n - 1$ is the variance.

Measurement x_i	Deviation $(x_i - \bar{x})$	(Deviation)^2 $(x_i - \bar{x})^2$
10.325 m	0.005 m	0.000025 m^2
10.320	0.000	0.000000
10.315	−0.005	0.000025
10.313	−0.007	0.000049
10.327	0.007	0.000049

$\Sigma x_i = 51.600$ m

$\bar{x} = 10.320$ m

0.000148 m$^2 = \Sigma(x_i - \bar{x})^2$

$$\text{Standard deviation} = s = \sqrt{\frac{\Sigma(x_i - \bar{x})^2}{n - 1}} = \sqrt{\frac{0.000148}{4}} = 0.0061 \text{ m}$$

In order to make the value of s useful for other calculations, it is customary to write it to one more decimal place than the least significant figure of the measurements; we shall do the same for $\bar{x}$ *when it is used in the calculation of s, but at no other time.* If we did not adopt such a convention, we would frequently find, for a good set of data, that a standard deviation rounded off to the decimal place corresponding to the least significant figure of the measurements would have a value of 0. This would be a useless and misleading result because variations in the measurements actually exist. A good compromise is to *write the average to the correct number of significant figures* as discussed on p 46, but to *write the error statement to one more decimal place* than corresponds to the least significant figure in the average. An error statement is the average deviation, the standard deviation, or one of the various confidence intervals to be discussed on pp 56–57. Thus, in the example just shown, you would write

$$x = 10.320 \pm 0.0048 \text{ m (for average deviation)}$$
$$x = 10.320 \pm 0.0061 \text{ m (for standard deviation)}$$

For a very poor set of data, with a very large standard deviation, it is foolish to show the error statements to one more decimal place than the least significant figure. If the room measurements just cited actually had a standard deviation of ± 0.0756 m, it means that you were silly to think of measuring to the nearest 0.001 m, and in any case you should round the standard deviation to ± 0.076 m. The decision as to when you do, or do not, write the error statement to one more decimal place than the least significant figure is arbitrary. (Some professors say that you should use the extra decimal place if the standard deviation is less than 0.4% of the average of the measurements.)

Many calculators permit you to determine $\bar{x}$ and s directly without the need of setting out the calculations as described in the last problem. After entering each measurement through the keyboard, you press the $\Sigma+$ key. After entering all of the numbers in this fashion, you press the $\bar{x}$ key (or keys) to get the average and then the s key (or keys) to get the standard deviation. The average—which may be displayed to, say, eight decimal places—must of course be rounded off to the proper number of significant figures.

A small complication may arise from the use of calculators that determine $\bar{x}$ and s directly. With these calculators, the standard deviation is calculated from the average deviation that includes *all* of the digits in the display. The resulting value probably will differ slightly from the one you would obtain if you previously rounded the average to one more decimal place than the least significant figure. This difference usually is small and can be ignored. Thus you need not worry about somewhat different answers obtained with different types of calculators.

Probability Distribution for Large Numbers of Measurements

In order to appreciate the usefulness of the standard deviation as a measure of reliability, we must take a closer look at the curves in Figures 5-1 and 5-2. The mathematical equation for these curves (when they represent a very large number of measurements) is called the Gaussian distribution equation. In one of its forms, the frequency of occurrence, F, is given by

$$F = \left[\frac{1}{\sigma(2\pi)^{\frac{1}{2}}} \right] e^{-\frac{(x_i - \mu)^2}{2\sigma^2}} \qquad (5\text{-}4)$$

where $\sigma = s$ (the standard deviation) when n is very large, just as $\mu = \bar{x}$ when n is very large. We use these special symbols for s and $\bar{x}$ to emphasize that this Gaussian equation does not apply to the curves obtained with only a small number of measurements. Note that Equation 5-4 is written with the base e rather than the base 10 (see p 17 for a discussion of natural logarithms). We next discuss some important characteristics of this distribution curve.

1. *Size of σ and the shape of the curve.* At the peak of the curve, $\bar{x} = \mu$ and

$$F = \frac{1}{\sigma(2\pi)^{\frac{1}{2}}}$$

   In other words, the maximum height of the Gaussian curve is determined solely by the value of σ and the constant 2π. If σ is small because the errors are relatively small, then F is large and the curve is tall (Figure 5-1). If σ is large because of relatively large errors, then F is small and the curve is short (Figure 5-2). For any other value of x than $\bar{x}$, the exponent $\frac{(x_i - \mu)^2}{2\sigma^2}$ is larger for a smaller value of σ, and the sides of the curve thus fall off faster for small σ (as in Figure 5-1) than for large σ (as in Figure 5-2).

2. *Significance of the area under the curve.* Consider the very small blackened area in Figure 5-1. Its width is the infinitesimal distance dx; its height is the value of F for the value of x we have chosen. Because dx is infinitesimal, we can regard this area as a rectangle. The area $F dx$ of the rectangle represents the number of measurements of x that lie between x and $x + dx$. If we take the consecutive sum of all such small areas from one end of the curve to the other, we have the total area under the curve, and we have included all our measurements. Because of the way that F is defined, the total area under the curve is 1.000, regardless of the size of σ. All *probability distributions* share this characteristic that the total area under the curve equals unity. The area under a portion of the curve represents the number of measurements of x lying between the limiting values of x that bound the portion. For example, if the area under a portion of the curve is 0.200, then that portion of the curve represents 20.0 percent of the measurements.

3. *Relationship between σ and area under the curve.* If we express the values of x in terms of σ, we find that the normal distribution curve always has the same shape, regardless of the size of σ. That is, the area under the curve for a particular multiple of σ on either side of μ is the same for any normal distribution curve.

Values of x between	*Area under this portion of the curve*
$\mu - \sigma$ and $\mu + \sigma$	0.683 (shaded area in Figure 5-1)
$\mu - 2\sigma$ and $\mu + 2\sigma$	0.954 (shaded area in Figure 5-2)
$\mu - 3\sigma$ and $\mu + 3\sigma$	0.997

Thus, whenever you have a very large number of measurements, the probability is that 68.3% of them (about two-thirds) will have values within the range $\mu \pm \sigma$ (that is, within one standard deviation of the average value). Similarly, 95.4% of the measurements (about 19 out of 20) probably will have values within the range $\mu \pm 2\sigma$, and only 0.3% of them (3 in 1000) are likely to have values *outside* the range $\mu \pm 3\sigma$. This also means that the probability is only 0.3% (3 times out of 1000 measurements) that any single measurement will yield a value differing by more than 3σ from the value μ.

4. *Confidence interval and confidence level.* We have seen that 3 out of 1000 measurements probably will have values outside the range $\mu \pm 3\sigma$. The range of values obtained in a particular large set of measurements will depend on the particular extreme values that happen to be obtained. Because the normal distribution curve is so regular, it is useful to express the average result in a form that reflects some particular percentage of the measurements, rather than listing the particular extremes obtained. For example, you might wish to report the result as $\mu \pm t\sigma$, where t is chosen so that the range will include some particular percentage of the measurements. For example, we have seen that a choice of $t = 2$ will yield a range that includes 95.4% of the measurements. Suppose you wish to report a range that includes 80% of the measurements; in this case, you will need to consult a t table such as Table 5-1 to find the appropriate value of t. We wish to find the value of t corresponding to a confidence level of 80%; the *confidence level* is the probability that any measurement picked at random will fall within the range $\mu \pm t\sigma$. Using the bottom line of the table (representing an infinite number of measurements), we see that the desired value of t is 1.282. Therefore, we can say that there is a probability of 80% that any random measurement will fall within the range $\mu \pm 1.282\sigma$; this range is called the *confidence interval*. When a result is reported with a confidence interval, the corresponding confidence level should be stated to make the range meaningful. For example, a result might be reported as 25.342 ± 0.003 with a confidence level of 80%.

TABLE 5-1
The t Values for Various Sample Sizes and Confidence Levels

Sample* size (n)	Confidence level (percentage)						
	50	60	70	80	90	95	99
2	1.000	1.376	1.963	3.078	6.314	12.706	63.657
3	0.816	1.061	1.386	1.886	2.920	4.303	9.925
4	0.765	0.978	1.250	1.638	2.353	3.182	5.841
5	0.741	0.941	1.190	1.533	2.132	2.776	4.604
6	0.727	0.920	1.156	1.476	2.015	2.571	4.032
7	0.718	0.906	1.134	1.440	1.943	2.447	3.707
8	0.711	0.896	1.119	1.415	1.895	2.365	3.499
9	0.706	0.889	1.108	1.397	1.860	2.306	3.355
10	0.703	0.883	1.100	1.383	1.833	2.262	3.250
20	0.688	0.861	1.066	1.328	1.729	2.093	2.861
30	0.683	0.854	1.055	1.311	1.699	2.045	2.756
40	0.681	0.851	1.050	1.303	1.684	2.021	2.704
50	0.680	0.849	1.048	1.299	1.676	2.008	2.678
60	0.679	0.848	1.046	1.296	1.671	2.000	2.660
120	0.677	0.845	1.041	1.289	1.658	1.980	2.617
∞	0.674	0.842	1.036	1.282	1.645	1.968	2.576

* Statistical manuals usually list "degrees of freedom" in this column, with values that are equal to the sample size minus one.

PROBLEM:
The average result of a set of 1000 measurements is to be reported with a confidence interval representing a confidence level of 80%. The average (the true value) is 2756, and the standard deviation is 13.0. Find the confidence interval.

SOLUTION:
The set of 1000 measurements is so large that the value of t differs negligibly from that for an infinitely large set of measurements, so we use the bottom row of Table 5-1. Looking in the column for an 80% confidence level, we find the desired value of t to be 1.282. Therefore, the desired confidence interval is

$$\mu \pm t\sigma = 2756 \pm (1.282)(13.0) = 2756 \pm 17$$

Either of the following two statements could be made.
1. The probability is 80% that any value taken at random from the 1000 measurements will lie within the interval 2756 ± 17.
2. Of the 1000 measurements, 800 (or 80%) lie within the interval 2756 ± 17.

Probability Distributions for Small Numbers of Measurements

In many cases, it is not practical to make large numbers of measurements. In chemical analyses, there often is only enough material to make a few measurements. The desire usually is to find a true value for the measured quantity, but the average of a small number of measurements is unlikely to represent the true value. You can see this if you make several small sets of measurements of the same quantity; the averages of the various groups are likely to differ somewhat from one another. Most people have the intuitive feeling that the average will be closer to the true value as the number of measurements increases. The formal mathematical statements of probability theory reflect this same viewpoint. It is possible to make some statements about the results of a small number of measurements, but these statements must be made with less confidence than we have in statements resulting from large numbers of measurements.

For this discussion we use a distribution curve as before, but this time it is not represented by an equation as simple as the Gaussian equation. In fact, there is not just one curve; there are many, one for each size of sample (different number of measurements). It isn't practical to draw a different curve for each size of sample, but we can describe the changing nature of the distribution curves: as the size of the sample gets smaller, the corresponding distribution curve becomes shorter and broader than the ones shown in Figure 5-1 for the same value of s. The same statement is true for Figure 5-2, where a larger value of s applies. This changing nature of the distribution curve is taken into account in the t table (Table 5-1), which can be used in place of the curves.

The Precision of a Single Measurement

The calculation of standard deviation (Equation 5-3) is the same whether you have many measurements or only a few; sample size affects only the selection of the t value. For a single measurement taken at random from a small number (n) of measurements, the confidence interval for the desired confidence level is

$$\bar{x} \pm ts \qquad\qquad (5\text{-}5)$$

This is a statement of the precision of a single measurement.

Note that you don't know the true value (you didn't take many measurements). You must use the average ($\bar{x}$) of your measurements as the best measure available as a substitute for the true value. Note also that the value of t you choose will depend on the sample size (n) as well as on the confidence level you desire. The confidence interval will get larger as the number of determinations gets smaller, or as the confidence level increases.

The Precision of the Mean

The main objective of making a series of measurements usually is to find the true value, and we would like to indicate the confidence we may have in the average of our values. Because the average of many measurements is more likely to be correct than the average of a few, our confidence interval should get smaller as we increase the number of measurements. This is commonly expressed quantitatively by using the *standard deviation of the mean,* defined as

$$\frac{s}{\sqrt{n}} = \text{standard deviation of the mean}$$

Note that the standard deviation of the mean decreases with the square root of n, not the first power. Thus, making 100 measurements rather than 4 does not improve the precision by a factor of $\frac{100}{4} = 25$, but only by a factor of $\sqrt{25}$, or 5. The useful statement we can make with the standard deviation of the mean is the following: for a series of n measurements and a specified confidence level, the *true value* of x will lie in the interval

$$\bar{x} \pm t \left(\frac{s}{\sqrt{n}} \right) \tag{5-6}$$

This is a statement of the precision of the mean.

PROBLEM:
The density of a liquid is measured by filling a 50 ml flask as close as possible to the index mark and weighing. In successive trials the weight of the liquid is found to be 45.736 g, 45.740 g, 45.705 g, and 45.720 g. For these weights calculate the average deviation, the standard deviation, the 95% confidence interval for a single value, and the 95% confidence interval for the mean.

SOLUTION:
Because the weighings are all for the same measured volume, we first average the weights. Let x refer to the weight measurement.

Weight x_i	Deviation $\lvert x_i - \bar{x} \rvert$	(Deviation)^2 $(x_i - \bar{x})^2$
45.736 g	0.0107 g	0.000114 g^2
45.740	0.0147	0.000216
45.705	0.0203	0.000412
45.720	0.0053	0.000028
$\Sigma x_i = 182.901$ g	$\Sigma\lvert x_i - \bar{x}\rvert = 0.0510$ g	$\Sigma(x_i - \bar{x})^2 = 0.000770$ g^2

$$\bar{x} = \frac{\Sigma x_i}{4} \qquad \text{Average deviation} = \frac{\Sigma\lvert x_i - \bar{x}\rvert}{4} \qquad s = \sqrt{\frac{\Sigma(x_i - \bar{x})^2}{3}}$$

$$= 45.7253 \text{ g} \qquad\qquad\qquad = 0.0128 \text{ g} \qquad\qquad\qquad = 0.0160 \text{ g}$$

From the t table, the t value of 3.182 is found in the row for sample size of 4 and in the column for 95% confidence level. The precision of a single value is therefore

$$45.725 \pm (3.182)(0.0160) = 45.725 \pm 0.0509 \text{ g}$$

There is a 95% probability that any weight value picked at random has a value that lies within 0.0509 g of the average. The precision of the mean is given by

$$45.725 \pm (3.182)\frac{(0.0160)}{\sqrt{4}} = 45.725 \pm 0.0255 \text{ g}$$

That is, there is a 95% probability that the true value of the weight of 50 ml of this liquid lies within the interval 45.725 ± 0.0255 g.

Relative Error

It frequently is convenient to express the degree of error on a relative basis, rather than on an absolute basis as above. A relative basis has the advantage of making a statement independent of the size of the measurements that were made. For example, the statement that a solid contains 10% silver is a *relative* statement; it says that one-tenth of the solid is silver, and it is understood that a large sample of the solid would contain more grams of silver than a small one. Percentage is "parts per hundred," and it is found for this example by multiplying the fraction of the sample that is silver (in this case, 0.1) by 100. It would be as correct to multiply the fraction by 1000 and call it "100 parts per thousand," or to multiply the fraction by 1,000,000 and call it "100,000 parts per million." The choice of parts per hundred (percentage) or parts per thousand or million is determined by convenience. If the fraction were very small, say 0.00005, it would be more convenient to call it 50 parts per million than to call it 0.005 parts per hundred or 0.005 per cent.

PROBLEM:
Express the 95% confidence level of the standard deviation of the mean obtained in the previous problem as percentage, as parts per thousand, and as parts per million.

SOLUTION:
The 95% confidence level of the standard deviation of the mean was found to be ±0.0255 g, where the weight itself was (on the average) 45.725 g. The *fraction* of the total weight that might be error is

$$\frac{0.0255 \text{ g}}{45.725 \text{ g}} = 0.000558$$

The relative error thus can be written as

$$(0.000558)(100) = 0.0558 \text{ parts per hundred} = 0.0558 \text{ percent}$$
$$(0.000558)(10^3) = 0.558 \text{ parts per thousand (ppt.)}$$
$$(0.000558)(10^6) = 558 \text{ parts per million (ppm.)}$$

Calculator Tips

Equation 5-3, which defines the standard deviation, can be rewritten in the form

$$s = \sqrt{\frac{n \sum x_i^2 - (\sum x_i)^2}{n(n-1)}} \qquad (5\text{-}7)$$

The advantage of this form is that you need not find the average before proceeding to find the deviations. If you use a calculator that doesn't have a built-in program for finding s and $\bar{x}$ directly, you can easily enter each of your measurements in a way that accumulates $\sum x_i$ in one memory storage and $\sum x_i^2$ in another. After entering all of the data, you can use $\sum x_i$ and $\sum x_i^2$ in Equation 5-7 to get s, and furthermore you can divide $\sum x_i$ by n to obtain $\bar{x}$, even though you didn't have to find it to calculate s.

EPILOGUE

In an age requiring closer looks at the factors that affect our health, safety, environment, and life style, it will become more and more important to examine carefully and intelligently the statistical significance of the data on which important decisions are based. The *intelligent* use of your calculator can be of enormous help in these kinds of evaluations. You must always resist the urge to include far more figures in your reported results than the data justify. Calculations and computers cannot improve experimental reliability, and it is *your responsibility* to round off the final answer to the proper number of significant figures.

PROBLEMS A

1. State the number of significant figures in each of the following measurements.
   (a) 374; (b) 0.0374; (c) 3074; (d) 0.0030740; (e) 3740 (f) 3.74×10^5; (g) 75 million; (h) 21 thousand; (i) 6 thousandths; (j) 2 hundredths.

2. Express the answer in each of the following calculations to the proper number of significant figures (assume that the numbers represent measurements).
   (a) $3.196 + 0.0825 + 12.32 + 0.0013$
   (b) $721.56 - 0.394$
   (c) $525.3 + 326.0 + 127.12 + 330.0$
   (d) $5.23 \times 10^{-2} + 6.01 \times 10^{-3} + 8 \times 10^{-3} + 3.273 \times 10^{-2}$
   (e) $\dfrac{3.21 \times 432 \times 650}{563}$
   (f) $\dfrac{8.57 \times 10^{-2} \times 6.02 \times 10^{23} \times 2.543}{361 \times 907}$

(g) $\dfrac{4.265 \times (3081)^2 \times 8.275 \times 10^{-8}}{0.9820 \times 1.0035}$

(h) $\dfrac{6.327 \times 10^{-5} \times 7.056 \times 10^{-7} \times 9.0038 \times 10^{-9}}{6.022 \times 10^{23} \times 27.00 \times 10^{-2}}$

3. Water analysts often report trace impurities in water as "parts per million"— that is, parts by weight of impurity per million parts by weight of water. In an analysis, 2.5 liters of a water sample are evaporated to a very small volume in a platinum dish; the residue is treated with a sensitive reagent that develops a red color, whose intensity is a measure of the amount of nickel present. The amount of nickel present is found to be 0.41 mg. How many parts per million of nickel were present in the original sample of water? (Assume that the density of water is 1 g/ml.)

4. State the precision, both in parts per thousand and in percentage, with which each of the following measurements is made.
   (a) 578 with a standard deviation of 2.0
   (b) 0.0578 with a standard deviation of 0.00020
   (c) 5078 with a standard deviation of 2.0
   (d) 0.005078 with a standard deviation of 0.000030
   (e) 0.0050780 with a standard deviation of 0.00000010
   (f) 5078 with a standard deviation of 50
   (g) 5.78×10^5 with a standard deviation of 2.0×10^3

5. A radioactive sample shows the following counts for one-minute intervals: 2642; 2650; 2649; 2641; 2641; 2637; 2651; 2636. Find the average deviation, the standard deviation, and the 90% confidence interval for a single value and for the mean.

6. A student wishes to calibrate a pipet by weighing the water it delivers. A succession of such measurements gives the following weights: 5.013 g; 5.023 g; 5.017 g; 5.019 g; 5.010 g; 5.018 g; 5.021 g. Calculate the average deviation and the 95% confidence interval for a single value.

7. In determining the viscosity of a liquid by measuring the time required for 5.00 ml of the liquid to pass through a capillary, a student records the following periods: 3 min 35.2 sec; 3 min 34.8 sec; 3 min 35.5 sec; 3 min 35.6 sec; 3 min 34.9 sec; 3 min 35.3 sec; 3 min 35.2 sec. Find the average deviation and the 70% confidence interval for a single value, expressing both as a percentage.

8. A student wishes to determine the mole weight of a gas by measuring the time required for a given amount of the gas to escape through a pinhole. He observes the following time intervals: 97.2 sec; 96.6 sec; 96.5 sec; 97.4 sec; 97.6 sec; 97.1 sec; 96.9 sec; 96.4 sec; 97.3 sec; 97.0 sec. Find the average deviation and the 99% confidence interval for the mean, expressing both in parts per thousand.

9. The height (h) to which a liquid will rise in a capillary tube is determined by the force of gravity (g), the radius (r) of the tube, and the surface tension (γ) and density (d) of the liquid at the temperature of the experiment. The relationship is $\gamma = \frac{1}{2}hdgr$. A student decides to determine the radius of a capillary

by measuring the height to which water rises at 25.0°C. Several attempts yield the following heights: 75.7 mm; 75.6 mm; 75.3 mm; 75.8 mm; 75.2 mm. Find the average deviation of these heights and the 80% confidence interval for the mean, expressing both in parts per thousand.

10. At 25.00°C, the density and surface tension of water are 0.997044 g/ml and 71.97 dynes/cm, respectively. What actual values for these properties should be used with the data of Problem 9 to determine the radius of the capillary?

11. A sample of a copper alloy is to be analyzed for copper by first dissolving the sample in acid and then plating out the copper electrolytically. The weight of copper plated is to be measured on a balance that is sensitive to 0.1 mg. The alloy is approximately 5% copper. What size of sample should be taken for analysis so that the error in determining the weight of copper plated out does not exceed one part per thousand? (Remember that two weighings are needed in order to find the weight of copper.)

PROBLEMS B

12. State the number of significant figures in each of the following measurements. (a) 6822; (b) 6.822×10^{-3}; (c) 6.82; (d) 682; (e) 0.006820; (f) 6.82×10^6; (g) 0.0682; (h) 34 thousandths; (i) 167 million; (j) 62 hundredths.

13. Express the answer in each of the following calculations to the proper number of significant figures (assume that the numbers represent measurements).
    (a) $0.0657 + 23.77 + 5.369 + 0.0052$
    (b) $365.72 - 0.583$
    (c) $365.2 + 27.3 + 968.45 + 5.62$
    (d) $4.27 \times 10^{-5} + 1.05 \times 10^{-6} + 5 \times 10^{-6} + 1.234 \times 10^{-5}$
    (e) $\dfrac{65.4 \times 1.23 \times 464}{231}$
    (f) $\dfrac{6.55 \times 10^{-27} \times 2.045 \times 7.34 \times 10^5}{565 \times 432}$
    (g) $\dfrac{5280 \times (2885)^3 \times 6.570 \times 10^{-12}}{4.6295 \times 0.8888}$
    (h) $\dfrac{96.08 \times 4.712 \times 10^{-5} \times 7.308 \times 10^{-3}}{6.547 \times 10^{-27} \times 6.022 \times 10^{23}}$

14. Water analysts often report trace impurities in water as "parts per million"— that is, parts by weight of impurity per million parts by weight of water. A swimming pool whose dimensions are $20 \times 50 \times 9$ m has 14 lb of chlorine added as a disinfectant. How many parts per million of chlorine are present in this swimming pool? (Assume that the density of water is 1.00 g/ml.)

15. State the precision, in both percentage and parts per thousand, with which each of the following measurements is made.
    (a) 6822 with a standard deviation of 4.0
    (b) 6.822×10^{-3} with a standard deviation of 0.00040

(c) 6.82 with a standard deviation of 4.0×10^{-3}

(d) 682 with a standard deviation of 4.0

(e) 0.00682 with a standard deviation of 0.0000040

(f) 6.82×10^6 with a standard deviation of 4000

(g) 0.0682 with a standard deviation of 0.004

16. A radioactive sample shows the following counts for one-minute intervals: 3262; 3257; 3255; 3265; 3257; 3264; 3259. Calculate the average deviation and the 60% confidence interval for the mean.

17. A student wishes to calibrate a pipet by weighing the water it delivers. A succession of such determinations gives the following weights in grams: 4.993; 4.999; 4.991; 4.994; 4.995; 4.995. Find the average deviation, the standard deviation, and the 70% confidence interval for a single value.

18. If the viscosity of a given liquid is known, the viscosity of another may be determined by comparing the time required for equal volumes of the two liquids to pass through a capillary. To do this, a student makes the following observations of time intervals: 4 min 9.6 sec; 4 min 8.8 sec; 4 min 10.2 sec; 4 min 9.8 sec; 4 min 9.0 sec. Find the average deviation and the 90% confidence interval for the mean, expressing both in parts per thousand.

19. The time required for a given amount of gas to effuse through a pinhole under prescribed conditions is a measure of its molecular weight. A student making this determination observes the following effusion times: 1 min 37.3 sec; 1 min 38.5 sec; 1 min 36.9 sec; 1 min 37.2 sec; 1 min 36.5 sec; 1 min 38.7 sec; 1 min 37.0 sec. Find the average deviation and the 99% confidence interval for a single value, expressing both as percentages.

20. The surface tension of a liquid may be determined by measuring the height to which it will rise in a capillary of known radius. A student makes the following observations of capillary rise with an unusual liquid that he has just prepared in the laboratory: 63.2 mm; 63.5 mm; 62.9 mm; 62.8 mm; 63.7 mm; 63.4 mm. Find the average deviation of these heights and the 90% confidence interval for the mean, expressing both in parts per thousand.

21. How accurately should the values of liquid density and capillary radius be known if all of the figures in the measurements in Problem 20 are to be considered significant?

22. A graduated tube arranged to deliver variable volumes of a liquid is called a buret. If a buret can be read to the nearest 0.01 ml, what total volume should be withdrawn so that the volume will be known to a precision of 3 parts per thousand? (Remember that two readings of the buret must be made for every volume of liquid withdrawn.)

23. At some time or another, nearly everyone must decide whether to reject a suspicious-looking result or to include it in the average of all the other results. There is no agreement on what criteria should be used but, lacking information about errors made in the experimental procedure, a rejection may be made on the following *statistical* basis. If d and r are, respectively, the differences

between the questionable result and the values closest to it and farthest from it, then there is a 90% probability that the questionable result is grossly in error and should be rejected if

$$\text{for 5 values, } d/r > 0.64$$

$$\text{for 4 values, } d/r > 0.76$$

$$\text{for 3 values, } d/r > 0.94$$

Which, if any, result should be rejected from the following series of measurements?

(a) 9.35, 9.30, 9.48, 9.40, 9.28   (c) 2534, 2429, 2486

(b) 9.35, 9.30, 9.48, 9.32, 9.28   (d) 2534, 2429, 2520

6

Graphical Representation

Measurements made on chemical and physical systems often are presented in graphical form. The graph may show vividly how different variables are related to each other, or how the change in one variable affects the change in another. Another reason for graphical representation is that an important *derived* property can be obtained from, say, the slope or intercept of a straight-line graph. In this chapter, we discuss only those kinds of graphs and graph papers that are normally used at an elementary level in chemistry.

CONSTRUCTING A GRAPH

If the following guidelines are followed, the resulting graph generally will have maximal usefulness.

1. Use a scale large enough to cover as much as possible of the full page of graph paper. Do not cramp a miserable little graph into the corner of a large piece of graph paper.

2. Use a convenient scale, to simplify the plotting of the data and the reading of the graph.

3. Do not place the "zero" or origin of the coordinate system at one corner of the graph paper if doing so would make a small, cramped, inconvenient scale.

4. Label the axes (the vertical axis is the ordinate, the horizontal axis the abscissa) with both units and dimensions.

5. When possible and desirable, simplify the scale units in order to use simple figures. For example, if you wanted to plot as scale units 1000 min, 2000 min, 3000 min, 4000 min, etc., it would be simpler to use the figures 1, 2, 3, 4, etc. and then label the axis as min $\times$ 10^{-3}. Such labeling states that the *actual* figures (in minutes) have been multiplied by 10^{-3} in order to give the simple figures shown along the axis.

6. Draw a smooth curve that best represents all the points; such a curve may not necessarily pass through any of the points. Straight-line segments should not be drawn between consecutive points, unless there is a reason to believe that discontinuities (angles) in the curve really do occur at the experimental points; *such reasons almost never exist.*

PROPERTIES OF A STRAIGHT-LINE GRAPH

Whenever possible, cast data into such a form that a straight-line graph results from their plotting. A straight line is much easier to draw accurately than a curved one; often one can obtain important information from the slope or intercept of the straight line. If the two variables under discussion are x and y (the convention is to plot x as the abscissa and y as the ordinate), and if they are *linearly* related (i.e., if the graph is a straight line), the form of the mathematical equation that represents this line is

$$y = mx + b \qquad\qquad (6\text{-}1)$$

No matter what the value of m, when $x = 0$, then $y = b$. It is for this reason that b is called the "y intercept," the point at which the line intersects the y axis (see Figure 6-1).

If two arbitrary points (x_1, y_1) and (x_2, y_2) are selected from this line, both sets of points must satisfy the general equation for the line. Consequently, we have two specific equations:

$$y_2 = mx_2 + b \qquad\qquad (6\text{-}2)$$

and

$$y_1 = mx_1 + b \qquad\qquad (6\text{-}3)$$

If we subtract the second equation from the first, we obtain

$$y_2 - y_1 = mx_2 - mx_1 = m(x_2 - x_1) \qquad\qquad (6\text{-}4)$$

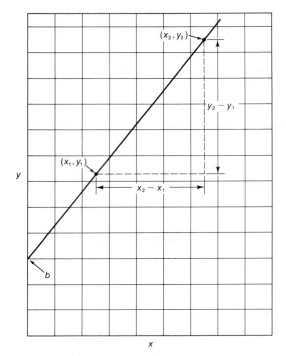

FIGURE 6-1

which, when rearranged, gives

$$m = \frac{y_2 - y_1}{x_2 - x_1} \tag{6-5}$$

By looking at Figure 6-1, you can see that $(y_2 - y_1)$ and $(x_2 - x_1)$ are the two sides of the right triangle made by connecting the two arbitrary points. The *ratio* of the sides, $\frac{y_2 - y_1}{x_2 - x_1}$ is the slope of the curve; this ratio has the same value whether the two arbitrary points are taken close together or far apart. It is for this reason that m is said to be the "slope" of the line.

PROBLEM:

The junction of two wires, each made of a different metal, constitutes a *thermocouple*. Some pairs of metals can generate a significant electrical voltage that varies substantially with the temperature of the junction. The following values of voltage (in millivolts) were observed at the temperatures (°C) shown for a junction of two alloys, chromel and alumel.

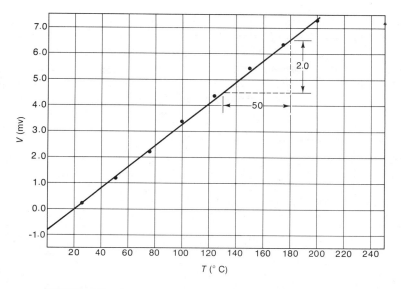

FIGURE 6-2

T (temperature, °C)	25.0	50.0	75.0	100	125	150	175	200
V (voltage, mv)	0.23	1.20	2.24	3.32	4.34	5.35	6.33	7.31

Plot these values of millivolts and temperature. From the resulting graph, determine the mathematical equation that describes V as a function of T.

SOLUTION:
The graph of these values of V and T is shown in Figure 6-2. Note that the best line passes through only one of the experimental points. In order to include the y intercept, one must take some negative values along the V axis. The value of the y intercept then can be taken directly from the graph; it is -0.80. The slope of the line is calculated by selecting two convenient points *on* this line, as shown, then taking the ratio of the two sides of the triangle formed from these points; in the figure, it is the ratio $2.0/50 = 0.040$. Knowing the slope and the y intercept, we can directly write the mathematical equation for the relationship between V and T as

$$V = 0.040T - 0.80$$

Students often select two of the *experimental* points in order to calculate the slope of a straight-line function; doing so usually is bad practice, because there are experimental errors inherent in the individual data points. The straight line that best represents *all* the points minimizes the experimental errors, and the slope calculated from this best straight line thus will (usually) be more reliable than one calculated from two randomly selected experimental points.

Frequently, in order to use a convenient scale and to construct a graph of a reasonable size, one must place the origin of the x-axis away from the lower left corner of the graph paper. When you have such a graph, you cannot obtain the value of b, the y intercept, directly from the graph. This situation need cause no difficulty, however, because m (the slope) still can be obtained from the graph. Once obtained, it can be used, along with any arbitrarily chosen point on the line, to *solve* for the value of b. The following problem illustrates this.

PROBLEM:
The following data show how the density D of mercury varies with the Kelvin temperature T. Plot these data. From the resulting graph, determine the mathematical equation that shows how D varies with T.

T (°K)	263	273	283	293	303	313
D (g/ml)	13.6201	13.5955	13.5708	13.5462	13.5217	13.4971

SOLUTION:
The graph of these values of D and T is shown in Figure 6-3. Note that a greatly expanded scale is needed along the y axis to show in a reasonable way the small changes in D with temperature; also note how far from absolute zero the scale along the x axis starts. The slope m is calculated by taking two convenient points *on* the line, as shown, then taking the ratio of the two sides of the triangle formed from these points; in the figure it is the ratio $0.0550/-22.5 = -0.00244 = m$. Note also that the sign of the slope is *negative,* as it must always be when the line slopes downward from left to right—that is, when y decreases with increasing values of x. Because the mathematical equation for this line must be of the form

$$D = mT + b$$

we can take any point on the line (300 and 13.530 for convenience) and use it with the value of $m = -0.00244$ to solve for the value of b:

$$b = D - mT$$
$$= 13.530 - (-0.00244)(300) = 13.530 + 0.732$$
$$= 14.262$$

The equation that relates D to T for mercury over this range of temperatures thus is

$$D = -0.00244T + 14.262$$

It often is desirable to recast data into a different form before making a graph, in order to obtain a straight-line graph. The data in Table 11-1 can be used to illustrate this procedure. With the data plotted just as they are given in the table (vapor pressure of water expressed in torr, and the corresponding temperature in °C), we obtain the graph shown in Figure 6-4.

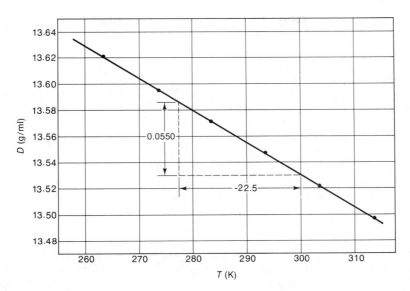

FIGURE 6-3

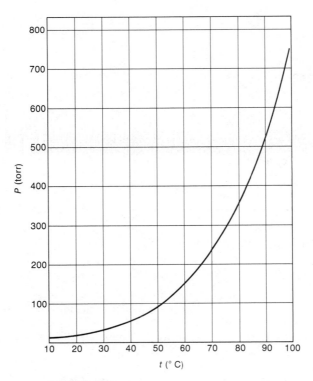

FIGURE 6-4

TABLE 6-1
Temperature and Pressure Data

t	P	1/T	log P
20	17.5	0.003412	1.243
30	31.8	0.003301	1.502
40	55.3	0.003195	1.743
50	92.5	0.003095	1.966
60	149.4	0.003004	2.174
70	233.7	0.002915	2.369
80	355.1	0.002832	2.550
90	525.8	0.002755	2.721
100	760	0.002681	2.881

We know from theoretical principles, however, (and your text may explain this) that the vapor pressure of a liquid is related to its heat of vaporization (ΔH_v), which is a physical constant characteristic of the liquid, to the gas constant ($R = 1.987$ cal/mole deg), and to the Kelvin temperature (T), by the equation

$$\log P = -\frac{\Delta H_v}{2.303R} \times \frac{1}{T} + B \qquad (6\text{-}6)$$

From this expression, it follows that a straight-line graph should be obtained if we let $y = \log P$ and $x = 1/T$; then the equation will be of the form

$$y = mx + b \qquad (6\text{-}1)$$

with the slope composed of a collection of constants, $m = -\dfrac{\Delta H_v}{2.303R}$, and the intercept, $b = B$.

PROBLEM:
Using the data of Table 11-1, plot a graph from which you can calculate the heat of vaporization of water, ΔH_v.

SOLUTION:
Table 6-1 shows the data given in Table 11-1, along with the values of $\log P$ and $1/T$ calculated from the data. If you need help with the logarithms, see pp 13–15. (K = °C + 273.2.)

Figure 6-5 shows the plot of $\log P$ against $1/T$. The slope of this straight-line graph is

$$m = \frac{0.500}{-0.000225} = -2.22 \times 10^3 \text{ deg}$$

Because $\qquad m = -\dfrac{\Delta H_v}{2.303R}$,

$$\Delta H_v = -(m)(2.303)(R) = -(-2.22 \times 10^3 \text{ deg})(2.303)\left(1.987 \frac{\text{cal}}{\text{mole deg}}\right)$$

$$= +1.02 \times 10^4 \text{ cal/mole}$$

If you also want to find B, then choose a simple pressure, such as $P = 100$ torr (so that $\log P = 2.000$) and the corresponding value of $1/T$ (it turns out to be 0.003075), and substitute them into the basic equation along with the calculated value of the slope (-2.22×10^3), to give:

$$\log P = -\frac{\Delta H_v}{2.303R} \times \frac{1}{T} + B$$

$$2.000 = (-2.22 \times 10^3)(0.003075) + B$$

$$B = 2.000 + 6.826 = 8.83$$

The general equation showing how the vapor pressure of water varies with temperature thus is

$$\log P = -\frac{2220}{T} + 8.83$$

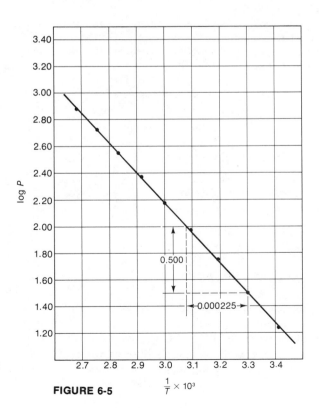

FIGURE 6-5 $\dfrac{1}{T} \times 10^3$

In a plot such as that in Figure 6-5, one can eliminate the extra labor of converting all the pressure values to the corresponding logarithms by using semilog graph paper, which has a logarithmic scale in one direction and the usual linear scale in the other. Semilog paper is available in one-cycle, two-cycle, or three-cycle forms (and so on), with each cycle able to handle a 10-fold spread of data. For example, one-cycle paper can handle data from 0.1 to 1.0, or 1.0 to 10.0, or 10^4 to 10^5, and so on. Two-cycle paper can handle data from 10^{-4} to 10^{-2}, or 0.1 to 10, or 10 to 1000, and so on. Always choose the paper with the smallest number of cycles that will do the job, in order to have the largest possible graph on the paper. The use of this kind of paper is illustrated in the following problem.

PROBLEM:
Calculate the heat of vaporization of water, using the data of Table 11-1 in a semilog plot.

SOLUTION:
It still is necessary to convert the Celsius temperature values to reciprocal degrees Kelvin, as in the previous problem, but the pressure values can be used directly, as shown in Figure 6-6.

At this point, take great care in calculating the slope of the line because, although a logarithmic *scale* was used, the *values* shown along the ordinate are *not* logarithmic values. In calculating the slope, therefore, select two convenient points on the line, but convert the two pressure values selected into their logarithms *before* making the calculations. For example, in Figure 6-6 when $1/T \times 10^3$ is 2.85, the value of P is 323, and $\log P$ is 2.509. Similarly, when $1/T \times 10^3$ is 3.00, P is 150, and $\log P$ is 2.176. Thus, the slope is given by

$$ m = \frac{2.176 - 2.509}{0.00300 - 0.00285} = -2.22 \times 10^3 $$

As before,

$$ \Delta H_v = -(m)(2.303)(R) = -(-2.22 \times 10^3)(2.303)(1.987) $$
$$ = +1.02 \times 10^4 \text{ cal/mole} $$

THE METHOD OF LEAST SQUARES

"Draw a smooth curve that best represents all the points; such a curve may not necessarily pass through any of the points." This is the sixth guideline given for curve construction on p 65. In all of the figures of this chapter, we have assumed that this has indeed been done. Nevertheless, curve drawing is a rather subjective process, prone to many abuses, unless some objective rules are

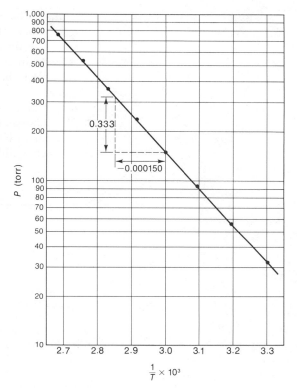

FIGURE 6-6

followed. There is universal acceptance of the "method of least squares" as the proper way to draw the best-fit line for a linear (straight-line) relationship. The ready availability of calculators and computers makes it feasible for almost anyone to use this method easily.

This method takes pairs of experimental points (x_1, y_1), (x_2, y_2), (x_3, y_3), ..., (x_n, y_n) and seeks to find the values of m (the slope) and b (the y intercept) that best represent the whole collection of pairs of points, referred to in general as (x_i, y_i), in the linear relationship

$$y = mx + b \tag{6-1}$$

If we substitute each value of x_i into this equation we can *calculate* what we might expect for the corresponding value of y_i—call it $(y_i)_{\text{calc}}$. For example,

$$(y_1)_{\text{calc}} = mx_1 + b \tag{6-7}$$

$$(y_2)_{\text{calc}} = mx_2 + b \tag{6-8}$$

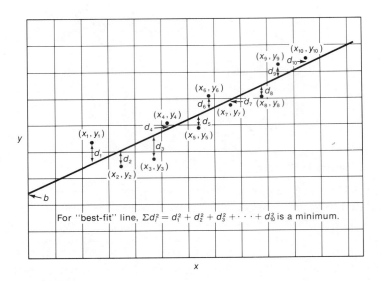

For "best-fit" line, $\Sigma d_i^2 = d_1^2 + d_2^2 + d_3^2 + \cdots + d_{10}^2$ is a minimum.

FIGURE 6-7

We could then compare the values of $(y_i)_{\text{calc}}$ with the corresponding *observed* values of y_i by taking the difference (d_i) between them, as follows:

$$d_i = (y_i)_{\text{calc}} - y_i \qquad (6\text{-}9)$$

or

$$d_i = (mx_i + b) - y_i \qquad (6\text{-}10)$$

If all of the experimental points lie on the line, then of course $(y_i)_{\text{calc}} = y_i$, and the differences (d_i) all equal zero. However, this rarely happens. A more typical situation is shown in Figure 6-7, where there is a fair scatter of points, and none of the differences equals zero.

The principle of least squares assumes that the "best-fit" line is the one for which the sum of the squares of the differences (d_i) is a minimum. Note that this assumption considers all experimental error to be associated with y and none to be associated with x. In finding a best-fit line, therefore, it is important to let x represent the variable that is known most accurately. The sum of the squares of the differences, taken for all values of i (from $i = 1$ up to and including $i = n$) is

$$\text{sum} = \Sigma d_i^2 = \Sigma(mx_i + b - y_i)^2 \qquad (6\text{-}11)$$

We expand the righthand side of Equation 6-11 and then treat it by the methods of calculus, so as to find the values of m and b that yield the *smallest* value for the sum, Σd_i^2. We thus obtain the two relations that we need.

$$\text{Slope} = m = \frac{\Sigma(x_i - \bar{x})(y_i - \bar{y})}{\Sigma(x_i - \bar{x})^2} \qquad \text{(6-12a)}$$

$$= \frac{n\Sigma x_i y_i - \Sigma x_i \Sigma y_i}{n\Sigma x_i^2 - (\Sigma x_i)^2} \qquad \text{(6-12b)}$$

$$y \text{ Intercept} = b = \bar{y} - m\bar{x} \qquad \text{(6-13a)}$$

$$= \frac{\Sigma y_i}{n} - m\frac{\Sigma x_i}{n} \qquad \text{(6-13b)}$$

Use of Calculators in the Method of Least Squares

Equations 6-12a and 6-13a show the slope (m) and intercept (b) in terms more clearly related to the raw data and to Figure 6-7, but Equations 6-12b and 6-13b are in a form that is more ideally suited for use with computers and many calculators. The "b" equations are related to the "a" equations in the same way that Equation 5-7 is related to Equation 5-3 for the calculation of standard deviation (s). Some calculators have built-in programs that require nothing more than the entry of x_i and y_i through the keyboard, followed by pressing the "least-square" keys. All programmable calculators can be arranged to accomplish the same thing. Nonprogrammable calculators must have at least four storage registers to accumulate each of the different kinds of sums, and then they must be operated with care. It is likely that in the near future most hand calculators will have built-in least-squares programs. If at all possible, you should always treat your data (to the extent they can be resolved into a linear form) by the method of least squares. The following problem illustrates the application of the method.

PROBLEM:
Using the data of the problem on p 67 that relates thermocouple voltages to temperature, find the equation of the best-fit line using the method of least squares. Draw a graph that shows the experimental points and the best-fit line.

SOLUTION:
Using your calculator, enter all of the data pairs (25, 0.23), (50, 1.20), and so on, considering T to be x and V to be y. If your calculator has a built-in program or is programmable, you will have to know the proper procedure for your brand of calculator—that is, which keys to press to enter the data pairs, and how to obtain a display of m and b. If your calculator doesn't have the aforementioned capability but does have at least four storage registers, you can use it to accumulate Σx_i, Σx_i^2, Σy_i, and $\Sigma x_i y_i$, and then use Equations 6-12b and 6-13b to calculate m and b. You will have to calculate m before b. Note that in this problem the number of data pairs (n) is 8. Note also that Σx_i^2 is *not* the same as $(\Sigma x_i)^2$. Your calculator will show that $m = 0.04074$ and $b = -0.794$, giving the equation for the best-fit line as

$$V = 0.04074T - 0.794$$

Note that the method of least squares will give you values of m and b more accurate than those you could obtain directly from the graph. Also, you now have a simple and unequivocal basis for drawing the best-fit line. The easiest approach is to locate two points accurately, then draw a line through them. For one point use b, the y intercept corresponding to $T = 0°C$. For the other point use some simple value of T, such as $100°C$, which corresponds to $V = 4.074 - 0.794 = 3.280$. Drawing the line through these two points, $(0, -0.794)$ and $(100, 3.280)$, will give the graph shown in Figure 6-2.

Usually it is a good idea to plot the experimental points first (but *not* draw a line that represents them) so that, if there should happen to be a really bad point that clearly doesn't represent the experiment (due probably to a serious experimental error), you could omit this point when using the method of least squares. You should still show the bad point on the graph, but with a note that it was not included in determining the equation of the best-fit line that is drawn.

Correlation Coefficient

Whenever you get the equation for the best-fit line, or draw its graphical representation, there is the question of how well the equation represents the data, or how good the correlation is between x and y. We can find the answer to this question as follows.

The choice of which variable is x and which is y is arbitrary but, in the method of least squares, whichever is chosen as y is assumed to possess all of the error, and x is assumed to possess none. If the variables were interchanged (that is, if x and y were plotted the other way around), this assumption would be reversed, and the slope of the line would be given by

$$\text{slope} = m' = \frac{\Sigma(x_i - \bar{x})(y_i - \bar{y})}{\Sigma(y_i - \bar{y})^2} \tag{6-14a}$$

$$= \frac{n\Sigma x_i y_i - \Sigma x_i \Sigma y_i}{n\Sigma y_i^2 - (\Sigma y_i)^2} \tag{6-14b}$$

Then, x intercept $= b' = \bar{x} - m'\bar{y}$ $\tag{6-15a}$

$$= \frac{\Sigma x_i}{n} - m' \frac{\Sigma y_i}{n} \tag{6-15b}$$

If there is a perfect correlation between x and y, then the slope of the best-fit plot of x versus y should be just the reciprocal of the slope of the best-fit plot of y versus x, and the product of the slopes (mm') should equal 1.00000000. The degree to which this product does not equal unity is considered to be the fraction of the variation in a set of measurements that can be explained by the linear dependence of one variable on another; it is called the *coefficient of determination, r^2.*

$$r^2 = mm' \tag{6-16}$$

The more commonly used term is r itself, which is called the *correlation coefficient:*

$$r = (mm')^{\frac{1}{2}} \tag{6-17}$$

Calculators that have built-in least-squares programs, and programmable calculators that perform the same function, always provide for the simultaneous calculation of r, because exactly the same sums are needed for its calculation as for the calculation of m and b. For a nonprogrammable calculator (with at least five storage registers), there would be the same effort to calculate m' as to find m; then Equation 6-17 would be used to calculate r.

PROBLEM:
Determine the correlation coefficient r for the best-fit equation obtained in the problem on p 75 involving thermocouple voltage versus temperature.

SOLUTION:
Work the problem exactly as you did previously, but with the additional knowledge of how to display r with your make of calculator, or with the program you use. If you use a nonprogrammable calculator with at least five storage registers, first accumulate Σx_i, Σx_i^2, Σy_i, Σy_i^2, and $\Sigma x_i y_i$. Then calculate m with Equation 6-12b, m' with Equation 6-14b, and finally r with Equation 6-17. The answer will be $r = 0.9999$, a very good correlation indeed.

Reliability of Slope and the y Intercept

It is a common practice to derive some important physical or chemical characteristics from the slope or the y intercept of a graph constructed from experimental data points. These data points have, of course, some error associated with them and, as a consequence, even the "best-fit" line must have some uncertainty associated with it. Just how good is a value derived from the slope and intercept of a best-fit line?

In the same way that one uses the variance and standard deviation to describe the scatter of points around their average, one can also use the variance and standard deviation to describe the scatter, in the vertical (y) direction, of the points about the best-fit line. Statisticians have shown that the variance is given by

$$s_{y/x} = \left[\frac{\Sigma(y_i - \bar{y})^2 - m^2\Sigma(x_i - \bar{x})^2}{n - 2} \right]^{\frac{1}{2}} \tag{6-18a}$$

$$= \left[\frac{n - 1}{n - 2} \right]^{\frac{1}{2}} (s_y^2 - m^2 s_x^2)^{\frac{1}{2}} \tag{6-18b}$$

where s_x^2 and s_y^2 are the variances of x and y, as calculated by Equation 5-2 in the last chapter. The variance $s_{y/x}$ is called the *standard error of estimate*.

Then, in the same way that one talks about a certain (percentage) confidence interval for a given series of measurements (see pp 54–57), one can also talk about the (percentage) confidence intervals for the slope and y intercept of a best-fit line. They are related to the standard error of estimate $s_{y/x}$ and, for the desired level of confidence, the t value that corresponds to *one less* than the number of data pairs, as follows. For the slope,

$$\text{the confidence interval is } m \pm \frac{ts_{y/x}}{s_x \sqrt{n - 1}} \tag{6-19}$$

For the y intercept,

$$\text{the confidence interval is } b \pm ts_{y/x} \left[\frac{1}{n} + \frac{\bar{x}^2}{(n - 1)s_x^2} \right]^{\frac{1}{2}} \tag{6-20}$$

PROBLEM:
Find the 95% confidence interval of the slope and intercept of the best-fit equation obtained in the problem on p 75 involving thermocouple voltage versus temperature.

SOLUTION:
You will need to use Equations 6-18b, 6-19, and 6-20, as well as Equation 5-7 from the last chapter. You will also need the values of $m = 0.04074$ and $b = 0.7936$ already obtained. Also, $n = 8$ and $\bar{x} = 112.50$. With Equation 5-7 you find that $s_x = 61.23724$ and $s_y = 2.49523$. Substitution into Equation 6-18b gives

$$s_{y/x} = \left[\frac{8 - 1}{8 - 2} \right]^{\frac{1}{2}} [(2.49523)^2 - (0.04074)^2(61.23724)^2]^{\frac{1}{2}}$$

$$= 0.038097$$

From Table 5-1, we find that $t = 2.447$ for $n = 7$ (*one less* than the number of data pairs) at the 95% confidence level. Then, using Equations 6-19 and 6-20, we find

$$95\% \text{ confidence interval of the slope} = 0.04074 \pm \frac{(2.447)(0.038097)}{(61.23724)(7)^{\frac{1}{2}}}$$

$$= 0.04074 \pm 0.000576 \frac{\text{volts}}{\text{deg}}$$

95% confidence interval of the y intercept

$$= -0.794 \pm (2.447)(0.038097)\left[\frac{1}{8} + \frac{(112.50)^2}{(7)(61.23724)^2}\right]^{\frac{1}{2}}$$

$$= -0.794 \pm 0.0727 \text{ volts}$$

Note: Problems 3 through 7 and 10 through 14 are, in part, given at two levels of sophistication and expectation. The "best" answers involve doing the part marked with an asterisk (*) *instead of* the immediately preceding part. The parts with an * involve the use of the method of least squares, correlation coefficients, and confidence intervals for the slopes and intercepts; the preceding parts do not.

PROBLEMS A

1. Construct graphs for each of the following functions, plotting q as ordinate and p as abscissa.
   (a) $q = 6p$ (d) $q = 10p^2 + 5$
   (b) $q = 6p + 10$ (e) $q = 5.0 \times 10^{-3}p$
   (c) $qp = 20$

2. Rearrange those equations in Problem 1 that do not give a straight-line relationship in such a way that, when plotted in a different fashion, they will yield a straight line. Plot each of these new equations, showing which function extends along each axis.

3. (a) Plot (as points) the solubility of Pb $(NO_3)_2$ in water as a function of temperature; the solubility S is given in g per 100 g of H_2O. The experimental data are the following.

t (°C)	20.0	40.0	60.0	80.0	100.0
S (g/100 g H_2O)	56.9	74.5	93.4	114.1	131.3

   (b) Draw a graph (add a line) on the plot made in (a), and determine the mathematical equation for S as a function of t.
   *(c) (*i*) Using the method of least squares, find the equation of the line that best fits the given data. (*ii*) Find the correlation coefficient and the 95% confidence intervals for both the slope and the y intercept of the best-fit equation. (*iii*) On the plot made in (a), draw the line that corresponds to the best-fit equation.

4. In colorimetric analysis, it is customary to use the fraction of light absorbed by a given dissolved substance as a measure of the concentration of the substance present in solution; monochromatic light must be used, and the length of the absorbing light path must be known or must always remain the same. The incident light intensity is I_0, and the transmitted light intensity is I; the fraction of light transmitted is I/I_0.
   (a) The following data are obtained for the absorption of light by MnO_4^- ion at a wavelength of 525 nm in a cell with a 1.00 cm light path. Construct a graph

of I/I_0 against C (as abscissa), where C is the concentration expressed in mg of Mn per 100 ml of solution.

C	1.00	2.00	3.00	4.00
I/I_0	0.418	0.149	0.058	0.026

(b) Replot the given data on semilog paper, again plotting concentration along the abscissa, but plotting I_0/I along the ordinate.

(c) From the graph made in (b), find the mathematical equation for the relationship between I_0/I and C.

*(d) (i) Using the method of least squares, find the equation of the best-fit line for the data as plotted in (b) above. (ii) Find the correlation coefficient and the 95% confidence interval for the slope of the best-fit equation. (iii) On the plot made in (b), draw the line that corresponds to the best-fit equation.

5. When radioactive isotopes disintegrate, they obey a rate law that may be expressed as $\log N = -kt + K$, where t is time, and N is the number of radioactive atoms (or something proportional to it) present at time t. A common way to describe the number of radioactive atoms present in a given sample is in terms of the number of disintegrations observed per minute on a Geiger counter (this is referred to as "the number of counts per minute," or simply as cpm).

(a) The following cpm are obtained for a sample of an unknown isotope at 10-minute intervals (beginning with $t = 0$): 10,000; 8166; 7583; 6464; 5381; 5023; 4466; 3622; 2981; 2690; 2239; 2141; 1775; 1603; 1348; 1114; 1048. Plot these data, using semilog graph paper.

(b) Construct a graph on the plot made in part (a), and from this graph evaluate the constants k and K to give the equation of the rate law.

*(c) (i) Using the method of least squares, find the rate-law equation of the best-fit line for the data as plotted in (a) above. (ii) Find the correlation coefficient and the 95% confidence intervals for both the slope and the y intercept of the best-fit equation. (iii) On the plot made in (a), draw the line that corresponds to the best-fit equation.

(d) What is the physical significance of K?

6. When the rates of chemical reaction are studied, it is common to determine the "rate constant" k, which is characteristic of a given reaction at a given temperature. Still further information can be obtained by determining the value of k at several different temperatures, because these values of k are related to the Kelvin temperature (T) at which they were measured according to the equation

$$\log k = - \frac{\Delta H_a}{2.3R} \times \frac{1}{T} + Q$$

In this equation ΔH_a is the so-called "energy of activation," Q is a constant characteristic of the reaction, and R is the gas constant of 1.987 cal mole^{-1} K^{-1}.

(a) The following data are obtained for the reaction

$$N_2O_5 \rightarrow 2\ NO_2 + \frac{1}{2} O_2$$

t (°C)	25.0	35.0	45.0	55.0	65.0
k	3.46×10^5	1.35×10^6	4.98×10^6	1.50×10^7	4.87×10^7

Plot these data on semilog graph paper.
   (b) Construct a graph on the plot made in (a), and determine the mathematical equation for the relationship between k and T.
*(c) (i) Using the method of least squares, find the equation of the best-fit line for the data as plotted in (a) above.   (ii) Find the correlation coefficient and the 95% confidence intervals for both the slope and the y intercept of the best-fit equation.   (iii) On the plot made in (a), draw the line that corresponds to the best-fit equation.
   (d) Calculate the value of the activation energy for this reaction.
   (e) What is the physical significance of the constant Q?

7. Regardless of how fast a chemical reaction takes place, it usually reaches an equilibrium position at which there appears to be no further change, because the reactants are being reformed from the products at the same rate at which they are reacting to form the products. This position of equilibrium commonly is characterized at a given temperature by a constant K_e, called the equilibrium constant (see Chapter 16). The equilibrium constant commonly is measured at several different temperatures for a given reaction, because these values of K_e are related to the Kelvin temperature (T) at which they are measured; the relationship is

$$\log K_e = -\frac{\Delta H}{2.3R} \times \frac{1}{T} + Z$$

In this equation, ΔH is the so-called "energy (or enthalpy) of reaction," Z is a constant characteristic of the reaction, and R is the ideal gas constant, 1.987 cal mole^{-1} K^{-1}.
   (a) The following data are obtained for the reaction

$$H_2 + I_2 \rightarrow 2\ HI$$

t (°C)	340	360	380	400	420	440	460
K_e	70.8	66.0	61.9	57.7	53.7	50.5	46.8

Plot these data on semilog graph paper.
   (b) Construct a graph on the plot made in (a), and determine the mathematical equation for the relationship between K_e and T.
*(c) (i) Using the method of least squares, find the equation of the best-fit line for the data as plotted in (a) above.   (ii) Find the correlation coefficient and the 95% confidence intervals for both the slope and the y intercept of the best-fit equation.   (iii) On the plot made in (a), draw the line that corresponds to the best-fit equation.

(d) Calculate the value of the energy of the reaction for this reaction.

(e) What is the physical significance of the constant Z?

PROBLEMS B

8. Construct graphs for each of the following functions, plotting m as ordinate and n as abscissa.

(a) $m = 2.5\pi n$

(b) $m = 20 - 4n$

(c) $mn = 1/4$

(d) $m = 4(n - 1)^2$

(e) $m = 4e^{-9.2n}$

9. Rearrange those equations in Problem 8 that do not give a straight-line relationship in such a way that, when plotted in a different fashion, they will yield a straight line. Plot each of these new equations, showing which function extends along each axis.

10. (a) Plot the solubility of K_2SO_4 in water as a function of temperature; the solubility S is given as g per 100 g of H_2O. The experimental data are the following.

t (°C)	20.0	40.0	60.0	80.0	100.0
S	9.8	14.2	18.3	21.6	25.9

(b) Draw a graph on the plot made in (a), and determine the mathematical equation for S as a function of t.

*(c) (*i*) Using the method of least squares, find the equation of the line that best fits the given data. (*ii*) Find the correlation coefficient and the 95% confidence intervals for both slope and the y intercept of the best-fit equation. (*iii*) On the plot made in (a), draw a line that corresponds to the best-fit equation.

11. (a) The nature of colorimetric analysis is described in Problem 4. With this in mind, use the following data (obtained for the absorption of light by CrO_4^{2-} ions at a wavelength of 3660 Å in a cell with a 1.00 cm light path) to construct a graph of I/I_0 against C (as abscissa), where C is the concentration of CrO_4^{2-} in moles of CrO_4^{2-} per liter of solution.

C	0.80×10^{-4}	1.20×10^{-4}	1.60×10^{-4}	2.00×10^{-4}
I/I_0	0.410	0.276	0.174	0.111

(b) Replot the given data on semilog paper, again plotting concentration along the abscissa, but plotting I_0/I along the ordinate.

(c) From the graph made in (b), find the mathematical equation for the relationship between I_0/I and C.

*(d) (*i*) Using the method of least squares, find the equation of the best-fit line for the data as plotted in (b) above. (*ii*) Find the correlation coefficient and the 95% confidence intervals for both the slope and the y intercept of

the best-fit equation. (*iii*) On the plot made in (b), draw the line that corresponds to the best-fit equation.

12. (a) The nature of the radioactive disintegration of unstable isotopes is described in Problem 5. Bearing this in mind, use the following data (obtained for a given sample of an unknown radioactive isotope, and taken at five-minute intervals, beginning with $t = 0$): 4500; 3703; 2895; 2304; 1507; 1198; 970; 752; 603; 496; 400; 309; 250; 199. Plot these data, using semilog graph paper.

   (b) Construct a graph on the plot made in (a), and from this graph evaluate the constants k and K to give the equation of the rate law.

   *(c) (*i*) Using the method of least squares, find the rate-law equation of the best-fit line for the data as plotted in (a) above. (*ii*) Find the correlation coefficient and the 95% confidence intervals for both the slope and the y intercept of the best-fit equation. (*iii*) On the plot made in (a), draw the line that corresponds to the best-fit equation.

   (d) What is the physical significance of K?

13. Pertinent statements about rate constants for chemical reactions are given in Problem 6. Keeping these statements in mind, use the following data obtained for the reaction

$$CO + NO_2 \rightarrow CO_2 + NO$$

t (°C)	267	319	365	402	454
k	0.00160	0.0210	0.120	0.630	2.70

   (a) Plot these data on semilog graph paper.

   (b) Construct a graph on the plot made in (a), and determine the mathematical equation for the relationship between k and T.

   *(c) (*i*) Using the method of least squares, find the equation of the best-fit line for the data as plotted in (a) above. (*ii*) Find the correlation coefficient and the 95% confidence intervals for both the slope and the y intercept of the best-fit equation. (*iii*) On the plot made in (a), draw the line that corresponds to the best-fit equation.

   (d) Calculate the value of the activation energy for this reaction.

   (e) What is the physical significance of the constant Q?

14. (a) Some fundamental statements are made about chemical equilibria in Problem 7. Keeping these statements in mind, use the following data for the reaction

$$H_2 + CO_2 \rightarrow CO + H_2O$$

t (°C)	600	700	800	900	1000
K_e	0.39	0.64	0.95	1.30	1.76

   Plot these data on semilog graph paper.

   (b) Construct a graph on the plot made in (a), and determine the mathematical equation for the relationship between K_e and T.

*(c) (*i*) Using the method of least squares, find the equation of the best-fit line for the data as plotted in (a) above. (*ii*) Find the correlation coefficient and the 95% confidence intervals for both the slope and the y intercept of the best-fit equation. (*iii*) On the plot made in (a), draw the line that corresponds to the best-fit equation.

(d) Calculate the value of the "energy of reaction" for this reaction.

(e) What is the physical significance of the constant Z?

Density and Buoyancy

Either directly or indirectly, the concept of density plays an important role in a myriad of scientific operations: construction of equipment, preparation of solutions, determination of volumes, accurate weighings, measuring buoyancy of objects, studying properties of gases, and so on. Density is defined as the mass per unit volume, or

$$\text{density} = \frac{\text{mass}}{\text{volume}} \tag{7-1}$$

In scientific work, the densities of solids and liquids usually are expressed in grams per cubic centimeter or grams per milliliter, whereas the densities of gases usually are expressed in grams per liter. In engineering work, densities customarily are expressed in pounds per cubic foot.

DENSITY OF A LIQUID

The simplest way to determine the density of a liquid is to weigh an empty vessel of known volume and then weigh it again when it is filled with the liquid. An approximate value may be determined with a simple graduated cylinder weighed on a triple-beam balance. Only a crude value can be obtained because the balance can be read only to the nearest 0.1 g and the cylinder only to the nearest 0.1 ml.

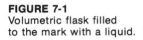

FIGURE 7-1
Volumetric flask filled
to the mark with a liquid.

PROBLEM:

A clean dry 10-ml graduated cylinder weighs 37.6 g empty; it weighs 53.2 g when filled to the 7.4 ml mark with an unknown liquid. Calculate the density of the liquid.

SOLUTION:

The weight of the liquid is 53.2 g − 37.6 g = 15.6 g.
The volume of the liquid is 7.4 ml.

The density of the liquid $= \dfrac{15.6 \text{ g}}{7.4 \text{ ml}} = 2.1 \dfrac{\text{g}}{\text{ml}}$.

A more accurate value, using the same method, involves a volumetric flask as the vessel and an analytical balance for the weighings. The volumetric flask has a narrow neck that makes accurate measurement easy; liquid is added until the bottom of the curved liquid surface (the meniscus) appears to just touch the mark that is etched on the neck (Figure 7-1).

PROBLEM:

A clean dry volumetric flask, known to contain exactly 10.000 ml when it is properly filled to the mark, weighs 12.754 g when empty. When filled to the mark with a liquid at 23°C, it weighs 33.671 g. Calculate the density of the liquid.

SOLUTION:

$$\text{Density} = \frac{33.671 \text{ g} - 12.754 \text{ g}}{10.000 \text{ ml}} = 2.092 \, \frac{\text{g}}{\text{ml}} \text{ at } 23°C$$

TABLE 7-1
Density of Water at Various
Temperatures

t (°C)	Density (g/ml)	t (°C)	Density (g/ml)
15	0.9991	23	0.9975
16	0.9989	24	0.9973
17	0.9988	25	0.9970
18	0.9986	26	0.9968
19	0.9984	27	0.9965
20	0.9982	28	0.9962
21	0.9980	29	0.9959
22	0.9978	30	0.9956

The previous problem draws attention to an important property that must be taken into account for accurate measurements. Almost every material expands with an increase in temperature. In the last problem, if the measurement had been made at 25°C (instead of 23°C), the liquid would have expanded, and a smaller amount (weight) would have been required to adjust the meniscus to the mark. The flask, being a solid, would have undergone a negligible expansion, so its volume remains unchanged. As a result, the measured liquid density would be smaller at the higher temperature. For this reason, it is always necessary to report the temperature at which an accurate density measurement is made. The most common liquid, water, has had its density measured with great accuracy over its entire liquid range. Table 7-1 gives a few values for the density of water near room temperature.

Once the density of a liquid, such as water, is known with great accuracy as a function of temperature, it provides a very useful means of determining the accurate volumes of vessels. Volumetric flasks are purchased with a *nominal* (approximate) value of the volume printed on their walls. The *accurate* volume can be determined by calibration with water, as illustrated in the next problem. Once calibrated, the flask can be used over and over again for other accurate measurements.

PROBLEM:
A 25 ml volumetric flask is calibrated by weighing it filled to the mark with distilled water at 26°C; it weighs 48.4636 g. When empty and dry, the flask weighs 23.5671 g. Assume that the weights have been corrected for buoyancy (see pp 92–95). Determine the accurate volume of the flask.

SOLUTION:

$$\text{Volume of flask} = \text{volume of water} = \frac{\text{mass of water}}{\text{density of water at 26°C}}$$

Look up the density of water at 26°C in Table 7-1; it is 0.9968 g/ml. Then substitute, to give

$$\text{volume} = \frac{48.4636 \text{ g} - 23.5671 \text{ g}}{0.9968 \frac{\text{g}}{\text{ml}}}$$

$$= 24.976 \text{ ml}$$

DENSITY OF A GAS

The density of gases can be determined experimentally in the same way as for liquids, but the values so obtained are not particularly reliable because the total weight of gas in a flask of a reasonable size for weighing is very small—quite likely less than 100 mg. Furthermore, a very special flask would have to be used that would be absolutely leakproof and capable either of being flushed out with the gas to be studied, or of being evacuated before filling. Moreover, provision would have to be made for measuring the pressure of the gas in the flask as well as its temperature.

If the molecular weight of a gas is known, it is possible to *calculate* its density for a given temperature and pressure using principles described in Chapter 11. We discuss gas densities in that chapter.

DENSITY OF A SOLID

If a solid has a regular geometric form, the density may be computed from its weight and volume.

PROBLEM:
A cylindrical rod weighing 45.0 g is 2.00 cm in diameter and 15.0 cm in length. Find the density.

SOLUTION:
Mass = 45.0 g.
Volume = $\pi r^2 L$ = (3.14)(1.00 cm)^2(15.0 cm) = 47.1 cm^3.
Density = $\dfrac{\text{mass}}{\text{volume}}$ = $\dfrac{45.0 \text{ g}}{47.1 \text{ cm}^3}$ = 0.955 g/cm^3 or 0.955 $\dfrac{\text{g}}{\text{ml}}$.

When a solid is irregular in shape, it seldom is convenient or possible to find its volume by measurement of its dimensions. A convenient procedure is to immerse the object in a liquid, and then to determine its volume by measuring the volume of the liquid it displaces, taking care that air bubbles don't cling to the solid's surface and also displace liquid. The method will work only if the solid is insoluble in the liquid used. An approximate value, using a graduated cylinder, may be found as follows.

PROBLEM:
A 5.7 g sample of metal pellets is put into a graduated cylinder that contains 5.0 ml of water. After the pellets are added, the water level stands at 7.7 ml. Find the density of the pellets.

SOLUTION:
Mass of pellets = 5.7 g.
Volume of pellets = volume of water displaced = 7.7 − 5.0 = 2.7 ml.

$$\text{Density of pellets} = \frac{\text{mass}}{\text{volume}} = \frac{5.7 \text{ g}}{2.7 \text{ ml}} = 2.1 \ \frac{\text{g}}{\text{ml}}.$$

A variation of the previous method that gives much more accurate values utilizes a volumetric flask and an analytical balance, as illustrated in the following problem.

PROBLEM:
A volumetric flask was weighed empty, then filled with water to the mark and reweighed. After the flask was emptied and dried, some solid sample was added and the flask was weighed again. Finally, water was added to the sample in the flask until the meniscus was again at the mark, and the flask was weighed once again. The following data were obtained with this method (assume the weights to be corrected for buoyancy as on pp 92–95). Calculate the density of the solid.

    A. Weight of empty flask = 24.3251 g.

    B. Weight of flask filled to mark with water at 23°C = 74.2613 g.

    C. Weight of flask + sample = 55.7884 g.

    D. Weight of flask + sample + water (at 23°C) to the mark = 101.9931 g

SOLUTION:
To find the density of the sample, we must calculate its mass and its volume.

$$\text{Mass of sample} = 55.7884 \text{ g} - 24.3251 \text{ g} = 31.4633 \text{ g}$$

To find the volume of the sample, we first find the *weight* of the water displaced by the sample, then convert this weight to volume using the density of water at 23°C from Table 7-1. The volume of water displaced is just equal to the volume of the sample.

$$\text{Weight of water in flask (with } no \text{ sample)} = 74.2613 \text{ g} - 24.3251 \text{ g}$$
$$= 49.9362 \text{ g}$$
$$\text{Weight of water in flask (} with \text{ sample)} = 101.9931 \text{ g} - 55.7884 \text{ g}$$
$$= 46.2047 \text{ g}$$
$$Weight \text{ of water displaced by sample} = 49.9362 \text{ g} - 46.2047 \text{ g}$$
$$= 3.7315 \text{ g}$$
$$Volume \text{ of sample} = volume \text{ of water displaced} = \frac{3.7315 \text{ g}}{0.9975 \frac{\text{g}}{\text{ml}}}$$
$$= 3.7409 \text{ ml}$$
$$\text{Density of sample} = \frac{31.4633 \text{ g}}{3.7409 \text{ ml}} = 8.4106 \frac{\text{g}}{\text{ml}}$$

Five significant figures are justified on the basis of the one set of data given, but repeated measurements would show that you would be unable to reproducibly adjust the water meniscus with four-place accuracy, so a density value of 8.411 g/ml would be more reasonable.

BUOYANCY

Archimedes' Principle

There is still another method by which the density of insoluble solids can be determined. It is based on an ancient principle known as Archimedes' principle: *when an object is suspended in a fluid, it* APPEARS *to lose weight equal to the weight of the fluid displaced.* We say that the object is "buoyed up," and that the "buoyancy" is equal to the apparent weight loss. In equation form, Archimedes' principle could be stated as

(wt of object in air) − (apparent wt of object in fluid) = (wt of fluid displaced)

Another principle, which we might call the "common sense principle" for immersed objects, is one we've used in the last two problems:

(the volume of the object) = (the volume of fluid displaced)

The application of Archimedes' principle to the determination of densities of liquids and solids is illustrated in the next two problems.

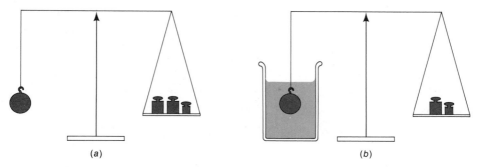

FIGURE 7-2
(a) Metal object weighs 25.0 g in air. **(b)** Same metal object appears
to weigh only 15.0 g when suspended in water.

PROBLEM:
A metal object suspended by a very fine wire from the arm of a balance weighs
25.0 g in air, but when it is suspended in water it appears to weigh only 15.0 g
(Figure 7-2). Find the density of the metal.

SOLUTION:

$$\text{Weight of metal object in air} = 25.0 \text{ g}$$

$$\text{Weight of metal object in water} = 15.0 \text{ g}$$

$$\text{Buoyancy} = \textit{apparent} \text{ weight loss} = 10.0 \text{ g}$$

By Archimedes' principle,

$$\text{weight of water displaced} = \text{apparent weight loss} = 10.0 \text{ g}$$

By the common sense principle,

$$\text{volume of object} = \text{volume of water displaced}$$

$$= \frac{\text{wt of water displaced}}{\text{density of water}}$$

$$= \frac{10.0 \text{ g}}{1.00 \ \frac{\text{g}}{\text{ml}}} = 10.0 \text{ ml}$$

We use 1.00 g/ml for the density of water here because the weighings were done
only to the nearest 0.1 g. More accurate weighings would have justified the use of
Table 7-1.

$$\text{Density of metal object} = \frac{\text{mass}}{\text{volume}} = \frac{25.0 \text{ g}}{10.0 \text{ ml}} = 2.50 \ \frac{\text{g}}{\text{ml}}$$

PROBLEM:

For routine liquid-density determinations, a glass bob fastened to a fine platinum wire is available for hanging on the end of an analytical balance arm. It can easily be weighed while suspended in a liquid. The glass bob has a density of 2.356 g/ml, and it weighs 11.780 g in air. When suspended in a liquid of unknown density, the bob appears to weigh only 7.530 g. Calculate the density of the liquid.

SOLUTION:

$$\text{Volume of the bob} = \frac{11.780 \text{ g}}{2.356 \text{ g}} = 5.000 \text{ ml}$$

$$\textit{Apparent} \text{ wt loss of bob} = 11.780 \text{ g} - 7.530 \text{ g} = 4.250 \text{ g}$$

By Archimedes' principle,

$$\text{wt of liquid displaced} = \text{apparent wt loss} = 4.250 \text{ g}$$

By the common sense principle,

$$\text{volume of liquid displaced} = \text{volume of bob} = 5.000 \text{ ml}$$

$$\text{Density of liquid} = \frac{\text{mass}}{\text{volume}} = \frac{4.250 \text{ g}}{5.000 \text{ ml}} = 0.850 \; \frac{\text{g}}{\text{ml}}$$

Buoyancy Correction for Weighing in Air

Air too is a fluid that exerts a small buoyant effect on any object it surrounds. At ordinary conditions the density of air is 1.2×10^{-3} g/ml, and in very accurate weighings it is necessary to take into account the buoyant effect of the displaced air; that is, we must calculate what the weight of an object would have been had the weighing been done *in vacuo* where there would be no buoyant effect. We can make this correction, which involves both the object and the weights, as follows (see Figure 7-3).

When a two-pan balance is "balanced," the total torque on the balance is zero, and you can set the clockwise torque equal to the counterclockwise torque because the lever principle,

$$F_1 \times L_1 = F_2 \times L_2 \tag{7-2}$$

states that at equilibrium, force #1 (F_1) times its distance (L_1) from the fulcrum is exactly equal to force #2 (F_2) times its distance (L_2) from the fulcrum. F_1 and F_2 are the products of the acceleration of gravity (g) and the effective masses (M_1 and M_2) at each pan, so that Equation 7-2 becomes

$$M_1gL_1 = M_2gL_2 \tag{7-3}$$

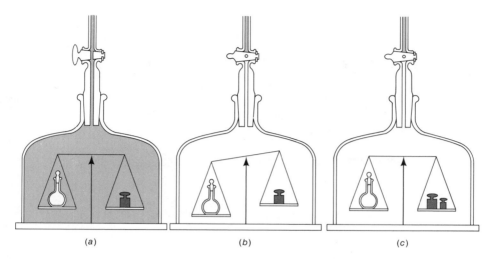

FIGURE 7-3
(a) A sealed flask whose volume is 100.0 ml weighs 100.0 g when weighed in air. Both the weights (volume = 12.5 ml) and the flask are buoyed up by the air. **(b)** When the air is pumped out, a greater buoyant support is withdrawn from the flask than from the weights, because the flask had previously displaced a larger volume of air; the position of balance is therefore lost and the left pan drops down. **(c)** The position of balance in the vacuum is restored by adding 0.105 g to the right pan. This increase in weight is equal to the difference between the two buoyant forces due to the air in part a. The *true* weight of the flask is 100.105 g; the *buoyancy correction factor* is 1.001050.

For a two-pan balance, the lever arms are equal ($L_1 = L_2$), the pan weights are equal, and the effect of gravity cancels, so that "at balance" the situation simply reduces to the fact that the effective masses of the object (o) and the weights (w) are equal. That is,

$$(M_o)_{\text{eff}} = (M_w)_{\text{eff}} \qquad (7\text{-}4)$$

In each case, the *effective* mass is the true mass (M_o or M_w, corresponding to weighing *in vacuo*) minus the buoyance (B_o for object, and B_w for weights) due to the mass of the air displaced. The true mass of the weights (M_w) is always known because this information is supplied by the manufacturer. Equation 7-4 can be rewritten as

$$M_o - B_o = M_w - B_w \qquad (7\text{-}5)$$

The buoyancies (the masses of the air displaced) are given by the products of the respective volumes and the density of air (d_a):

$$B_o = V_o d_a = \left(\frac{M_o}{d_o}\right) d_a \qquad (7\text{-}6)$$

$$B_w = V_w d_a = \left(\frac{M_w}{d_w}\right) d_a \tag{7-7}$$

We can substitute Equations 7-6 and 7-7 into Equation 7-5 and solve for M_o, the true mass of the object, to obtain

$$M_o = M_w \left(1 - \frac{d_a}{d_w}\right)\left(1 - \frac{d_a}{d_o}\right)^{-1} \tag{7-8}$$

A useful approximation can be made by expressing

$$\left(1 - \frac{d_a}{d_o}\right)^{-1}$$

as the series

$$\left[1 + \frac{d_a}{d_o} + \left(\frac{d_a}{d_o}\right)^2 + \left(\frac{d_a}{d_o}\right)^3 + \cdots\right]$$

and then, after multiplying by

$$\left(1 - \frac{d_a}{d_o}\right)$$

neglecting all those terms that possess d_a^2, d_a^3, d_a^4, and so on. This approximation is justified because d_a is so small (about $1.5 \times 10^{-4} \times d_w$) that these terms will be negligible compared to all the others. This approximation yields the simple formula

$$M_o = M_w \left[1 + d_a \left(\frac{1}{d_o} - \frac{1}{d_w}\right)\right] \tag{7-9}$$

This same expression can be derived for single-pan balances that use the method of substitution of weights. The fact that the lever arms are unequal and that there is a constant load on the balance does not alter the final expression.

The factor

$$\left[1 + d_a \left(\frac{1}{d_o} - \frac{1}{d_w}\right)\right]$$

in Equation 7-9 is called the *buoyancy correction factor*. It is a number—involving only the densities of air, the weights, and the object—by which you multiply the sum of the observed weights $(M)_w$ in order to get the true weight of the object

(M_o). You can see that, if the object and the weights had the same density, they would have the same volume; each would be buoyed by the same amount, and the correction factor would be 1.0000000 (that is, there would be no correction). If there is a big difference in densities of weights and object (as there is in the case of water as the object), then the correction is significant, as illustrated in the following problem.

PROBLEM:
A sample of water at 20°C is weighed in air with brass weights and found to weigh 99.8365 g. Calculate the true mass of the water. The density of brass is 8.0 g/ml and the density of air is 1.2×10^{-3} g/ml.

SOLUTION:

$$\text{The buoyancy correction factor} = 1 + d_a \left(\frac{1}{d_o} - \frac{1}{d_w} \right)$$

$$= 1 + (1.2 \times 10^{-3}) \left(\frac{1}{1.0} - \frac{1}{8.0} \right)$$

$$= 1.001050$$

The true weight $= M_o = M_w (1.001050)$

$$= (99.8365 \text{ g})(1.001050) = 99.9413 \text{ g}$$

There is no point in using the very accurate density for water from Table 7-1 because the other densities are given only to two significant figures; furthermore, the slight correction has almost no effect on the value of the factor or the true weight. If we used $d_o = 0.9982$ g/ml, the buoyancy factor would be 1.001052 and the true weight would be 99.9415 g. The error is only two parts in a million.

A huge fraction of all weighing is done by difference—that is, the weight of the object is found as the difference between the weight of the empty container and the weight of the container with the object. There is no point in making a buoyancy correction to both weighings, because the error in container weight is the same both times and cancels out when one weight is subtracted from the other. A buoyancy correction need be applied only to the "difference"—that is, only to the object itself.

If the object being weighed is small, and of some significant density, the need for buoyancy correction vanishes. The more nearly d_o is like d_w, the less important is buoyancy. For example, if you wanted to know the true weight of 0.5000 g AgCl weighed in air with brass weights ($d_w = 8.0$ g/ml), you would look up the density of AgCl in a handbook (it is 5.56 g/ml) and calculate the true value as

$$\text{true wt of AgCl} = (0.5000 \text{ g}) \left[1 + (1.2 \times 10^{-3}) \left(\frac{1}{5.56} - \frac{1}{8.0} \right) \right]$$

$$= 0.500033 \text{ g}$$

In reality there is no correction, because you could weigh only to four decimals in the first place; the true weight is the observed weight of 0.5000 g.

Finally, if the object has a greater density than the weights ($d_o > d_w$), the buoyancy correction factor will be *less* than 1.000000, because the smaller volume of the object displaces a smaller mass of air than the weights. Platinum, gold, and lead are metals with appreciably greater density than brass.

Calibration of Volumetric Flasks

The accurate calibration of volumetric glassware must also take buoyancy into account. For example, in the previous problem, if the observed weight of water is that needed to fill a 100 ml volumetric flask exactly to the mark, we could easily calculate the *true volume* of the volumetric flask just as we did on p 86:

$$\text{true volume} = \frac{\text{true mass}}{\text{density of water at } 20°C} = \frac{99.9415 \text{ g}}{0.9982 \frac{\text{g}}{\text{ml}}} = 100.12 \text{ ml}$$

If we had not made a buoyancy correction, we would have thought the volume to be $\frac{99.8365 \text{ g}}{0.9982 \text{ g/ml}} = 100.02$ ml; the error is 0.1%. Because the true volume of the flask is 100.12 ml, and the nominal volume is 100.00 ml, we would say that the calibration correction is +0.12 ml; that is, we must add 0.12 ml to the nominal value to get the true volume.

PROBLEMS A

Show units in all calculations.

1. The density of mercury is 13.54 g/ml. How many milliliters of mercury are needed to weigh 454 g?

2. The density of a sulfuric acid solution is 1.540 g/ml. How much does 1 liter of this solution weigh?

3. If 17.5 g of brass filings (density = 8.0 g/ml) are put into a dry 10 ml graduated cylinder, what volume of water is needed to complete the filling of the cylinder to the 10 ml mark?

4. A ring weighing 7.3256 g in air weighs 6.9465 g when suspended in water at 24°C. Is the ring made of gold (density = 19.3 g/ml) or brass (density = 8.0 g/ml)?

5. Find the weight of 1 cu ft of air at 21°C, assuming a density of 0.00120 g/ml at this temperature.

6. A glass bulb with a stopcock weighs 54.9762 g when evacuated, and 54.9845 g when filled with a gas at 25°C. The bulb will hold 50.0 ml of water. What is the density of the gas at 25°C?

7. Mercury (density = 13.54 g/ml) is sold by the "flask," which holds 76 lb of mercury. If the cost is $1130 per flask, how much does 1 ml of mercury cost?

8. (a) A column of mercury (density = 13.54 g/ml) 730 mm high in a tube of 8 mm inside diameter is needed to balance a gas pressure. What weight of mercury is in the tube?
   (b) If the same gas pressure were balanced by mercury in a tube of 16 mm inside diameter, what weight of mercury would be needed?

9. In normal whole blood there are about 5.4×10^9 red cells per milliliter. The volume of a red cell is about 90 μm^3, and the density of a red cell is 1.096 g/ml. How many pints of whole blood would we need in order to collect 8 oz (avoirdupois) of red cells?

10. How far (in centimeters) does a 1 cm cube of wood stick out of the water if its density is 0.85 g/ml?

11. A piece of Invar (density = 8.00 g/ml) weighs 15.4726 g in air and 13.9213 g when suspended in liquid nitrogen at a temperature of −196°C. What is the density of liquid nitrogen at that temperature? (Invar has a very small coefficient of thermal expansion, and its change in density with temperature may be neglected in this problem.)

12. The weight of a metal sample is measured by finding the increase in weight of a volumetric flask when the metal sample is placed in it. The volume of the metal sample is measured by finding how much less water the volumetric flask holds when it contains the metal sample. Compute the density of the metal sample from the following data, assuming that the density of water is 1.000 g/ml.

$$\text{Weight of empty flask} = 26.735 \text{ g}$$
$$\text{Weight of flask} + \text{sample} = 47.806 \text{ g}$$
$$\text{Weight of flask} + \text{sample} + \text{water} = 65.408 \text{ g}$$
$$\text{Weight of flask} + \text{water (no sample)} = 50.987 \text{ g}$$

13. (a) Repeat the calculations of Problem 12, using an accurate density value for water (Table 7-1) and assuming the measurements were made at 27°C.
   (b) In which decimal place does it make a significant difference if the accurate density for water is used instead of 1 g/ml?

14. A chemical is soluble in water but insoluble in benzene. As a consequence, benzene may be used to determine its density. The density of benzene is 0.879 g/ml at 20°C. From the following data (obtained as in Problem 12), compute the density of the sample.

Weight of empty flask = 31.862 g

Weight of flask + sample = 56.986 g

Weight of flask + sample + benzene to fill = 75.086 g

Weight of flask + benzene (without sample) = 52.175 g

15. A sample of benzene (D = 0.88 g/ml) weighs 25.3728 g in air with brass weights. What is the weight *in vacuo*?

16. A platinum crucible (D = 21.5 g/ml) weighs 56.3724 g in air with brass weights. What is the weight *in vacuo*?

17. A brass sample weighs 16.3428 g in air with brass weights. What is the weight *in vacuo*?

18. The density of water at 25°C is 0.9970 g/ml. What volume is occupied by 1.0000 g of water weighed in air with brass weights?

19. A chemical of density 2.50 g/ml is weighed in air with brass weights. The observed weight is 0.2547 g. Show by calculation that it is not necessary to make a correction of the weight to *vacuo* if the weighing is reliable only to the nearest 0.2 mg.

20. What weight of water at 25°C, weighed in air with brass weights, should be delivered by a 25.00 ml pipet if it is accurately graduated?

21. A spherical balloon 100 ft in diameter is filled with a gas having a density one-fifth that of air. The density of air is 1.20 g/liter. How many pounds, including its own weight, can the balloon lift? (The lift is the difference between the weight of the gas and that of an equal volume of air.)

PROBLEMS B

22. The density of a sodium hydroxide solution is 1.1589 g/ml. How much does 1 liter of this solution weigh?

23. The density of carbon tetrachloride is 1.595 g/ml. How many milliliters of carbon tetrachloride are needed to give 500.0 g?

24. The density of benzene is 0.879 g/ml. How many grams of benzene will be needed to fill a 25 ml graduated cylinder?

25. A glass bulb with a stopcock weighs 66.3915 g evacuated, and 66.6539 g when filled with xenon gas at 25°C. The bulb holds 50.0 ml of water. What is the density of xenon at 25°C?

26. If 20 g of magnalium lathe turnings of density 2.50 g/ml are put into a 25 ml graduated cylinder, what volume of water will be needed to complete the filling of the cylinder to the 25 ml mark?

27. What is the weight (in pounds) of 1 cu ft of aluminum (density = 2.70 g/ml)?

28. The density of Dowmetal, a magnesium alloy, is 1.78 g/ml. Find the weight (in grams) of a rod $\frac{1}{2}$ inch in diameter and 2 ft long.

29. Neptunium has a density of 17.7 g/ml. What would be the radius of a sphere of neptunium that weighed 500 g?

30. Gallium (density = 3.01 g/ml) may be purchased at the rate of $6.90 per gram. How much does 1 cu ft of gallium cost?

31. When 235 g of uranium-235 disintegrates by nuclear fission, 0.205 g is converted into 4.4×10^{12} calories (4.4 million million calories!) of energy. What volume of uranium is converted to energy if its density is 18.9 g/ml?

32. (a) Find a general factor by which pounds per cubic foot could be multiplied to convert to density in grams per milliliter.
    (b) Find a factor to convert grams per milliliter to pounds per cubic foot.

33. A metal earring weighs 2.6321 g when suspended in air. When immersed in water at 22°C, it weighs 2.3802 g. What is the density of the earring?

34. A thousandth of a milliliter is called a lambda (λ). A certain biochemical procedure calls for the addition of 5 λ of a 2.00% solution of sodium chloride (density = 1.012 g/ml). How many milligrams of sodium chloride will be added from the 5 λ pipet?

35. The weight of a metal sample is measured by finding the increase in weight of a volumetric flask when the metal sample is placed in it. The volume of the metal sample is measured by finding how much less water the volumetric flask holds when it contains the metal sample. Compute the density of the metal sample from the following data, assuming that the density of the water is 1.000 g/ml.

$$\text{Weight of empty flask} = 23.482 \text{ g}$$

$$\text{Weight of flask} + \text{sample} = 40.375 \text{ g}$$

$$\text{Weight of flask} + \text{sample} + \text{water} = 63.395 \text{ g}$$

$$\text{Weight of flask} + \text{water (without sample)} = 48.008 \text{ g}$$

36. (a) Repeat the calculations of Problem 35, using an exact density value for water (refer to Table 7-1) and assuming a temperature of 68°F.
    (b) In which decimal place does it make a significant difference if the exact density for water is used instead of the approximate value of 1 g/ml?

37. A chemical is soluble in water but insoluble in kerosene. As a consequence, kerosene may be used to determine its density. The density of kerosene is 0.735 g/ml at 22°C. From the following data (obtained as in Problem 35), compute the density of the sample.

$$\text{Weight of empty flask} = 28.176 \text{ g}$$

$$\text{Weight of flask} + \text{sample} = 40.247 \text{ g}$$

$$\text{Weight of flask} + \text{sample} + \text{kerosene to fill} = 51.805 \text{ g}$$

$$\text{Weight of flask} + \text{kerosene (without sample)} = 45.792 \text{ g}$$

38. A sample of liquid of density 0.85 g/ml weighs 32.3524 g in air with brass weights. What is the weight *in vacuo?*

39. A platinum dish (D = 21.5 g/ml) weighs 65.2364 g in air with brass weights. What is the weight *in vacuo?*

40. A crude brass weight of nominal value 100 g weighs 99.9986 g in air, with brass weights. What is its true weight *in vacuo?*

41. A buret is calibrated by filling with water to the zero mark, withdrawing to the 25 ml mark, and weighing the water withdrawn. From the following data, compute the correction that must be applied at the 25 ml mark (similar corrections at other intervals may be used to construct a correction curve for the entire range of the buret).

$$\text{Final reading} = 25.00 \text{ ml}$$
$$\text{Initial reading} = 0.03 \text{ ml}$$
$$\text{Weight of flask + water delivered} = 81.200 \text{ g}$$
$$\text{Weight of flask} = 56.330 \text{ g}$$
$$\text{Temperature of water} = 21.5°C$$

42. A geologist often measures the density of a mineral by mixing two dense liquids, carbon tetrachloride and acetylene tetrabromide, in such proportions that the mineral grains will *just* float. She then determines the density of the liquid mixture, which is equal to the density of the solid. When a sample of the mixture in which calcite (calcium carbonate) just floats is put in a special density bottle, the weight is 6.2753 g. When empty, the bottle weighs 2.4631 g, and when filled with water it weighs 3.5441 g. What is the density of this calcite sample? (The temperature of these measurements is 25°C.)

43. How many pounds, including its own weight, can a hydrogen-filled balloon lift in air if it is 50.0 ft in diameter? (The density of air is 1.205 g/liter, and the density of hydrogen is 0.0833 g/liter. The lift is the difference between the weight of the gas and that of an equal volume of air.)

44. What percentage of your body (density approximately 1.03 g/ml) would be out of the water while you floated on your back in Great Salt Lake (density of water approximately 1.19 g/ml)?

45. How far into a 2-inch cube of wood (density = 0.90 g/ml) must a 14 g iron screw, ¼ inch in diameter, be driven in order that the block will just float? (Assume that the block does not change in size when the screw is driven into it. The density of iron is 7.60 g/ml. Assume that the density of water is 1.00 g/ml.)

46. A "Cartesian diver" is a hollow sealed bulb made of thin glass, such that the overall density is just a little less than 1 g/ml. As a consequence, it floats in water. If pressure is applied to the gas over the water in which the bulb floats, the bulb collapses a little and then sinks. When the pressure is released, it expands and floats again. A glass (density = 2.20 g/ml) sphere made on this principle weighs 3.25 g.

(a) What is its radius?

(b) What is the thickness of the glass?

47. The density of mercury at several selected temperatures is known as follows

$t(°C)$	-10.00	0.00	10.00	20.00	30.00	40.00	50.00
d (g/ml)	13.6202	13.5955	13.5708	13.5462	13.5217	13.4973	13.4729

(a) Using the method of least squares, find the mathematical equation that best fits these data (see pp 72–75).

(b) Find the correlation coefficient and the 95% confidence intervals for both the slope and the intercept of the best-fit equation found in (a).

8

Formulas and Nomenclature

At first it may seem difficult to learn the formulas and names of the hundreds of chemical compounds used in the introductory chemistry course. Actually the job is not so hard if you start in a systematic way, by learning the atomic building blocks that make up the compounds and the rules for naming the compounds.*

One of the many ways to classify inorganic compounds is into electrolytes, nonelectrolytes, and weak electrolytes. When *electrolytes* are dissolved in water, the resulting solution is a good conductor of electricity; the water solutions of *nonelectrolytes* do not conduct electricity; the solutions of *weak electrolytes* are very poor conductors. Water itself is an extremely poor conductor of electricity. A flow of current is a movement of electrical charges caused by a difference in potential (voltage) between the two ends of the conductor.

In metals, electrons are the structural units that carry the electrical charge, a negative one. But because electrons cannot exist for any significant length of time as independent units in water, some other kind of charged structural unit must be present in solutions of electrolytes. The general term for this charged structural unit is *ion*. A negatively charged ion is an atom (or a group of atoms) carrying one or more extra electrons it has received from other atoms. And, naturally, those atoms or groups of atoms that gave up electrons are no longer

---

* Most of the rules for chemical nomenclature are discussed in this chapter. The approved system for naming complex ions is given on page 391, and that for naming *ortho, meta,* and *pyro* compounds on page 420.

electrically neutral; they are positively charged ions. At all times, the solutions or crystals that contain ions are electrically neutral, because the total number of negative charges gained by one group of atoms always exactly equals the total number of positive charges created in the groups from which the electrons came. Ions tend to stay in the vicinity of each other because of the attraction of opposite electrical charges. Ions that contain more than one atom in stable combinations often are called *radicals*. In addition to explaining the electrical conductivity of water solutions, ions are important because a tremendous number of chemical reactions take place between them.

Inorganic compounds also may be classified as acids, bases, and salts. This classification is particularly useful as a basis for naming the chemicals with which we shall deal.

The names of the elements (and many of the symbols used to represent them) are traditional, rather than part of a logical system. The electrical charges that the ions usually carry can be reasoned out, but chemists do not work through such reasoning every time they want to use the charges or talk about them; they simply know them as characteristic properties.

Once you have learned the symbols for the elements, you will easily recognize and understand formulas. A *formula* is the shorthand notation used to identify the composition of a molecule. It includes the symbol of each element in the molecule, with numerical subscripts to show how many atoms of each element are present if there are more than one. For example, the formula for sulfuric acid, H_2SO_4, shows that this molecule has 2 hydrogen atoms, 1 sulfur atom, and 4 oxygen atoms. Note that a molecular formula does not tell how the atoms are bound together, only the kinds and numbers of atoms.

Listed in the paragraphs that follow are the names, symbols, and usual·electrical charges for 30 common positive ions, and the names and formulas for 37 common acids that are frequently mentioned in this text. You should memorize these so as to have them at instant recall; the use of flash cards or other foreign-language learning aids is recommended (computer-generated drill programs also are helpful). Aside from the direct intrinsic value of these names and formulas, you can reason out from them the names and formulas of almost 50 bases and over 1600 different salts, *none* of which should be memorized.

POSITIVE IONS WHOSE CHARGES DO NOT VARY

The ions listed in Table 8-1 carry exactly the same names as the elements from which they are derived. For example, Na and Mg are sodium and magnesium atoms, whereas Na^+ and Mg^{2+} are sodium and magnesium ions. These elements do not normally form ions that have charges other than those shown.*

---

* Some of the elements listed here do exhibit other charges under very unusual conditions, but this occurs so infrequently that we need not worry about it here. The principal exception, H^-, is discussed on page 109.

TABLE 8-1
Positive Ions Whose Charges Do Not Vary

Single charge		Double charge		Triple charge	
Hydrogen	H^+	Beryllium	Be^{2+}	Aluminum	Al^{3+}
Lithium	Li^+	Magnesium	Mg^{2+}		
Sodium	Na^+	Calcium	Ca^{2+}		
Potassium	K^+	Strontium	Sr^{2+}		
Rubidium	Rb^+	Barium	Ba^{2+}		
Cesium	Cs^+	Zinc	Zn^{2+}		
Silver	Ag^+	Cadmium	Cd^{2+}		
Ammonium	NH_4^+				

POSITIVE IONS WHOSE CHARGES VARY

The atoms of some metals can lose different numbers of electrons under different conditions (Table 8-2). For these atoms it has been traditional to add the
suffix *-ous* to the atom's root name for the lower charge state, and the suffix *-ic*
for the higher charge state. Thus the aurous ion is Au^+ and the auric ion is Au^{3+}.
There is difficulty when an element has more than two charge states. To over-

TABLE 8-2
Positive Ions Whose Charges Vary

IUPAC names	Root	Traditional names	
		-ous ending	-ic ending
Copper(I) and (II)	Cupr-	Cu^+	Cu^{2+}
Gold(I) and (III)	Aur-	Au^+	Au^{3+}
Mercury(I) and (II)	Mercur-	Hg^+ (Hg_2^{2+})	Hg^{2+}
Chromium(II) and (III)	Chrom-	Cr^{2+}	Cr^{3+}
Manganese(II) and (III)	Mangan-	Mn^{2+}	Mn^{3+}
Iron(II) and (III)	Ferr-	Fe^{2+}	Fe^{3+}
Cobalt(II) and (III)	Cobalt-	Co^{2+}	Co^{3+}
Nickel(II) and (III)	Nickel-	Ni^{2+}	Ni^{3+}
Tin(II) and (IV)	Stann-	Sn^{2+}	Sn^{4+}
Lead(II) and (IV)	Plumb-	Pb^{2+}	Pb^{4+}
Cerium(III) and (IV)	Cer-	Ce^{3+}	Ce^{4+}
Arsenic(III) and (V)	Arsen-	As^{3+}	As^{5+}
Antimony(III) and (V)	Antimon-	Sb^{3+}	Sb^{5+}
Bismuth(III) and (V)	Bismuth-	Bi^{3+}	Bi^{5+}

come this difficulty and to clarify some subtler points, the International Union of Pure and Applied Chemistry (IUPAC) has in recent years recommended the adoption of a system that, when applied to a positive ion, requires that the name of the element be used and that it be followed immediately by its charge in Roman numerals within parentheses. For example, Au^+ would be gold(I), and Au^{3+} would be gold(III); they would be read as "gold one" and "gold three," respectively. Until there is unanimous adoption of the IUPAC system, chemists will have to use both systems. In any case, neither system relieves the student of the responsibility of knowing the usual charge states of the common elements.

BASES

We define a *base* as any substance that can accept or react with a hydrogen ion, H^+. This definition includes a wide variety of compounds, but for the present it is convenient to limit our discussion to one special type of base called a *hydroxide*. A hydroxide is any compound that has one or more replaceable hydroxide ions, OH^-. Any of the positive ions cited in the preceding sections might combine with the hydroxide ion, the principle being that the resulting compound must be electrically neutral. Naturally there must be as many OH^- ions as there are positive charges on the other ion. In naming, the word "hydroxide" is preceded by the name of the positive ion; for example,

NaOH	sodium hydroxide
$Co(OH)_2$	cobaltous hydroxide or cobalt(II) hydroxide
$Sn(OH)_4$	stannic hydroxide or tin(IV) hydroxide

ACIDS

We define an *acid* as any substance that has one or more replaceable hydrogen ions (H^+); we often say that an acid can donate one or more hydrogen ions to a base. No matter how many H atoms a molecule might have, if none of them can be replaced by some positive ion, then the molecule does not qualify as an acid. On the other hand, if a molecule does have several H atoms, of which only one is replaceable, that alone is enough to qualify it as an acid. In writing the formula of an acid, it is traditional to show the replaceable H ions at the beginning of the formula, separated from those that are not replaceable. For example, in acetic acid, $HC_2H_3O_2$, there is one replaceable H^+ and there are three H atoms that are not replaceable. Besides identifying the number of replaceable H atoms, this traditional method also provides a simple way of knowing the charge of the negative ions; for every H^+ that is removed from an acid there must be one negative charge left on the residue. For example, if HBr loses its H^+, the Br^- has a minus-one charge; if H_2SO_4 loses its two H^+, the SO_4^{2-} has a minus-two charge, and so on. The names and formulas of acids that you should know in this course are listed in two separate groups below, because the traditional method of naming them depends on whether or not the molecule contains oxygen.

Acids That Do Not Contain Oxygen

These acids are named by putting the prefix *hydro-* before the rest of the name of the characteristic element (or elements) and adding the suffix *-ic*.

HF	hydrofluoric acid	HCN	hydrocyanic acid
HCl	hydrochloric acid	H_2S	hydrosulfuric acid
HBr	hydrobromic acid	HN_3	hydrazoic acid
HI	hydriodic acid		

Acids That Contain Oxygen

If the characteristic element forms only one oxygen acid, the name is that of the characteristic element followed by the suffix *-ic*. Thus H_2CO_3 is carbonic acid. If the characteristic element forms two oxygen acids, the name of the one with the larger number of oxygen atoms ends in *-ic*, and the name of the one with the smaller number of oxygen atoms ends in *-ous*. Thus HNO_3 is nitric acid, and HNO_2 is nitrous acid.

If there are several oxygen acids, there is a systematic terminology to indicate more or less oxygen atoms than the number assigned to the acid whose name ends in *-ic*. This can be illustrated by the oxygen acids of chlorine:

$HClO_4$	perchloric acid	one more oxygen than the *-ic* acid
$HClO_3$	chloric acid	arbitrarily given the *-ic* ending
$HClO_2$	chlorous acid	one less oxygen than the *-ic* acid
HClO	hypochlorous acid	one less oxygen than the *-ous* acid

Table 8-3 includes the common acids that contain oxygen.

The prefix *thio-* indicates that sulfur is present, usually as a replacement for one or more oxygen atoms in a compound whose name is familiar. For example, in thiosulfuric acid, one O atom of sulfuric acid has been replaced by an S atom; in thiocyanic acid, the only O atom of cyanic acid has been replaced; in thioarsenic acid, all of the O atoms of arsenic acid have been replaced by S. The rules of naming do not normally show how many O atoms are replaced. HSCN is placed in Table 8-3 because of the many similarities between oxygen and sulfur.

TABLE 8-3
Acids That Contain Oxygen

Formula	Name	Formula	Name
H_2CO_3	Carbonic acid	$HClO_4$	Perchloric acid
H_3BO_3	Boric acid	(HIO_4 is similar to $HClO_4$)	
H_4SiO_4	Silicic acid	$HClO_3$	Chloric acid
HNO_3	Nitric acid	(HIO_3 and $HBrO_3$ similar to $HClO_3$)	
HNO_2	Nitrous acid	$HClO_2$	Chlorous acid
H_2SO_4	Sulfuric acid	$HClO$	Hypochlorous acid
H_2SO_3	Sulfurous acid	(HIO and $HBrO$ similar to $HClO$)	
$H_2S_2O_3$	Thiosulfuric acid	$HMnO_4$	Permanganic acid
H_2CrO_4	Chromic acid	$HOCN$	Cyanic acid
$H_2Cr_2O_7$	Dichromic acid	$HSCN$	Thiocyanic acid
H_3PO_4	Phosphoric acid	$H_2C_2O_4$	Oxalic acid
H_3PO_3	Phosphorous acid	$H_2C_8H_4O_4$	Phthalic acid
H_3AsO_4	Arsenic acid	$HC_2H_3O_2$	Acetic acid
H_3AsO_3	Arsenious acid	$H(NH_2)SO_3$	Sulfamic acid

SALTS

One general type of chemical reaction is that occurring when a hydroxide reacts with an acid. This reaction, like all chemical reactions, can be represented by a *chemical equation* in which the reactants are separated by "+" signs to indicate that they are mixed together, the products are separated by "+" signs to indicate that they are produced as a mixture, and the products are separated from the reactants by an arrow to show that the reactants are producing the products. In order to write a chemical equation for the general reaction between an acid and a hydroxide, we need to know first that in every case the acid donates an H^+ to each OH^- of the hydroxide to form water (H_2O), and second that the electrically neutral combination of the positive ions from the hydroxide and the negative ions from the acid is what constitutes a *salt*. If we express this in very general terms, this type of reaction can be written as

$$\text{hydroxide} + \text{acid} \rightarrow \text{salt} + \text{water}$$

Now that we know the names and formulas of some acids and bases, we can also write equations for specific examples of this type of reaction, using our shorthand notation. For example, if potassium hydroxide reacts with hydrobromic acid, we write

$$KOH + HBr \rightarrow KBr + H_2O$$

If cupric hydroxide reacts with perchloric acid, we write

$$Cu(OH)_2 + 2 \ HClO_4 \rightarrow Cu(ClO_4)_2 + 2 \ H_2O$$

And if aluminum hydroxide reacts with sulfuric acid, we write

$$2 \ Al(OH)_3 + 3 \ H_2SO_4 \rightarrow Al_2(SO_4)_3 + 6 \ H_2O$$

In addition to showing which chemicals react and which are produced, the chemical equation shows the relative numbers of molecules needed in the reaction. In the above reactions we had to use the proper numbers of acid molecules and base molecules so that the salt that was formed would be electrically neutral. In the third reaction, for example, the two Al^{3+} ions require three SO_4^{2-} ions in order to have electroneutrality. Besides showing the correct chemical formulas of all the reactants and products, a *balanced* chemical equation always abides by a principle of conservation, which might be stated as "atoms are never created or destroyed in a chemical reaction." In other words, no matter how drastically the atoms are rearranged in a chemical reaction, there must always be the same number of each kind of atom in the collection of products (the righthand side of the equation) as there is in the collection of reactants (the lefthand side of the equation).

The rules for naming a salt depend on whether or not the acid from which it is derived contains oxygen.

1. If the acid does not contain oxygen, the salt is named by replacing the acid prefix *hydro-* by the name of the positive ion and changing the suffix *-ic* to *-ide;* for example,

KBr	potassium bromide
Sb_2S_3	antimonous sulfide or antimony(III) sulfide
$Hg(CN)_2$	mercuric cyanide or mercury(II) cyanide

2. If the acid does contain oxygen, the salt is named by giving the name of the positive ion followed by the name of the acid, but changing the suffix *-ic* to *-ate* or the suffix *-ous* to *-ite;* for example,

$SrCO_3$	strontium carbonate
$Cu(ClO_4)_2$	cupric perchlorate or copper(II) perchlorate
$(NH_4)_2SO_3$	ammonium sulfite
$Co(BrO)_2$	cobaltous hypobromite or cobalt(II) hypobromite

By changing the proportions in which some acids and bases are mixed, it is possible to make salts in which only some of the replaceable hydrogen atoms are actually replaced. For example, with H_3PO_4 and NaOH, three different reactions are possible:

$$3 \text{ NaOH} + H_3PO_4 \rightarrow Na_3PO_4 \quad + 3 H_2O$$
$$2 \text{ NaOH} + H_3PO_4 \rightarrow Na_2HPO_4 + 2 H_2O$$
$$\text{NaOH} + H_3PO_4 \rightarrow NaH_2PO_4 + H_2O$$

Each of these reactions is possible, and each equation is balanced. Obviously, we need different names for the three different kinds of sodium phosphate salts. Of the several ways to accomplish this, the most highly recommended and unequivocal is to include as part of the name the number of H^+ ions that have *not* been replaced in salt formation. In the example above we would have

Na_3PO_4	sodium phosphate
Na_2HPO_4	sodium monohydrogen phosphate
NaH_2PO_4	sodium dihydrogen phosphate

If the acid contains only two H^+, then only two different salts are possible; they are most acceptably named by the method just described. However, another traditional method that will probably continue in use for many years is to use the prefix *bi-* for the salt of the acid with just half of the hydrogen atoms replaced, as illustrated for the sodium bisulfate that is produced by the reaction

$$\text{NaOH} + H_2SO_4 \rightarrow NaHSO_4 + H_2O$$

BINARY COMPOUNDS NOT DERIVED FROM ACIDS

The atoms listed in this section may combine with many metals to form binary compounds (compounds made up of two elements) that are saltlike in nature, but are not derived from acids. For purposes of naming, it is convenient to assign negative charges to these atoms. Except in the names of the oxides, the suffixes *-ous* and *-ic* are not used with metals forming compounds in this group. The names of all these compounds end in *-ide*. Only the metal oxides of this group are common.

H^-	LiH	lithium hydride
O^{2-}	FeO	ferrous oxide (Fe_2O_3 ferric oxide)
N^{3-}	Sn_3N_4	tin nitride
P^{3-}	Ba_3P_2	barium phosphide
As^{3-}	Na_3As	sodium arsenide
C^{4-}	Al_4C_3	aluminum carbide
Si^{4-}	Mg_2Si	magnesium silicide

BINARY COMPOUNDS COMPOSED OF TWO NONMETALS

Allowing for a bit of quibbling one way or the other, there are only 20 nonmetallic elements. Of these 20 elements, six are so unreactive that until recently they were thought never to combine with other elements. If, for practical considerations, we eliminate these six elements and hydrogen (which we've already dealt with in various forms above), this leaves just 13 nonmetallic elements (B, C, N, O, F, Si, P, S, Cl, Se, Br, Te, I) that can combine with each other. These nonmetallic binary compounds are designated by the names of the two elements followed by the ending -*ide*. Before the name of the second element there is a prefix to indicate how many atoms of it are in the molecule. Unfortunately, this is not usually done for the first element. The following are examples of some common compounds of this type:

CO	carbon monoxide	SO_3	sulfur trioxide
CO_2	carbon dioxide	CCl_4	carbon tetrachloride
Cl_2O	chlorine monoxide	PF_5	phosphorus pentafluoride
ClO_2	chlorine dioxide	SF_6	sulfur hexafluoride
Cl_2O_7	chlorine heptoxide	N_2O_4	nitrogen tetroxide

PROBLEMS A

Name the following compounds (proper spelling is required).

1. $Ca(OH)_2$	12. FeC_2O_4	23. $Al(ClO_4)_3$
2. Ag_3PO_4	13. Hg_2Cl_2	24. $Hg(C_2H_3O_2)_2$
3. $AgSCN$	14. $MnCO_3$	25. $CsClO_3$
4. $MgC_8H_4O_4$	15. $Mn(OH)_3$	26. $Sr(IO)_2$
5. $(NH_4)_2SO_4$	16. $Ni(ClO)_2$	27. Rb_3AsO_3
6. ZnS	17. $CrAsO_4$	28. Be_3N_2
7. $Cd(CN)_2$	18. $SnBr_4$	29. $Ca(HCO_3)_2$
8. $Ba(IO_3)_2$	19. CrF_2	30. $Sb(NO_3)_3$
9. $CuSO_3$	20. $Pb(MnO_4)_2$	31. PCl_3
10. CuI	21. Na_4SiO_4	32. $Bi(OCN)_3$
11. $Fe(NO_3)_3$	22. Bi_2O_5	33. $Al_2(S_2O_3)_3$

Without consulting a text, give the formulas for the following compounds.

34. aluminum bromate   36. bismuth(III) oxide

35. mercurous phosphate   37. strontium bicarbonate

38. aurous iodide

39. chromium(III) iodate

40. manganous hydroxide

41. lithium arsenide

42. arsenic(III) sulfate

43. stannic chloride

44. nickelous periodate

45. chlorine heptoxide

46. silver oxalate

47. chromium(II) borate

48. antimonous sulfide

49. aluminum acetate

50. calcium oxalate

51. sodium chlorite

52. tin(II) azide

53. mercury(II) cyanide

54. ammonium sulfite

55. cobalt(II) permanganate

56. plumbous carbonate

57. zinc phosphide

58. cupric silicate

59. barium hypoiodite

60. Given that selenium (Se) is similar in properties to sulfur (S), and that francium (Fr) is similar to sodium (Na), write the formulas for the following compounds.
    (a) zinc selenide
    (b) francium phosphate
    (c) cobalt(II) selenite
    (d) selenium dioxide
    (e) francium selenate
    (f) selenium hexafluoride
    (g) francium hydride

61. Write balanced chemical equations for the following reactions.
    (a) beryllium hydroxide + thiocyanic acid
    (b) periodic acid + antimonic hydroxide
    (c) mercurous hydroxide + acetic acid
    (d) arsenious acid + chromous hydroxide
    (e) ceric hydroxide + boric acid
    (f) hydrazoic acid + ferrous hydroxide
    (g) lithium hydroxide + hypobromous acid

PROBLEMS B

Name the following compounds (proper spelling is required).

62. Ag_3PO_4

63. $CoCl_3$

64. $Be(NO_2)_2$

65. $Fe(MnO_4)_3$

66. NH_4NO_2

67. Al_2S_3

68. $Zn(IO_4)_2$

69. $Pb_3(BO_3)_2$

70. $As(CN)_3$

71. $Ni_3(AsO_3)_2$

72. I_2O_7

73. Ba_2Si

74. $AuHSO_3$

75. $Ba(BrO)_2$

76. CaH_2

77. N_2O_5

78. Sb_2S_3

79. MgC_2O_4

80. ICl   83. Ca_3P_2 86. $CrBr_2$

81. Rb_4SiO_4 84. MnO_2 87. $BaCrO_4$

82. SF_6 85. $Cu(C_2H_3O_2)_2$ 88. $Cd(SCN)_2$

Without consulting a text, give the formulas for the following compounds.

89. potassium oxalate   102. iron(III) bromate

90. cupric arsenate   103. arsenic(V) perchlorate

91. bismuthous carbonate   104. magnesium monohydrogen borate

92. manganese(III) oxide   105. boron trifluoride

93. mercurous sulfate   106. strontium silicate

94. nitrogen tri-iodide   107. beryllium hydroxide

95. cobalt(II) borate   108. stannic oxide

96. cesium hypoiodite   109. gold(III) fluoride

97. boron nitride   110. ferric chromate

98. cadmium dichromate   111. iodine pentoxide

99. ammonium acetate   112. lithium thiocyanate

100. zinc cyanide   113. silver thiosulfate

101. tin(II) phosphate   114. antimonic permanganate

115. Given that astatine (At) is similar in properties to chlorine (Cl), and that
     gallium (Ga) is similar to aluminum (Al), write the formulas for the following
     compounds.
     (a) potassium astatate   (d) hydrastatic acid
     (b) barium astatide   (e) gallium thiocyanate
     (c) gallium sulfate   (f) gallium hypoastatite

116. Write balanced chemical equations for the following reactions.
     (a) hydrosulfuric acid + zinc hydroxide
     (b) ferric hydroxide + permanganic acid
     (c) oxalic acid + plumbous hydroxide
     (d) aluminum hydroxide + carbonic acid
     (e) bromic acid + cupric hydroxide
     (f) auric hydroxide + dichromic acid
     (g) sulfamic acid + strontium hydroxide

Sizes and Shapes of Molecules

A knowledge of molecular shapes and sizes is important to an understanding of chemical reactions. The shape of a molecule (and the bond types it possesses) has important implications for the manner in which it enters into chemical reactions. The shape and size of molecules also influence their packing in the crystalline state.

When atoms combine to produce molecules, they often do so in accord with the *octet rule*. Your text undoubtedly contains a fairly detailed discussion of the octet rule. In essence, it may be described as the tendency for an atom to lose, gain, or share electrons in order to achieve an s^2p^6 configuration in the outermost shell. The simplest atoms (H, Li, Be, and so on) tend to achieve a $1s^2$ configuration, according to what might be called the duet rule.

In Chapter 8, we emphasize the loss and gain of electrons, leading to the formation of electrically charged ions, such as Na^+ and Cl^-. When electrons are *shared,* a molecule is formed, and the atoms are connected by a *covalent bond*. In this chapter we emphasize the approximate shapes, interatomic distances, and bond energies of molecules and molecular ions that are held together by covalent bonds.

COVALENT BOND ENERGIES

The strengths of the bonds that hold the atoms together in a molecule can be determined in a variety of ways: for example, by direct calorimetric measurement, by dissociation equilibrium measurements, by absorption spectrum measurements, or by mass spectrometry.

We define *bond energy* as the energy change (ΔH) for the chemical process in which one mole of a given bond is broken, when both the reactants and the products are in the hypothetical ideal-gas state of 1 atm and 25°C. For a diatomic molecule, the bond energy is identical to the energy required to dissociate the *gaseous* molecule into its respective *gaseous* atoms. For the dissociation of Cl_2 gas this corresponds to the reaction

$$Cl_{2(g)} \rightleftarrows 2\ Cl_{(g)}$$

for which the Cl–Cl bond energy is 58.0 kcal. We say that $\Delta H_{Cl-Cl} = 58.0$ kcal.

For a polyatomic molecule of the type AB_n, which possesses n A–B bonds, our definition of bond energy implies that each bond is the same, and that it corresponds to $1/n$ of the total energy required to dissociate the gaseous AB_n molecule into $A + n$ B gaseous atoms. This is a useful definition except when studying the detailed steps of a chemical reaction. For example, the total binding energy in a CH_4 molecule is 397 kcal/mole and, by our definition of bond energy, the C–H bond energy = 397/4 = 99.3 kcal. Extensive, complicated, and detailed studies have shown, however, that each H atom is not equally easily removed from carbon in this molecule; it is estimated that the individual bond energies are 104 kcal for CH_3–H, 106 kcal for CH_2–H, 106 kcal for CH–H, and 81 kcal for C–H, with a total of 397 kcal. In most cases, such detailed information is not available; neither is it normally needed except in discussion of the individual steps involved in chemical reactions.

The atoms of some elements (such as C, N, and O) are able to share more than one pair of electrons between them, to form single, double, or triple bonds—depending on whether one, two, or three pairs of electrons are shared. In general, the bonding energy increases and the internuclear distance decreases as the number of bonds between a pair of atoms increases.

By studying the experimentally determined bond energies of hundreds of compounds, researchers have uncovered some useful generalizations, such as the following.

1. A single bond between two *identical* atoms has about the same strength (ΔH_{A-A}) in any molecule in which it occurs. Because the atoms are identical, the bond is purely covalent.

2. A strictly covalent bond between two *different* atoms (A and B) is about the same strength as the average of the bond strengths that would be observed if each atom were bonded to another like itself. That is,

$$\Delta H_{A-B} = \tfrac{1}{2}[\Delta H_{A-A} + \Delta H_{B-B}] \qquad (9\text{-}1)$$

TABLE 9-1
Covalent Bond Radii and Energies, and Electronegativities

Element	Radius (Å)	Bond energy ΔH_{A-A} (kcal/mole)	Electronegativity (ϵ) of element
H—H	0.30	104.2	2.1
B—B	0.88	62.7	2.0
C—C	0.772	83.1	2.5
C=C	0.667	147	
C≡C	0.603	194	
Si—Si	1.17	42.2	1.8
Si=Si	1.07		
Ge—Ge	1.22	37.6	1.8
Sn—Sn	1.40	34.2	1.8
N—N	0.70	38.4	3.0
N=N	0.60	100	
N≡N	0.55	226.2	
P—P	1.10	51.3	2.1
P=P	1.00	117	
As—As	1.21	32.1	2.0
Sb—Sb	1.41	30.2	1.9
Bi—Bi	1.52	25	1.9
O—O	0.66	33.2	3.5
O=O	0.55		
S—S	1.04	50.9	2.5
S=S	0.94		
Se—Se	1.17	44.0	2.4
Se=Se	1.07		
Te—Te	1.37	33.0	2.1
F—F	0.64	36.6	4.0
Cl—Cl	0.99	58.0	3.0
Br—Br	1.14	46.1	2.8
I—I	1.33	36.1	2.5

These two generalizations mean that to each element there can be assigned a "covalent bond energy" (ΔH_{A-A}), which can be used to estimate the strength of a bond between any two atoms that form a strictly covalent bond. Table 9-1 lists a few of these covalent bond energy assignments. Carbon and iodine form a purely covalent bond so, by our second generalization, its approximate bond energy would be

$$\Delta H_{C-I} = \tfrac{1}{2}[\Delta H_{C-C} + \Delta H_{I-I}]$$

$$= \tfrac{1}{2}[83.1 + 36.1] = 59.6 \text{ kcal}$$

ELECTRONEGATIVITY

Unfortunately, as in so many sharing processes, the pair of electrons in a covalent bond often is not shared equally by the two atoms. The atom with the greater electron affinity will hold the pair closer to its nucleus, with the result that its end of the bond (and its end of the molecule) will be somewhat more negatively charged than the other end. When this happens, we say that the bond is partially ionic and, because opposite charges attract each other, this partially ionic bond will be stronger than it would have been with equal sharing.

Linus Pauling made a careful study of a tremendous number of partially ionic covalent bonds and came to the conclusion that, as a measure of its electron affinity, each element could be assigned an *electronegativity value* (ϵ) that would make it possible to estimate the energy of these bonds. To calculate the approximate bond energy (ΔH_{A-B}) between the atoms A and B, his formula requires knowledge of the A–A bond energy (ΔH_{A-A}), the B–B bond energy (ΔH_{B-B}), and the electronegativities of A and B (ϵ_A and ϵ_B). Table 9-1 lists some electronegativity values along with covalent bond energies. Pauling's finding, which we might call generalization #3, is the following.

3. The energy of a partially ionic covalent bond between two atoms (A and B) is equal to the energy expected from a strictly covalent bond between the two atoms *plus* an amount of energy related to the square of the difference in their electronegativities. In equation form, Pauling's formula (with units in kcal) is

$$\Delta H_{A-B} = \tfrac{1}{2}[\Delta H_{A-A} + \Delta H_{B-B}] + 23.06(\epsilon_A - \epsilon_B)^2 \qquad (9\text{-}2)$$

You can see that, if A and B have equal electronegativities, then ΔH_{A-B} is simply the average of the two covalent bond energies. The greater the difference in electronegativites, the greater the percentage of ionic character and the stronger the bond. If the difference in electronegativity becomes too great, the bond is essentially ionic, and the atoms are held together by electrostatic forces as in an ionic crystal; the concept of a molecule disappears.

From this brief discussion you can see that more often than not most covalent bonds will be "partially ionic," and most ionic bonds will be "partially covalent." We shall describe a bond as being ionic or covalent according to its predominant characteristic. Some chemists like to say that a bond possesses a certain "percent ionic character." One way of calculating an *approximate* value for this quantity is with the expression

$$\log F_c = -\left(\frac{\epsilon_A - \epsilon_B}{3}\right)^2 \qquad (9\text{-}3)$$

where F_c is the fraction of *covalent* character. The fraction of *ionic* character is $F_i = 1 - F_c$.

PROBLEM:
Calculate approximate values for the bond energy and the percent ionic character for the C–Cl bond in CCl_4.

SOLUTION:
From Table 9-1 we find

$$\Delta H_{C-C} = 83.1 \text{ kcal/mole}$$
$$\Delta H_{Cl-Cl} = 58.0 \text{ kcal/mole}$$

and

$$\epsilon_C = 2.5$$
$$\epsilon_{Cl} = 3.0$$

With the Pauling equation, we obtain

$$\Delta H_{C-Cl} = \tfrac{1}{2}[83.1 + 58.0] + (23.06)(3.0 - 2.5)^2$$
$$= 70.6 + 5.8 = 76.4 \text{ kcal/mole}$$
$$\log F_c = -\left(\frac{3.0 - 2.5}{3}\right)^2 = -0.027778$$
$$F_c = 0.938$$
$$F_i = 1 - 0.938 = 0.062$$

Percent ionic character = 6.2%

COVALENT BOND RADII

Before considering shape principles, we should take a look at the sizes of molecules, particularly their internuclear distances. Electron diffraction studies of molecules in the gas phase are especially useful for the determination of these distances (and the angles that the atom pairs form with one another). An interatomic distance is defined as the *average* internuclear distance between two atoms bonded together. Because these two atoms vibrate, the distance between them alternately lengthens and shortens in rapid succession but, if the distance is averaged for a period of time, the atoms appear to be separated by some fixed distance. A fourth generalization can be made about these distances.

4. The atoms of a given element always appear to have essentially the same radius (R), regardless of the kinds of atoms to which they are bound (or their electronegativities). That is, covalent radii are additive,

and the approximate interatomic distance (D) between two atoms (A and B) will be given by

$$D_{A-B} = R_A + R_B$$

PROBLEM:
Calculate the C–Cl distance in the CCl_4 molecule.

SOLUTION:
The interatomic distance is the sum of the covalent radii so, with data from Table 9-1,

$$D_{C-Cl} = R_C + R_{Cl} = 0.77 + 0.99 = 1.76 \text{ Å}$$

We have already observed in the preceding problem that the C–Cl bond energy is about 76.4 kcal/mole, and that it possesses about 6% ionic character.

SHAPES OF MOLECULES

To predict the shape of a molecule, we can apply a set of simple principles. Before we outline them, however, you should realize that they have little predictive value when you try to answer the question, "What molecule is formed when two or more particular elements react?" From the same elements, but under differing conditions, it is possible to prepare many different molecules, not just one. Instead, our principles will answer the question, "What is the shape of the molecule with a particular formula?"

The concept of the *central atom* is convenient to use in discussing the shapes of molecules. In a simple molecule, one of the atoms usually is "central" to the whole molecule. For example, in CCl_4 the central atom is C, the one to which all the other atoms are attached.

We shall use the term *ligand* (L) in its broadest sense to refer to those atoms that are attached to the central atom. What we mean by the shape of a molecule or ion is the geometrical arrangement of the ligands about the central atom.

General Predictive Principles

In broad outline, our shape-prediction approach will assume that the valence electrons of the central atom (M) are all spin-paired (that is, their axes are parallel but spinning in opposite directions), and that these pairs of electrons will repel each other in such a way that they occupy positions of minimum repulsion (that is, positions of minimum potential energy). The electron pairs will try to get as far away from each other as they can and still stay in the molecule. The central atom's valence electron pairs will fall into one of two categories: (1) *bond pairs* that are used to bind ligands to the central atom, or (2) *lone pairs* that are held by the central atom but not used for bonds. Each pair, regardless of the type, will occupy a *molecular* orbital, not an atomic orbital. The key to success in prediction hinges on finding the correct number of valence electrons associated with M.

A summary of *all* the procedural steps for shape prediction is given here because it will be useful for later reference; the application of the steps is illustrated in detail on the following pages.

1. *Find the number of valence electrons around the central atom* by drawing an "electron-dot formula" for the molecule or ion whose shape is desired. See pp 120–121. Follow these rules.
   (a) The central atom must be the least electronegative atom in the molecule or ion. H can never be a central atom; it can form only one bond.
   (b) *Every* ligand must obey the octet rule (or duet rule for H).
   (c) If the shape involves an ion (but *not* a complex ion), take account of the charge on the ion by adding one valence electron to M for each negative charge, or subtracting one valence electron from M for each positive charge.
   (d) If possible, also make M obey the octet rule. Manipulation will be possible only if one or more *ligands* can form multiple bonds. The four or six electrons involved in double or triple bonds count as belonging to both the ligand *and* the central atom, just as do the two electrons in a single bond. See pp 121 and 132.
   (e) For complex ions you can *ignore* the electrons originally possessed by M (usually a metal ion) and the charge on the ion. You do not need to draw an electron-dot formula. You need to know only that each ligand contributes a bond pair to M; there are no lone pairs. The number of ligands therefore directly tells you the number of pairs of valence electrons around M. See p 141.

2. *Calculate the number P of electron pairs around the central atom* (M), as follows.
   (a) If the electron-dot formula shows no multiple bonds (double or triple bonds) between M and the ligands, divide the total number of electrons around M by 2 to get the number P of electron pairs.
   (b) If the electron-dot formula shows one or more multiple bonds between M and the ligands, *only two* electrons are counted for *each* multiple bond (double *or* triple). Add these electrons to those involved in single bonds or lone pairs, then divide the total by 2 to get the number P of electron pairs around M. See pp 121, 132, and 138.
   (c) If the number of valence electrons in (a) or (b) is an *odd* number, then (*for the purpose of this calculation only*) increase the number of electrons by one to make it even before proceeding as in (a) or (b).

This is equivalent to treating the one odd electron as a lone pair, though it is less effective in repulsion than a true lone pair. See p 140.

3. *Determine the electron-pair geometry* of the molecule or ion, using the guidelines in Table 9-2.

4. *Determine the molecular geometry*, taking into consideration the bond pairs and lone pairs involved in the electron-pair geometry. See pp 121–132.

Electron-Dot Formulas

The first thing you must be able to do in order to predict molecular shapes is to draw an electron-dot formula, so we'll tackle that subject first. Including H, there are 16 active nonmetals for which you should know the numbers of valence electrons in the uncombined atoms. Except for H (which has only one s electron), these elements are all found to the right of the diagonal in the p block of the periodic table (see inside front cover). Each atom has two s electrons in its valence shell; the number of p electrons is different for different atoms. (Basically, we are uninterested in metals here; metals rarely form predominantly covalent bonds, but tend to form ionic bonds. Except for Xe, we also can ignore the noble gases; with an already filled s^2p^6 configuration, they are unreactive.)

It will pay you to know (without having to look in the periodic table or tables of electron configurations) that the halogens (F, Cl, Br, I, At) all have seven valence electrons, that the oxygen family (O, S, Se, Te) all have six, that the nitrogen family (N, P, As) have five, that the carbon family (C, Si) have four, and that the boron family (B) have three. It will also pay you to know that electronegativities decrease from right to left in a row, or from top to bottom in a column, in the periodic table.

Figure 9-1 shows electron-dot formulas in which each ligand possesses eight electrons as a result of sharing the needed number from the central atom (except H, which needs only two electrons to fill its valence shell). In each case, the least electronegative atom is central, except for NH_4^+ where H cannot be central. In the case of the NH_4^+ and SO_4^{2-} ions, the number of dots reflects the loss or gain of electrons as required by the charge on the ions. In three cases (BCl_3, ClF_3, and PCl_5) the octet rule has been violated for the central atom, but it is not possible to rectify this because the ligands (all halogens in this case) are not able to form multiple bonds. If both the ligands and the central atom never violated the octet rule, there would always be just four pairs of electrons around the central atom, and all molecular structures would be tetrahedral. It's because there are so many exceptions to the octet rule for the central atom that there are so many different shapes of molecules.

The halogens and boron can form only single bonds. The oxygen-family

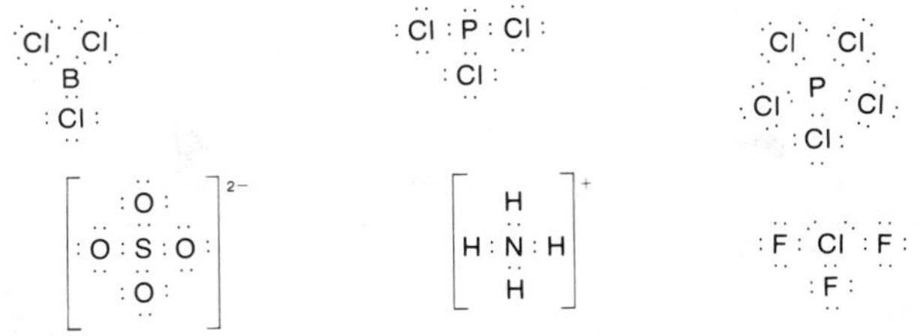

```
 :Cl  :Cl           :Cl : P : Cl :          :Cl  :Cl
   B                    : Cl :              :Cl  P  Cl:
 : Cl :                                        :Cl:


      :O:  ]2-            H  ]+
   :O : S : O :        H : N : H         :F : Cl : F:
      :O:                 H                   :F:
```

FIGURE 9-1
Electron-dot structures.

atoms may form two single bonds or one double bond. The nitrogen-family atoms may form three single bonds, a single and a double bond, or a triple bond. The carbon-family atoms can form four single bonds, two singles and a double, one single and a triple, or two double bonds. For our first examples, we exclude molecules or ions in which multiple bonds can be invoked in order to make the central atom obey the octet rule.

Electron-Pair Geometries

As we discuss the electron-pair geometries for the molecular systems with two or more pairs of electrons, imagine each pair of electrons to be attached to M by a weightless string that permits free movement within the confines of the tether. Under these conditions it would be natural to expect that, with only two such pairs of electrons, the pairs would be diametrically opposite each other with M in the center. Within the confines of the string, any other position would bring them closer together, in a more repulsive condition. Thus, when there are only two electron pairs around M (that is, when $P = 2$), the electron-pair geometry is always linear (that is, the angle, p–M–p, is always 180°). The letter p refers to a single pair of electrons.

When $P = 3$, the two most likely electron-pair geometries are a $\triangle$-pyramid with M at the apex, and a $\triangle$-coplanar structure in which M lies at the center of an equilateral triangle and in the same plane as the electron pairs, which lie at the corners of the triangle. A little reflection quickly leads you to the $\triangle$-coplanar structure as the one with less electron-pair repulsion, for the electron pairs are farther apart in this configuration. The p–M–p angle is 120°.

When $P = 4$, the two most likely electron-pair geometries are a $\square$-coplanar structure, in which M lies at the center of the square and in the same plane as the pairs of electrons, which lie at each corner of the square; and a tetrahedral structure with M at the center of the tetrahedron outlined by the four pairs of electrons, one pair at each apex of the tetrahedron. Again, for a given length of string, the pairs of electrons will be farther from each other in the tetrahedron, in which the p–M–p angle is 109°28′, than in the $\square$-coplanar structure, where

TABLE 9-2
Guidelines for Determining Electron-Pair Geometry

Number (P) of electron pairs	Electron-pair geometry (see pp 121–132)	Hybrid orbitals (see pp 135–136)
2	Linear	sp
3	△-Coplanar	sp^2
4	Tetrahedral	sp^3
5	△-Bipyramidal	sp^3d
6	Octahedral (same as □-bipyramidal)	d^2sp^3 or sp^3d^2
7	Pentagonal bipyramidal	sp^3d^3
8	□-Antiprism (or distorted dodecahedral)	d^4sp^3

the p–M–p angle is 90°; the tetrahedral electron-pair geometry thus will always be expected with four pairs of electrons.

When $P = 5$, there come to mind the possibilities of (1) a pentagonal-coplanar structure or (2) a △-bipyramid formed by placing two triangle-based pyramids base-to-base, with M centered in the plane where the two bases come together. Again, for a given length of string, the repulsion between the electrons will be greater in the coplanar pentagon, where the p–M–p angle is 72°, than in the △-bipyramid, where the p–M–p angle is either 90°, 120°, or 180°, depending on which two pairs of electrons are under consideration. The △-bipyramid will always be the expected electron-pair geometry.

At least three reasonable electron-pair geometries might occur to you for the $P = 6$ case. They are (1) a coplanar hexagon (with M centered in the plane), (2) a triangular prism (two triangular pyramids, apex-to-apex, with base edges parallel to each other and M located at the apex-to-apex contact), and (3) an octahedron (a □-bipyramid formed by placing two square-based pyramids base-to-base with M centered in this base plane). The same arguments used above lead to the conclusion that, when $P = 6$, the expected electron-pair geometry will always be the octahedron. Note that all the p–M–p angles between adjacent p positions are the same, 90°, which means that all the electron-pair positions are equivalent in the octahedron, in contrast to the nonequivalent positions in the △-bipyramid.

Having established the electron-pair geometry for the common numbers of electron pairs, we note that, in the more complicated case when $P = 7$, the expected electron-pair geometry is a pentagonal bipyramid with M centered in the common pentagonal base plane. And for $P = 8$, three electron-pair geometries are possible: a cube with M at the center; a square antiprism (two square-based pyramids, apex-to-apex, with the base of one rotated 45° relative to the other, and M located at the apex-to-apex contact); and a distorted dodecahedral arrangement. The square antiprism and the distorted dodecahedron are about equally probable and involve less electron-pair repulsion than the simple cubic. The number of compounds in which M has $P = 7$ or $P = 8$ is actually rather limited.

Molecular Geometries

Now, let us look at molecular geometries or shapes. These are determined by the number (BP) of electron pairs that are used as *bond* pairs, for these bond pairs will lie in molecular orbitals between M and L (the ligands) in positions determined by P, as we have just discussed. The total number of pairs often (usually) will be equal to the number of ligands attached to M. When this is true (that is, when $P = BP$), the molecular geometry will be identical to the electron-pair geometry. In the sketches, the electron-pair geometry is shown by shaded planes, and bond pairs of electrons by solid lines. The lone pairs (LP) of electrons are shown as dots.

When $P = 1$, we have a trivial case (such as HCl, in which H is considered to be the central atom), in which the only pair is of necessity a bond pair or there would be no molecule at all. The shape is typical of all diatomic molecules regardless of the number of electron pairs—it is a "dumbbell" molecule.

When $P = 2$, as in $BeCl_2$ (Figure 9-2), the molecular geometry is *linear*, because both pairs are bond pairs repelling each other at 180°.

Linear

FIGURE 9-2
Linear molecule.

The compound $SnCl_2$, which appears superficially to be similar to $BeCl_2$, actually is different because $P = 3$, which leads to a triangular coplanar electron-pair geometry. Here, however, $BP = 2$ and $LP = 1$, and the net result is that $SnCl_2$ is an *angular* molecule (Figure 9-3) rather than a linear one.

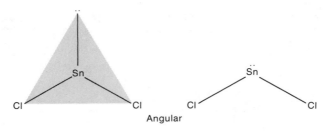

Angular

FIGURE 9-3
Angular molecule.

A triatomic molecule is never called triangular or planar, even though it always is both; it always should be called either linear or angular. In either case, the three atoms will lie in the same plane, because any three noncollinear points always determine a plane.

Boron trichloride also has three electron pairs ($P = 3$), but all three are bond pairs and the molecule is triangular-coplanar (Figure 9-4).

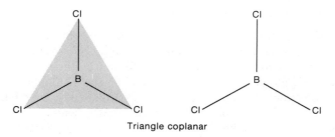

FIGURE 9-4
Triangular-coplanar molecule.

For CCl_4, with $P = 4$, the molecular structure is tetrahedral with C at the center, because BP also is 4 (Figure 9-5).

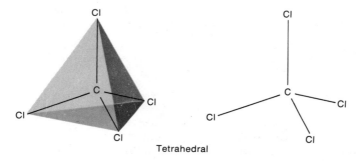

FIGURE 9-5
Tetrahedral molecule.

The ammonia molecule is another in which $P = 4$, but the molecule is called *pyramidal,* not tetrahedral, because there are only three bond pairs (Figure 9-6). At the apex of the pyramid is the N atom, and the lone pair of electrons

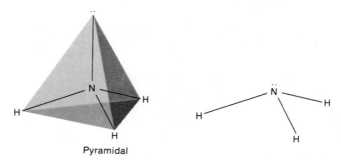

FIGURE 9-6
Pyramidal molecule.

occupies a tetrahedral position in the electron-pair geometry. The term "pyramid" is applied to *irregular* tetrahedra, the term "tetrahedron" only to regular (equal-edged) tetrahedra.

Still another type of molecule exists for $P = 4$; water is an example (Figure 9-7). Here, $BP = 2$, and $LP = 2$. The net result is an angular molecule with the two lone pairs occupying tetrahedral positions in the electron-pair geometry.

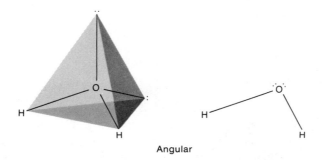

Angular

FIGURE 9-7
Angular molecule.

Note the change in *molecular* geometry that occurs when a proton (H^+ with no electrons) is bonded to an NH_3 or an H_2O molecule through a coordinate covalent bond to form NH_4^+ or H_3O^+; in each, P still equals 4, but the number of *bond* pairs has increased (Figure 9-8).

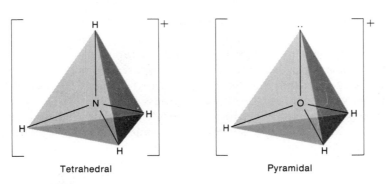

Tetrahedral   Pyramidal

FIGURE 9-8
Tetrahedral molecule (*left*) and pyramidal molecule (*right*).

A common type of molecule is exemplified by PCl_5, in which $P = 5$. Because all of the pairs are bond pairs, it follows that the molecular geometry will be the same as the electron-pair geometry, a $\triangle$-bipyramid (Figure 9-9). Note that all of the P–Cl bond distances are the same, but that the Cl–Cl distances (*not* bonds) are greater between any two Cl atoms in the plane than between an apical Cl

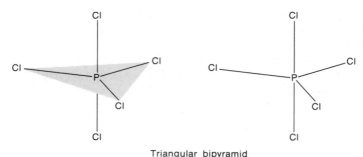

Triangular bipyramid

FIGURE 9-9
Triangular-bipyramidal molecule.

and a Cl atom in the plane. Up to this point, it has made no difference to molecular shape which pair of electrons in a given set of electron pairs was used as a lone pair. But when the electron-pair positions are not equivalent, then the lone-pair positions are all-important in determining the molecular shape.

A bond pair of electrons has a less effective negative charge than a lone pair, because the former's charge is more reduced by its lying between two positive nuclei than the latter's charge is by its being attached to only one nucleus. As a result, we would expect a lone-pair–lone-pair repulsion to be greater than a lone-pair–bond-pair repulsion, and this in turn to be greater than a bond-pair–bond-pair repulsion. That is,

$$LP–LP > LP–BP > BP–BP$$

To simplify the application of these differences in repulsion to the determination of molecular shapes, we can, for practical purposes, ignore the repulsions between pairs of electrons that lie at angles greater than 90° to each other in comparison with those that lie at angles of less than 90°.

Let us apply these principles to the molecule $TeCl_4$, in which $P = 5$, $BP = 4$, and $LP = 1$. Two molecular shapes are possible (Figure 9-10). The one that actually exists is the one with the least repulsion between the electron pairs.

Model (a) has 3 LP–BP repulsions at 90° and
3 BP–BP repulsions at 90°;
Model (b) has 2 LP–BP repulsions at 90° and
4 BP–BP repulsions at 90°.

In both models, all other repulsions are at angles greater than 90° and can be ignored. Because LP–BP repulsion is greater than BP–BP repulsion, it follows that (b) has less electron-pair repulsion than (a), and $TeCl_4$ has the shape of a seesaw, as observed.

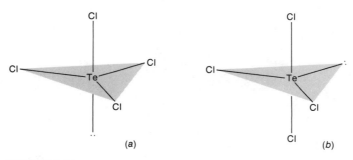

FIGURE 9-10
Two possible shapes of TeCl₄.

In ClF₃, where $P = 5$, we have to make a more complicated decision involving three possible structures (Figure 9-11). We apply the same principles just used for TeCl₄.

> Model (a) has 6 LP–BP repulsions at 90°;
> Model (b) has 1 LP–LP repulsion at 90°,
>   3 LP–BP repulsions at 90°, and
>   2 BP–BP repulsions at 90°.
> Model (c) has 4 LP–BP repulsions at 90° and
>   2 BP–BP repulsions at 90°.

Models (a) and (b) both have more electron-pair repulsion than model (c); hence we would choose the structure of ClF₃ to be a T-shaped molecule as in (c), which in fact it is.

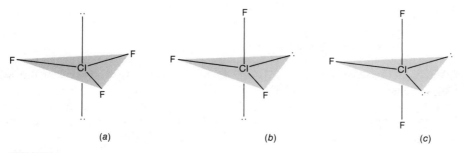

FIGURE 9-11
Three possible shapes of ClF₃.

When $P = 6$, as in SF₆, the structure is simple, because all the pairs are bond pairs, and the molecular geometry is the same as the electron-pair geometry; it is octahedral (Figure 9-12).

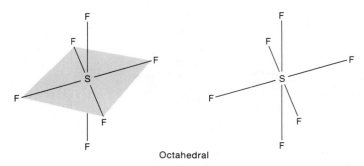

FIGURE 9-12
Octahedral molecule.

The problem is not significantly more difficult for a molecule such as IF$_5$, where P also equals 6, but $BP = 5$ and $LP = 1$. With an electron-pair geometry that is octahedral, all positions are equivalent and it does not matter which position is occupied by the one lone pair. No matter how you look at it, IF$_5$ is a □-based pyramid with the I atom centered in the base (Figure 9-13).

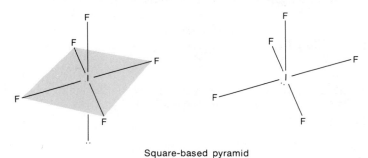

Square-based pyramid

FIGURE 9-13
Square-based pyramidal molecule.

Again, because all octahedral positions are equivalent, the shape of the ion ICl$_4^-$ can easily be determined. Here, $P = 6$, $BP = 4$, and $LP = 2$. Note that one electron was contributed to the central I atom by the negative charge on the ion. The two possible structures are shown in Figure 9-14. We evaluate the minimum repulsion as follows.

> Model (a) has 8 LP–BP repulsions at 90° and
> 4 BP–BP repulsions at 90°;
> Model (b) has 1 LP–LP repulsion at 90°,
> 6 LP–BP repulsions at 90°, and
> 5 BP–BP repulsions at 90°.

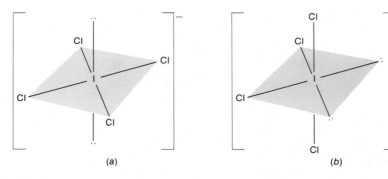

FIGURE 9-14
Two possible shapes of ICl_4^-.

Offhand, the minimum repulsion in the two structures might seem similar, but the decrease in repulsion by having the very strong LP–LP repulsion go from 90° as in (b) to 180° as in (a) is so great that it more than compensates for the smaller simultaneous increase in repulsion caused by having one BP–BP repulsion at 90° as in (b) go to a LP–BP repulsion also at 90° as in (a). As a consequence, ICl_4^- is a □-coplanar molecule as shown in (a).

A good example of a pentagonal-bipyramidal molecule is IF_7, where $P = 7$. All the pairs are bond pairs. Figure 9-15 shows the structure.

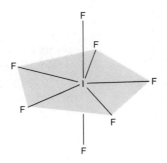

Pentagonal bipyramid

FIGURE 9-15
Pentagonal-bipyramidal molecule.

$SbBr_6^{3-}$ has an especially irregular shape, which can be deduced from the fact that $P = 7$ with $BP = 6$ and $LP = 1$. Note here that Sb has 5 valence electrons of its own, and that three more have been contributed to it by the charge on the ion. Of the two possible shapes in Figure 9-16, the irregular octahedron (b) is more likely, as may be deduced from the following considerations.

Model (a) has 5 LP–BP repulsions at 90°,
5 BP–BP repulsions at 72°, and
5 BP–BP repulsions at 90°.
Model (b) has 2 LP–BP repulsions at 72°,
2 LP–BP repulsions at 90°,
3 BP–BP repulsions at 72°, and
8 BP–BP repulsions at 90°.

The difference between the two models comes down to this:

Model (a) has 3 LP–BP repulsions at 90° and
2 BP–BP repulsions at 72°;
Model (b) has 2 LP–BP repulsions at 72° and
3 BP–BP repulsions at 90°.

Qualitatively, the difference is that model (a) has a partial LP–BP repulsion at 90°, and model (b) has a partial BP–BP repulsion at 90°. Because the LP–BP repulsion is greater, we choose model (b).

TaF_8^{3-} is a molecule that illustrates the situation for $P = 8$ and $BP = 8$. The electron-pair and molecular geometries are the same, corresponding to a square antiprism. A □-antiprism can be thought of as two □-based pyramids, apex-to-apex, with the central atom (Ta in this case) located at the junction of the apices, and the base edges not parallel. The simplest way to visualize this structure is to imagine the central atom at the center of a cube, with the ligands at the eight corners. Any two opposite faces can be thought of as the bases of the two □-based pyramids. Choose two opposite faces and then rotate one of them 45° with respect to the other, keeping them parallel. The resulting figure is a □-antiprism in which the ligands (and therefore the bond pairs) are farthest away from each other. Figure 9-17 illustrates the structure for TaF_8^{3-}. The

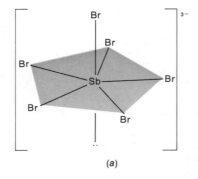
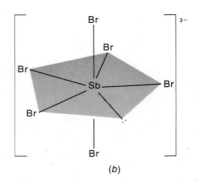

FIGURE 9-16
Two possible shapes of $SbBr_6^{3-}$.

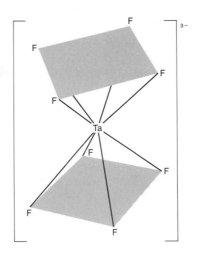

FIGURE 9-17
Square antiprism molecule.

distorted dodecahedron that sometimes occurs for $P = 8$ is too difficult to illustrate in a clear manner, so we ignore it here.

Another consequence of the unequal repulsion between lone pairs and bond pairs of electrons is the distortion of the molecule from the perfectly regular geometric shapes we have discussed. In CCl_4, the Cl–C–Cl angle is the tetrahedral value 109°28′, as expected. In NH_3, however, the H–N–H angle is only 106°45′; the distortion is caused in part by the stronger LP–BP repulsion forcing the H atoms closer together against a weaker BP–BP repulsion. The same effect can be seen in

$SnCl_2$, where the Cl–Sn–Cl angle is less than 120°;

H_2O, where the angle is only 104°27′ (instead of 109°28′);

$TeCl_4$, where the two apical Cl atoms are not quite linear with respect to Te, and the two Cl atoms in the plane have a Cl–Te–Cl angle of a little less than 120°;

ClF_3, where the T-shaped molecule has a bent cross to the T;

IF_5 molecule, where the I atom is slightly below the plane of the four F atoms; and so on.

Not all of these distortion effects can be ascribed to differences in LP–BP and BP–BP repulsions. Some distortion is due to differences in electronegativity between M and the ligands. If the two have equal electronegativity, then of course the bond pairs are shared equally between M and L. If M is more electronegative than L, however, the bond pairs will be held more closely to M. But as they are drawn closer to M, the bond pairs also are drawn closer to each other and, in an effort to reduce this added repulsion, they tend to widen the angle between them. The net result is a distortion. Compare the following sets of bond angles, which reflect both the differences in LP–BP and BP–BP repulsions, and the differences in electronegativity between M and L (H–M–H angles are cited):

$$NH_3, \ 106°45'; \quad PH_3, \ 93°50'; \quad AsH_3, \ 91°35'; \quad SbH_3, \ 91°30';$$
$$H_2O, \ 104°27'; \quad H_2S, \ 92°20'.$$

If the ligands are more electronegative than M, then the bond pairs are drawn farther from M and away from each other, a situation that assists the lone pair in making the L–M–L angle smaller as it operates against this weaker BP–BP repulsion. For example, compare NH_3 (106°45′) with NF_3 (102°9′), and H_2O (104°27′) with OF_2 (101°30′).

Multiple Bonds

If you were asked to draw the electron-dot formula for CH_2O you might be tempted to draw the structure shown in Figure 9-18(a), which would have the △-coplanar structure shown in Figure 9-18(b). This is an *in*correct structure

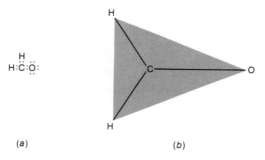

(a)   (b)

FIGURE 9-18
An *incorrect* structure for CH_2O.

because the predicted C–O distance is 1.43 Å (compared to the observed value of 1.23 Å), and because C does not obey the octet rule in this structure, whereas it *could* satisfy the rule by forming a double bond with O. One of the lone pairs on O can become a bond pair, as shown in Figure 9-19(a). The four electrons in the C=O double bond count as belonging to both C and O, so O still obeys the octet rule and now C does also. According to prediction rule #2(b) (p 119), a multiple bond counts as only *one* pair of electrons, so the structure in Figure 9-19(b) is still △-coplanar (not tetrahedral), but the C=O distance is now predicted to be 1.22 Å, in close agreement with fact.

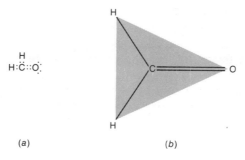

(a)   (b)

FIGURE 9-19
A *correct* structure for CH₂O.

In general, you should always follow the procedure just described when the central atom does not obey the octet rule and the ligands are able to form multiple bonds (see p 121).

PROPERTIES OF MOLECULAR ORBITALS

Resonance

A surprising thing is noted when we examine interatomic distances for certain ions and molecules, such as CO_3^{2-}, NO_2^-, or C_6H_6. If we draw the usual electron-dot formulas for these, we get the structures shown in Figure 9-20.

Using Table 9-1, we would say that in CO_3^{2-} there are one C=O bond of length 1.22 Å and two C–O bonds of length 1.43 Å; in NO_2^-, one N–O bond of length 1.36 Å and one N=O bond of length 1.15 Å; in C_6H_6, three C–C bonds of length 1.54 Å and three C=C bonds of length 1.34 Å. *In actual fact,* all the bond lengths in any one of these three molecules are the same; for CO_3^{2-} all lengths are 1.30 Å; for NO_2^- they are 1.24 Å; and for C_6H_6 they are 1.40 Å. These values are very close to the weighted averages of single and double bond lengths. It appears that, in such molecules or ions, there is no real preference for single or double

FIGURE 9-20
Electron-dot formulas.

bonds to be located between any specific atoms. How should this situation be described? Pauling suggested that the actual state of the molecule is a *resonance hybrid* of all the separate forms that can be written in the classical way. The principal resonance forms for our three examples would be those shown in Figure 9-21.

It must be emphasized that none of these resonance forms actually exists, but the superposition of all of them for a given molecule serves as a good representation, and shows that each bond possesses both single-bond and double-bond characteristics.

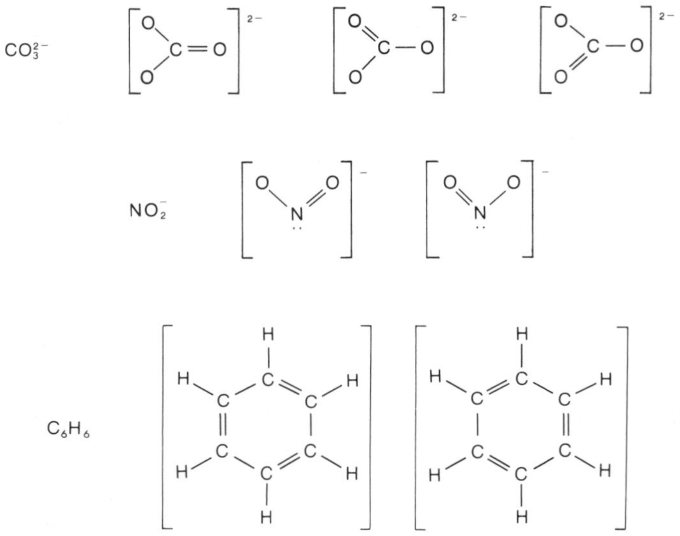

FIGURE 9-21
Resonance forms.

A Molecular Orbital Description (Sigma Bonds)

So far we have looked at molecular orbitals in a simplified way as electron pairs that try to seek locations of minimum potential energy. Now that double bonds are under consideration, it will pay us to examine in greater detail the characteristics of both atomic and molecular orbitals. You recall that an *atomic* orbital is a volume element oriented with respect to the nucleus of the atom where there is a high probability of finding (*at the most*) two electrons that are identical in quantum numbers except for direction of spin. The orbital of each type of electron (*s, p, d,* etc.) has a characteristic shape and orientation.

One important type of *molecular* orbital is regarded as being formed from the "end-overlap" of two atomic orbitals (one from each atom). If two electrons (of opposite spin) fill this molecular orbital (one from each atom, or two from one

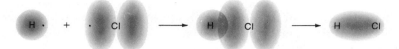

FIGURE 9-22
HCl molecular orbital.

and none from the other), a bond will be formed. In HCl the molecular orbital could be regarded as the end-overlap of the half-filled s orbital of H with the one half-filled p orbital of Cl (Figure 9-22). In ICl, the molecular orbital is the end-overlap of two half-filled p orbitals (Figure 9-23). In making an NH_4^+ ion

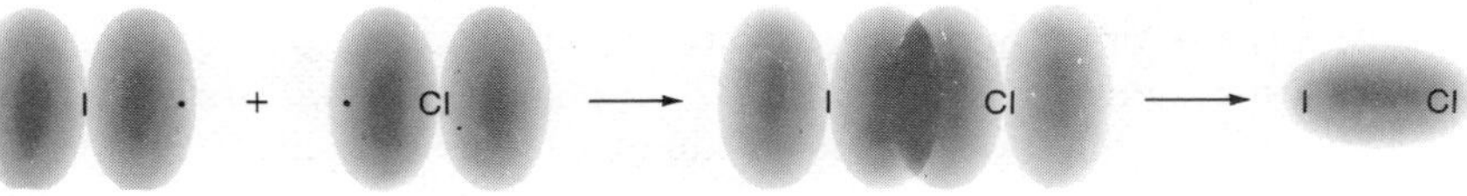

FIGURE 9-23
ICl molecular orbital.

from H^+ and NH_3, the new bond is the molecular orbital resulting from the end-overlap of the empty s orbital of H^+ and the filled (lone-pair) molecular orbital of NH_3 (Figure 9-24). These end-overlap molecular orbitals are called σ (sigma) bonds.

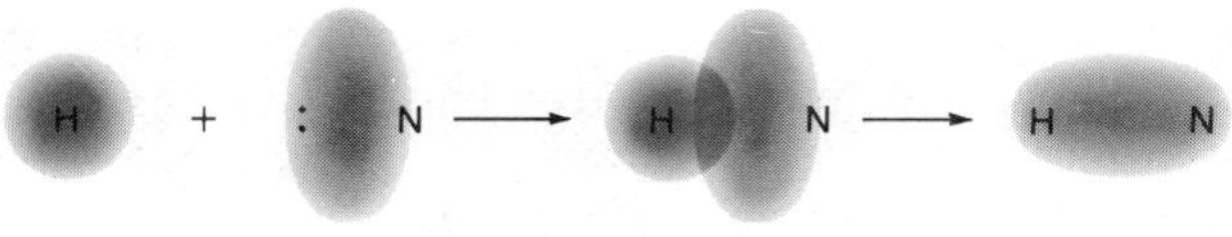

FIGURE 9-24
NH_4^+ molecular orbital.

Hybridization

There is an alternative to resonance as the explanation of why all the bond lengths are the same in molecules such as CO_3^{2-}, NO_2^-, and C_6H_6; it is known as *hybridization of atomic orbitals* to give molecular orbitals. Basically, hybridization is a mathematical operation that considers molecular orbitals to be made of atomic orbitals in whatever way is needed to give the minimum potential energy (the minimum electron-pair repulsion and the most stable arrangement of atoms). The mathematical result leads to a definite geometrical arrangement of *molecular* orbitals that are identical from the standpoint of bond length and bond strength. We shall illustrate the method qualitatively by showing, for a given atom, three different ways to hybridize its s and p atomic orbitals of the same principal quantum number to give three different kinds of molecular orbitals. A basic principle involved in all hybridization processes is that you will always get the same number of molecular orbitals as the number of atomic orbitals that were used. At the moment we shall hybridize atomic orbitals without regard to the element involved or the number of electrons that are available.

In each of the following hybridizations we shall deal with one s and three p atomic orbitals. When we don't hybridize all of them, we must consider how the resulting molecular orbitals are related to the atomic orbitals we don't use.

1. Hybridize one s and three p atomic orbitals to get
   (a) four identical sp^3 tetrahedral *molecular* orbitals

2. Hybridize one s and two p atomic orbitals to get
   (a) three identical sp^2 trigonal ($\triangle$-coplanar) *molecular* orbitals, and
   (b) one *atomic* p orbital perpendicular to the $\triangle$ plane.

3. Hybridize one s and one p atomic orbital to get
   (a) two identical sp digonal (linear) *molecular* orbitals, and
   (b) two mutually perpendicular *atomic* p orbitals that are perpendicular to the line of molecular orbitals.

If you refer back to the earlier figures in this chapter, you can see that, whenever $P = 4$ (corresponding to tetrahedral electron-pair geometry), you were actually using sp^3 molecular orbitals of the central atom to hold lone pairs and/or to end-overlap with ligand orbitals so as to make the molecular orbitals needed to hold bond pairs. Similarly, $\triangle$-coplanar electron-pair geometry uses sp^2 molecular orbitals, and linear electron-pair geometry uses sp molecular orbitals. These orbitals often are described as sp^3, sp^2, and sp without further description because the symbolism itself implies hybridization (and the number and kind of each atomic orbital used) as well as the geometry involved. These are summarized in Table 9-2.

The other electron-pair geometries that are listed in Table 9-2 are also related to specific hybrid molecular orbitals, but they are more complicated because they involve d orbitals as well as s and p. In every case, the s and p orbitals are of the same principal quantum number. However, if the d symbol is listed first (as in d^2sp^3), the d orbitals used in hybridization are of principal quantum number one less than that of s and p. If the d symbol is listed at the end (as in sp^3d^2), all orbitals are of the same principal quantum number.

A Molecular Orbital Description (Pi Bonds)

When the unhybridized p atomic orbitals that are associated with sp^2 and sp molecular orbitals contain no electrons, we need not worry about them. But every time you draw an electron-dot formula that involves a double or a triple bond, you must take these atomic *p* orbitals into account because they contain electrons. Let's say that M and L form a σ bond by the end-overlap of two *sp*² molecular orbitals (one from each atom). This pair of electrons will be localized between M and L. Now, if the △ planes of M and L both lie in the same plane, the *p* orbitals that stick out perpendicularly above and below each of the △ planes will *side-overlap* and then merge to form a *delocalized* molecular orbital consisting of two "clouds" parallel to the plane, one above and one below (Figure 9-25). It is a delocalized orbital (called a π orbital) because, if any electrons are in this orbital, they will be spread out or delocalized with time over both M and L; they will no longer be localized on either of the atoms as *p* electrons. Any double bond can be considered as composed of one σ bond and one π bond—the σ bond localized between M and L, and the π bond delocalized with one electron on each side of the plane. The atoms at *both* ends of a double bond must use *sp*² orbitals.

A triple bond has an analogous interpretation. Using N_2 (: N ::: N :) as an example, first imagine the end-overlap of one *sp* digonal orbital from each N atom to form a σ bond localized between the atoms. Then, orient the pairs of atomic *p* orbitals on each atom so that they are parallel. These *p* orbitals will side-overlap and merge to give two sets of delocalized π orbitals mutually perpendicular to each other (Figure 9-26). Each N atom will hold its lone pair in the *sp* orbital that isn't used to make the σ bond, and the remaining four electrons will be located in the π clouds to form π bonds, with one electron in each of the four delocalized clouds. Any triple bond can be considered as being composed of one localized σ bond and two delocalized π bonds. The atoms at *both* ends of a triple bond must use *sp* orbitals.

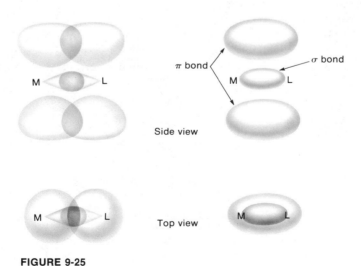

Side view

Top view

FIGURE 9-25
Double bond.

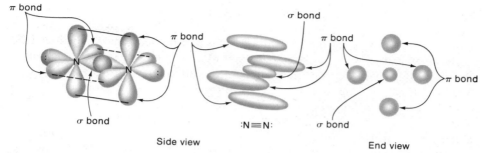

π bond

σ bond

π bond

π bond

σ bond

π bond

:N≡N:

σ bond

Side view

End view

FIGURE 9-26
Triple bond.

Because π bonds always are oriented parallel to the σ bonds with which they are associated, it follows that the molecular shape will be determined solely by the number of σ bonds that the central atom has. It is understandable, therefore, why predictive rule #2(b) on p 119 states that only two electrons of a multiple bond can be used to calculate the number of electron pairs around the central atom.

In those molecules or ions where resonance is involved (see Figure 9-21)—in CO_3^{2-}, for example—it isn't possible to say that *one* oxygen atom will be involved in a double bond and the other two in singles. We must treat all of them the same way, as though they all participate in double-bond formation. In other words, *all* of the O atoms must use sp^2 △-coplanar orbitals, using one to end-overlap with an sp^2 △-coplanar orbital from C to form a σ bond, and using the other two sp^2 orbitals to hold lone pairs. Because all four atoms are using sp^2 orbitals, each one has an atomic p orbital perpendicular to the plane that will side-overlap with the others and merge into two gigantic π clouds parallel to the plane of the CO_3^{2-}, one above the plane and one below (Figure 9-27). The planar CO_3^{2-} ion will be like the hamburger patty between two halves of a hamburger bun (the π clouds). The CO_3^{2-} ion has 24 valence electrons that must be ac-

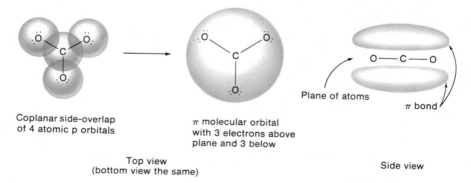

Coplanar side-overlap
of 4 atomic p orbitals

π molecular orbital
with 3 electrons above
plane and 3 below

Plane of atoms

π bond

Top view
(bottom view the same)

Side view

FIGURE 9-27
The π molecular orbital of CO_3^{2-}.

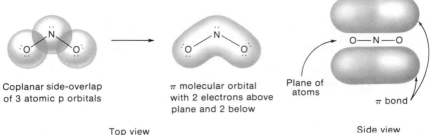

Coplanar side-overlap
of 3 atomic p orbitals

π molecular orbital
with 2 electrons above
plane and 2 below

Plane of
atoms

π bond

Top view
(bottom view the same)

Side view

FIGURE 9-28
The π molecular orbital of NO_2^-.

counted for (Figure 9-20). Six will be involved in the three C–O σ bonds, 12 will be held as lone pairs in sp^2 orbitals of the O atoms, and the remaining six will be delocalized in the π bond (three below the plane and three above).

Likewise for the NO_2^- ion, the two O atoms must be treated the same way as participating in double-bond formation. All three atoms must use sp^2 orbitals, with end-overlap between each O and N to form σ bonds. And of course the three p orbitals that rise perpendicularly from their common plane will side-overlap and merge to give a delocalized split-banana-shaped π molecular orbital, with one half-banana above the plane of the molecule and the other half below (Figure 9-28). Four of the 18 valence electrons are held in the two σ bonds, 10 are held as five lone pairs in the sp^2 orbitals (two on each O, and one on the N), and the remaining four are delocalized over the length of the molecule in the π bond (two above the plane and two below).

In the case of C_6H_6, every C atom must use sp^2 orbitals because each one is at the end of a double bond (see Figure 9-21). Each C will use two of its sp^2 orbitals to end-overlap with its neighboring C atoms to make σ bonds, and its third sp^2 orbital will be used to end-overlap with an s orbital of H to make a third σ bond. In addition, each C atom has a p atomic orbital that stands perpendicular to the plane of the molecule. These six atomic orbitals will side-overlap and merge to form a massive π orbital, again like a hamburger bun, with the planar C_6H_6 hamburger patty sandwiched in between (Figure 9-29). In accounting for the 30 valence electrons (Figure 9-20), we see that 24 of them are used to form 12 σ bonds (six C–C σ bonds, and six C–H) and the remaining six are delocalized over the entire molecule in the massive π molecular orbital (three above the plane and three below).

Some additional examples of double-bonded molecules are shown in Figure 9-30. You should realize that the NO_2 and O_3 molecules involve resonance (and therefore π bonds delocalized over the entire length of the molecules), for there is no reason to believe that one bond in each should be single and the other double. Both bond lengths in each molecule will be equal.

The N_2O molecule illustrates the situation that exists when one atom is involved in two double bonds, like the central N atom here. In order for the

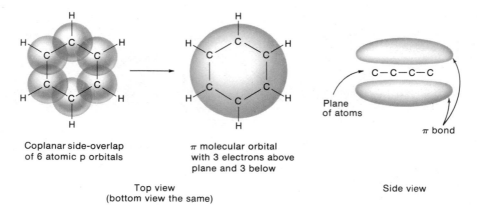

Coplanar side-overlap
of 6 atomic p orbitals

π molecular orbital
with 3 electrons above
plane and 3 below

Top view
(bottom view the same)

Side view

FIGURE 9-29
The π molecular orbital of C_6H_6.

central N atom to provide a p orbital for side-overlap for both the O and the other N, it will be necessary for it to use sp digonal orbitals. The sp orbitals will be used to make σ bonds, while the two remaining p orbitals will merge to make π bonds.

The NO_2 molecule illustrates what must be done with a single unpaired electron on the central atom [see predictive rule #2(c), p 119]. It must occupy a molecular orbital but, unlike the usual situation, the orbital won't be filled. In

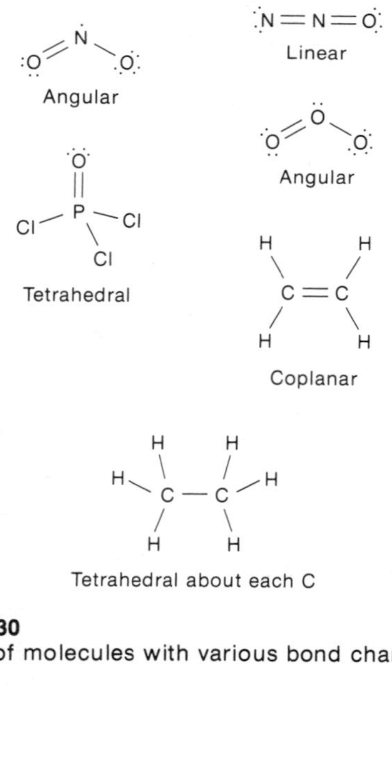

Angular

Linear

Tetrahedral

Angular

Coplanar

Tetrahedral about each C

FIGURE 9-30
Examples of molecules with various bond characteristics.

all respects the molecule will be identical to the NO_2^- ion illustrated in Figure 9-28, except that the sp^2 orbital on N that holds a lone pair in NO_2^- will have only a single electron in NO_2. This single electron will exert a repulsive effect on the other bond pairs, but much less so than a complete lone pair. You would expect the O–N–O angle (134° by experiment) in NO_2 to be greater than the O–N–O angle (115° by experiment) in NO_2^-, for example. You can also see why two NO_2 molecules will so readily react with each other to form N_2O_4; the two half-filled lone-pair orbitals will end-overlap to form an electron pair σ bond between the two N atoms. In addition, the π clouds of each NO_2 will side-overlap to make one gigantic π cloud delocalized over the whole plane of the N_2O_4 molecule.

Finally, we might comment on one other interesting property that is associated with a double or triple bond. The π bond keeps the two halves of a double bond from freely rotating about the σ bond axis that joins them. This property is of the utmost importance in explaining the shape and properties of many organic compounds. In C_2H_4, the two $=CH_2$ groups are unable to rotate freely with respect to each other. In C_2H_6, the two $-CH_3$ groups tend to arrange themselves to give an end-on view as in Figure 9-31, with each H atom at maximum possible distance from the others. The activation energy for rotation about the C–C bond is only 3 kcal/mole, so that rotation occurs relatively easily.

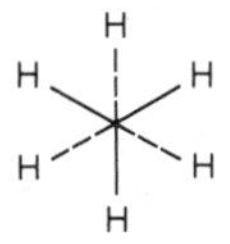

FIGURE 9-31
End-on view of C_2H_6.

Complex Ions

The complex ions that are studied in Chapter 25 consist of a metal ion acting as the "central atom," to which several ligands are attached. The metal ions frequently are transition metal ions, usually characterized as having lost their $4s^2$ (or $5s^2$ or $6s^2$) electrons and having a variable number of d electrons in the outermost (3rd, 4th, or 5th) shell. In these complex ions, the pairs of electrons by which the ligands are attached are *all* furnished by the ligands (thus each ligand must possess at least one lone pair before reacting), and there are *no lone pairs* on the central ion. Thus, the number P of electron pairs around the transition metal ion is just equal to the number of ligands attached; the molecular geometry is the same as the electron-pair geometry.

Although it makes no difference to the *shape* of the ion whether d^2sp^3 or sp^3d^2 orbitals are used (it is octahedral in either case), the *properties* of the resulting two ions may be enormously different (color, paramagnetic susceptibility, and

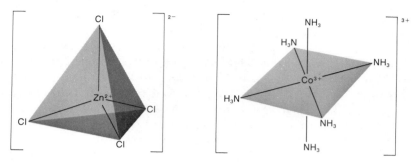

FIGURE 9-32
Molecular shapes of $ZnCl_4^{2-}$ and $Co(NH_3)_6^{3+}$.

strength of metal–ligand bond, for example). Just which of the two hybrid orbitals is used will be determined by how many d electrons the central ion possesses, and by the strength of the ligand field. Consult your textbook for the details. One major exception may occur for those complex ions that have only four ligands. If the central metal ion has eight d electrons, it is very likely that the four ligands will be arranged in a ☐-*coplanar* shape (not tetrahedral), using dsp^2 hybrid orbitals. Some ions with eight d electrons in their outermost shells are Ni^{2+}, Pd^{2+}, and Pt^{2+}.

If you seek to sketch the shape of the ions $ZnCl_4^{2-}$ and $Co(NH_3)_6^{3+}$, you must first recognize that Zn^{2+} and Co^{3+} are transition metal ions, and that $ZnCl_4^{2-}$ and $Co(NH_3)_6^{3+}$ are complex ions. Once this is established, you note that Zn^{2+} has four ligands and therefore four electron pairs with tetrahedral geometry, with each bond pair coming from a chloride ion,

$$:\ddot{Cl}:^-$$

You also note that Co^{3+} has six ligands and therefore six electron pairs with octahedral geometry, with each bond pair coming from the lone pair on the ammonia molecule,

$$
\begin{array}{c}
\ddot{N} \\
\diagup \mid \diagdown \\
H \quad \mid \quad H \\
H
\end{array}
$$

Figure 9-32 shows the shapes of the complex ions.

PROBLEMS A

1. Sketch each of the following molecules or ions, and give approximate values for bond angles, bond distances, and bond energies for those for which sufficient data are given in Table 9-1.

(a) CO_2	(f) IS_2	(k) NO_2Cl	(p) $S_2O_3^{2-}$	(u) ICl_3
(b) SiO_4^{4-}	(g) NO_2^+	(l) $Mo(CN)_8^{4-}$	(q) SbF_7^{2-}	(v) AsS_3^{3-}
(c) SO_3^{2-}	(h) NOF	(m) BrF_4^-	(r) N_2O_4	(w) SO_2Cl_2
(d) XeF_2	(i) $SeCl_4$	(n) AlF_6^{3-}	(s) $SiOF_2$	(x) C_6H_6
(e) SO_2	(j) SiF_6^{2-}	(o) OCN^-	(t) BrF_5	(y) C_2H_3Cl
				(z) $(SiO_3^{2-})_x$

PROBLEMS B

2. Sketch each of the following molecules or ions, and give approximate values for bond angles, bond distances and bond energies for those for which sufficient data are given in Table 9-1.

(a) F_2O	(f) ClO_2	(k) $PO_2F_2^-$	(p) BrF_3	(u) SO_3
(b) SO_4^{2-}	(g) SnS_3^{2-}	(l) XeF_4	(q) H_2O_2	(v) I_5^-
(c) IO_3^-	(h) $IF_2O_2^-$	(m) WF_8^{2-}	(r) SCN^-	(w) C_2H_2
(d) NO_3^-	(i) $COCl_2$	(n) HCN	(s) I_2Br^-	(x) ZrF_7^{3-}
(e) N_3^-	(j) POF_3	(o) H_5IO_6	(t) H_3BO_3	(y) Al_2Cl_6
				(z) $(PO_3^-)_x$

Stoichiometry I: Calculations Based on Formulas

A chemical formula tells the numbers and the kinds of atoms that make up a molecule of a compound. Because each atom is an entity with a characteristic mass, a formula also provides a means for computing the relative weights of each kind of atom in a compound. Calculations based on the numbers and masses of atoms in a compound, or the numbers and masses of molecules participating in a reaction, are designated *stoichiometric calculations*. These weight relationships are important because, although we may think of atoms and molecules in terms of their interactions as structural units, we often must deal with them in the lab in terms of their masses—with the analytical balance. In this chapter, we consider the stoichiometry of chemical formulas. In following chapters, we look at the stoichiometric relations involved in reactions and in solutions.

WEIGHTS OF ATOMS AND MOLECULES

The masses of the atoms are the basis for all stoichiometric calculations. Long before the actual masses of atoms were known, a *relative* scale of masses, called the *atomic weight scale,* was devised. Since 1961, this relative scale has been based on the assignment of the value 12.00000 to the most common isotope of carbon. The table on the inside back cover lists approximate values of the *relative* atomic masses. Because these relative masses represent ratios, they are dimensionless. The atomic masses of the different elements can be expressed in various units, but the ratios of the masses will always be the same as the ratios of these relative masses.

With an established scale of relative atomic weights, it is simple to compute the relative molecular weights on the same scale. All you have to do is take the sum of the relative masses of the atoms that make up the molecule. Thus, the relative molecular weight of H_2SO_4 is $(2 \times 1.0) + (1 \times 32.1) + (4 \times 16.0) = 98.1$. With more accurate values of the relative atomic weights, the value is found to be 98.07. Regardless of the mass units used, the masses of the H atom, the O_2 molecule, and the H_2SO_4 molecule will stand in the ratio of $1.0079 : 31.9984 : 98.07$.

Ever since instrumentation has been sophisticated enough to determine the masses of individual atoms in an indirect fashion, it has been possible to have a list of *absolute* values of the atomic masses. The mass of an atom is incredibly small; expressed in grams, these values are cumbersome to use in day-to-day work and conversation. In recent years, this difficulty has been surmounted by using a special mass unit called a *dalton* (d): $1 d = 1.6604 \times 10^{-24}$ g. This unit is chosen so that the mass of the most common isotope of the carbon atom equals 12.00000 d. Thus, the absolute atomic weight of H is 1.0079 d, and the absolute molecular weights of O_2 and H_2SO_4 are 31.9984 d and 98.07 d, respectively. The absolute atomic and molecular weights expressed in daltons (sometimes called the Dalton atomic and molecular weights) have the same numerical values as the relative atomic and molecular weights, but they represent the *actual masses* of the atoms and molecules involved; we can convert them to grams by multiplying by 1.6604×10^{-24} g/d.

THE MOLE

In the laboratory, we normally work with very large numbers of atoms or molecules, usually with weights conveniently expressed in grams. Suppose we have a certain weight of O_2 and another certain weight of H_2SO_4; if the ratio of these two weights is $31.9984 : 98.07$, we know that each weight contains the same number of molecules (because the weight ratio is the same as that of individual molecules from the relative molecular weights). It is convenient to use the relative atomic weight scale with values expressed in grams. Thus, the gram atomic weight of H is 1.0079 g; the gram molecular weight of O_2 is 31.9984 g; and the gram molecular weight of H_2SO_4 is 98.07 g. These gram atomic or molecular weights are also called *mole weights*. One mole weight of any substance must contain the same number of atoms or molecules as one mole weight of any other substance. This quantity is usually called one *mole* of the substance. Note that the same numerical scales are used for relative, Dalton, and gram molecular weights; only the units are different.

Molecule	Relative molecular weights	Dalton molecular weights	Gram molecular weights
H	1.0079	1.0079 d/atom	1.0079 g/mole
O_2	31.9984	31.9984 d/molecule	31.9984 g/mole
H_2SO_4	98.07	98.07 d/molecule	98.07 g/mole

If a chemical formula states that one molecule of oxygen will react with two molecules of hydrogen to form two molecules of water, we can immediately conclude that one mole of oxygen will react with two moles of hydrogen to form two moles of water. Thus we can readily determine the proportions of masses in grams that will react on the laboratory scale.

The Dalton scale is particularly useful when describing enormous and complicated biological structures, such as chromosomes, virus, mitochondria, and ribosomes. The mass of such a structure can be stated in daltons, even though the quantity of material available in the laboratory would normally be only a very tiny fraction of a mole.

AVOGADRO'S NUMBER

We have noted that one mole of a substance always contains a certain number of molecules (or atoms), regardless of the substance involved. This number is called *Avogadro's number* (in honor of the scientist who first suggested the concept, long before the value of the number could be determined). Using the definition of the dalton and the scale of Dalton atomic (or molecular) weights, we can readily determine the value of Avogadro's number. For example, we have seen that the absolute molecular weight of O_2 is 31.9984 d/molecule. Therefore,

$$\frac{31.9984 \text{ g/mole}}{31.9984 \text{ d/molecule} \times 1.6604 \times 10^{-24} \text{ g/d}} = 6.0225 \times 10^{23} \text{ molecules/mole}$$

This value, of course, was determined experimentally and was used to define the dalton; we have simply reversed that computation here. The value of Avogadro's number, 6.023×10^{23} molecules (or atoms) per mole, represents the fantastically large number of molecules you are dealing with every time you use a *gram* molecular weight (mole weight) of a substance. Note that Avogadro's number is numerically equal to the reciprocal of the dalton expressed in grams.

The term "mole" commonly is used to represent the number of molecules (or atoms) in a quantity of material; that is, one mole of molecules = Avogadro's number of molecules. In this sense, a mole is a dimensionless number, just as a dozen means 12. For example, you could talk about a mole of caterpillars, meaning 6.023×10^{23} caterpillars. However, the number is seldom useful except in talking about molecules or atoms. The metric prefixes commonly are used to give such units as millimoles, nanomoles, or picomoles.

PROBLEM:

How many molecules are there in 20.0 g of benzene, C_6H_6?

SOLUTION:

First find how many *moles* of C_6H_6 there are in 20.0 g, then use Avogadro's number to find the number of molecules.

Mole weight of C_6H_6 = $(6 \times 12.0 \text{ g/mole}) + (6 \times 1.0 \text{ g/mole}) = 78.0 \text{ g/mole}$

$$\text{Moles of } C_6H_6 = \frac{20.0 \text{ g}}{78.0 \text{ g/mole}} = 0.256 \text{ moles}$$

Number of molecules = $(0.256 \text{ moles})(6.023 \times 10^{23} \text{ molecules/mole})$

$$= 1.54 \times 10^{23} \text{ molecules}$$

PERCENTAGE COMPOSITION

A chemical formula may be used to compute the percentage composition of a compound; that is, the percentage by weight of each type of atom in the compound.

PROBLEM:

Calculate the percentages of oxygen and hydrogen in water, H_2O.

SOLUTION:

The formula shows that 1 mole of H_2O contains 2 moles of H and 1 mole of O. The mole weight is $(2 \times 1.0 \text{ g/mole}) + (1 \times 16.0 \text{ g/mole}) = 18.0 \text{ g/mole}$. This calculation shows that there are 2.0 g H and 16.0 g O in 18.0 g H_2O. Therefore,

$$\% \text{ H} = \frac{2.0 \text{ g}}{18.0 \text{ g}} \times 100 = 11.1\%$$

$$\% \text{ O} = \frac{16.0 \text{ g}}{18.0 \text{ g}} \times 100 = 88.9\%$$

PROBLEM:

Compute the percentages of K, Fe, C, N, and H_2O in $K_4Fe(CN)_6 \cdot 3H_2O$ crystals.

SOLUTION:

The dot in the formula indicates that 3 moles of water are combined with 1 mole of $K_4Fe(CN)_6$ in the crystalline compound. The mole weight is $(4 \times 39.1 \text{ g/mole}) + (1 \times 55.8 \text{ g/mole}) + (6 \times 12.0 \text{ g/mole}) + (6 \times 14.0 \text{ g/mole}) + (3 \times 18.0 \text{ g/mole}) = 422.2 \text{ g/mole}$. Each of these weights per mole divided by the weight of a mole will give the percentage, as follows.

$$\%K = \frac{4 \times 39.1 \text{ g}}{422.2 \text{ g}} \times 100 = 37.0\%$$

$$\%Fe = \frac{55.8 \text{ g}}{422.2 \text{ g}} \times 100 = 13.2\%$$

$$\%C = \frac{6 \times 12.0 \text{ g}}{422.2 \text{ g}} \times 100 = 17.1\%$$

$$\%N = \frac{6 \times 14.0 \text{ g}}{422.2 \text{ g}} \times 100 = 19.9\%$$

$$\%H_2O = \frac{3 \times 18.0 \text{ g}}{422.2 \text{ g}} \times 100 = 12.8\%$$

CALCULATION OF FORMULAS FROM CHEMICAL ANALYSIS

When a new chemical compound is prepared, we do not know its formula. To establish the formula, we find by experiment the weights of the various atoms in the compound, and from these weights we compute the relative number of each kind of atom in the molecule. The formula so computed is the simplest formula, not necessarily the true one. It is therefore called the *empirical* formula. For example, we would find the empirical formula for benzene to be CH, whereas the true formula is C_6H_6. To get the true formula from the empirical formula, we must also be able to determine the molecular weight. (This is accomplished by methods that we discuss later.)

PROBLEM:

A sample of chromium weighing 0.1600 g is heated with an excess of sulfur in a covered crucible. After the reaction is complete, the unused sulfur is vaporized by heating and allowed to burn away. The cooled residue remaining in the crucible weighs 0.3082 g. Find the empirical formula of the chromium–sulfur compound that formed.

SOLUTION:

This is a logical place to use Dalton atomic weights, because what we want to know is the number of atoms in a molecule. Instead of using grams, use daltons (d). Think of this experiment as having been performed with 1600 d of Cr to obtain

3082 d of residue; the increase in weight was due to $3082\ d - 1600\ d = 1482\ d$ of S combining with Cr. Then use the relative atomic weight scale in d/atom to get

$$\text{atoms of Cr in } 1600\ d = \frac{1600\ d}{52.0\ d/atom} = 30.8 \text{ atoms Cr}$$

$$\text{atoms of S in } 1482\ d = \frac{1482\ d}{32.1\ d/atom} = 46.2 \text{ atoms S}$$

We could write the formula as $Cr_{30.8}S_{46.2}$, but what we want is the *simplest* formula. The proportions will be the same if we just divide each of the numbers (subscripts) by the smallest, to get

$$Cr_{\frac{30.8}{30.8}} S_{\frac{46.2}{30.8}} = CrS_{1.5}$$

This formula is unacceptable because it erroneously implies that we can split atoms in chemical reactions. Therefore we multiply by 2 to get Cr_2S_3, the correct empirical formula.

We could have thought of the experiment as being done with 0.1600 d of Cr and 0.1482 d of S; the final result would have been the same. By taking the larger quantities we avoided the uneasy feeling you might have had when it looked as if fractions of atoms were combining (0.00308 atoms of Cr and 0.00462 atoms of S).

PROBLEM:
A compound contains 90.6% Pb and 9.4% O by weight. Find the empirical formula.

SOLUTION:
The results of analysis usually are given in percentages of the constituents, not in terms of the amounts actually weighed (as in the last problem). This permits comparison of results from different experiments. You recall that "percent" means "per 100." We can say, then, that if we have 100 g of the compound, 90.6 g is Pb and 9.4 g is O. We can also say that if we have 100 d of compound, 90.6 d is Pb and 9.4 d is O. Using Dalton atomic weights we can easily find the number of atoms in 100 d of compound, as follows.

$$\text{atoms of Pb in } 90.6\ d = \frac{90.6\ d}{207.2\ d/atom} = 0.437 \text{ atoms Pb}$$

$$\text{atoms of O in } 9.4\ d = \frac{9.4\ d}{16.0\ d/atom} = 0.587 \text{ atoms O}$$

As in the previous problem, we divide each of the numbers of atoms by the smaller number to give

$$Pb_{\frac{0.437}{0.437}} O_{\frac{0.587}{0.437}} = PbO_{1.34}$$

We know that the subscripts must be simple whole numbers, a situation readily obtained by multiplying through by 3 to give Pb_3O_4, the correct empirical formula.

Frequently the fractional numbers that must be multiplied through by simple integers do not yield *exactly* whole numbers (for example, $3 \times 1.34 = 4.02$, not

4.00). Do not worry about this. Instead, realize that there will always be a little experimental error that you will have to allow for. Normally, look for relatively simple numbers for subscripts.

PROBLEM:

Many crystalline compounds contain water of crystallization that is driven off when the compound is heated. The loss of weight in heating can be used to determine the formula. For example, a hydrate of barium chloride, $BaCl_2 \cdot xH_2O$, weighing 1.222 g is heated until all the combined water is expelled. The dry powder remaining weighs 1.042 g. Compute the formula for the hydrate.

SOLUTION:

In solving this problem we seek the value of x whose units must be moles of H_2O per mole of $BaCl_2$. Our objective, therefore, is to find the number of moles of each in the given sample. The 1.042 g residue is anhydrous $BaCl_2$ and the loss in weight is H_2O.

$$\text{Weight of } H_2O = 1.222 \text{ g} - 1.042 \text{ g} = 0.180 \text{ g } H_2O$$

From these weights we calculate the moles.

$$\text{Moles of } BaCl_2 = \frac{1.042 \text{ g } BaCl_2}{208.4 \dfrac{\text{g } BaCl_2}{\text{mole } BaCl_2}} = 0.00500 \text{ moles } BaCl_2$$

$$\text{Moles of } H_2O = \frac{0.180 \text{ g } H_2O}{18.0 \dfrac{\text{g } H_2O}{\text{mole } H_2O}} = 0.0100 \text{ moles } H_2O$$

We obtain x, the moles of H_2O per mole of $BaCl_2$, by dividing the moles of H_2O by the moles of $BaCl_2$ to which the water is attached, to give

$$x = \frac{0.0100 \text{ moles } H_2O}{0.00500 \text{ moles } BaCl_2} = 2 \frac{\text{moles } H_2O}{\text{mole } BaCl_2}$$

and the formula for the hydrate is $BaCl_2 \cdot 2H_2O$.

DETERMINATION OF ATOMIC WEIGHTS

There is a simple way by which you can find the *approximate* atomic weight of an element. It is described on pp 211–212. With this method, you determine the specific heat of the element experimentally, then divide it into 6.2 (using the rule of Dulong and Petit); the result is the approximate atomic weight. Even if you didn't know the name of the element, you could use this approximate value along with an accurate chemical analysis to find an *accurate* value of the atomic weight.

PROBLEM:

From a specific heat measurement, the approximate atomic weight of a metal (M) is found to be 135. A 0.2341 g sample of M is heated to constant weight in air to convert it to the oxide. The weight of the residue is 0.2745 g. Find the true atomic weight of the metal (and therefore its identity), and determine the formula of the metal oxide.

SOLUTION:

Using the approximate atomic weight, we can try to find the empirical formula for the metal oxide. We can say that 2341 d of M combine with 2745 d − 2341 d = 404 d of O. The number of atoms of each is

$$\text{atoms of M} = \frac{2341 \text{ d}}{135 \text{ d/atom}} = 17.3 \text{ atoms of M}$$

$$\text{atoms of O} = \frac{404 \text{ d}}{16 \text{ d/atom}} = 25.3 \text{ atoms of O}$$

from which we derive the empirical formula to be

$$M_{17.3}O_{25.3} = M_{\frac{17.3}{17.3}}O_{\frac{25.3}{17.3}} = MO_{1.46} = M_2O_{2.9}$$

We find that our *approximate* atomic weight value yields a formula of $M_2O_{2.9}$, which we know cannot be correct. We also know that $M_{20}O_{29}$ is unreasonable. What we perceive is that the formula is undoubtedly M_2O_3, and that the apparent error is undoubtedly caused by the approximate atomic weight. Assuming that M_2O_3 is correct and that the analysis is good, we can calculate the correct atomic weight (X) as follows:

$$\text{wt. fraction of M in } M_2O_3 = \frac{0.2341 \text{ g}}{0.2745 \text{ g}} = \frac{2 \times X \text{ g/mole}}{(2 \times X \text{ g/mole}) + (3 \times 16.0 \text{ g/mole})}$$

$$= \frac{2X}{2X + 48.0} = 0.853$$

$$X = 0.853X + (0.853)(24.0)$$

$$X = \frac{(0.853)(24.0)}{1 - 0.853} = 139 \frac{\text{g}}{\text{mole}}$$

The element must be lanthanum—not cesium or barium, as seemed likely from the approximate atomic weight. This is further confirmed by the formula, which would have been Cs_2O for Cs, or BaO for Ba. La_2O_3 agrees with the fact that La has a valence of +3.

ISOTOPES AND ATOMIC WEIGHT

There is the implication in the first part of this chapter that all of the atoms of a given element are the same and have the same mass. Although their electronic structures and reactivities are the same, their masses actually may vary. These different mass forms are called *isotopes*. It is because the percentage of each isotope of an element is always the same throughout nature that the mass of each atom of that element *appears* to be the same for all. The mass spectrometer provides a means of accurately finding the percentage of each isotope and its actual mass. For carbon, for example, two isotopes are found: 98.892% of one isotope whose mass is 12.00000 d/atom, and 1.108% of the other whose mass is 13.00335 d/atom. If we should take a million carbon atoms at random, then 988,920 of them would each weigh 12.00000 d and 11,080 of them would each weigh 13.00335 d. The total weight would be

$$(988,920 \text{ atoms})(12.00000 \text{ d/atom}) = 11,867,040 \text{ d}$$

$$(11,080 \text{ atoms})(13.00335 \text{ d/atom}) = \underline{\quad 144,080 \text{ d}}$$

$$\text{Total weight of } 10^6 \text{ atoms} = 12,011,120 \text{ d}$$

$$\text{Average weight of one atom} = \frac{12,011,120 \text{ d}}{10^6 \text{ atoms}} = 12.01112 \frac{\text{d}}{\text{atom}}$$

This average weight is the weight that *all* carbon atoms *appear* to have as they are dealt with by chemists. The average atomic weights for all the elements have been determined in a similar way, and it is these averages that are listed in the relative atomic weight scale inside the back cover.

PROBLEMS A

1. Find the percentage composition of (the percentage by weight of each element in) each of the following compounds.
   - (a) N_2O
   - (b) NO
   - (c) NO_2
   - (d) Na_2SO_4
   - (e) $Na_2S_2O_3$
   - (f) $Na_2SO_4 \cdot 10H_2O$
   - (g) $Na_2S_2O_3 \cdot 5H_2O$
   - (h) $Ca(CN)_2$
   - (i) $(NH_4)_2CO_3$
   - (j) $UO_2(NO_3)_2 \cdot 6H_2O$
   - (k) Penicillin, $C_{16}H_{26}O_4N_2S$

2. What is the weight of 1.00 mole of each compound in Problem 1?

3. How many moles are in 1.00 lb of each compound in Problem 1?

4. Find the number of molecules in
   - (a) 25.0 g H_2O
   - (b) 1.00 oz of sugar, $C_{12}H_{22}O_{11}$
   - (c) 1.00 microgram of NH_3
   - (d) 5.00 ml of CCl_4 whose density is 1.594 g/ml

NOTE: Problems concerning the determination of approximate atomic weights by the rule of Dulong and Petit may be found in Chapter 14.

(e) An aluminum rod 20.0 cm long and 1.00 cm in diameter. The density of aluminum is 2.70 g/ml.

5. In Friedrich Wöhler's *Grundriss der Chemie*, published in 1823, the atomic weight of oxygen is given as 100. On this basis calculate the molecular weight of NH_4Cl.

6. From the following analytical results (percentage by weight), determine the empirical formulas for the compounds analyzed.
   (a) 77.7% Fe, 22.3% O
   (b) 70.0% Fe, 30.0% O
   (c) 72.4% Fe, 27.6% O
   (d) 40.2% K, 26.9% Cr, 32.9% O
   (e) 26.6% K, 35.4% Cr, 38.0% O
   (f) 92.4% C, 7.6% H
   (g) 75.0% C, 25.0% H
   (h) 21.8% Mg, 27.9% P, 50.3% O
   (i) 66.8% Ag, 15.9% V, 17.3% O
   (j) 52.8% Sn, 12.4% Fe, 16.0% C, 18.8% N

7. Weighed samples of the following hydrates are heated to drive off the water, and then the cooled residues are weighed. From the data given, find the formulas of the hydrates.
   (a) 0.695 g of $CuSO_4 \cdot x H_2O$ gave a residue of 0.445 g
   (b) 0.573 g of $Hg(NO_3)_2 \cdot x H_2O$ gave a residue of 0.558 g
   (c) 1.205 g of $Pb(C_2H_3O_2)_2 \cdot x H_2O$ gave a residue of 1.032 g
   (d) 0.809 g of $CoCl_2 \cdot x H_2O$ gave a residue of 0.442 g
   (e) 2.515 g of $CaSO_4 \cdot x H_2O$ gave a residue of 1.990 g

8. Weighed samples of the following metals are completely converted to other compounds by heating them in the presence of other elements, and then are reweighed to find the increase in weight. The excess of the nonmetal is easily removed in each case. From the data given, find the formulas of the compounds formed.
   (a) 0.527 g of Cu gave a 0.659 g residue with S
   (b) 0.273 g of Mg gave a 0.378 g residue with N_2
   (c) 0.406 g of Li gave a 0.465 g residue with H_2
   (d) 0.875 g of Al gave a 4.325 g residue with Cl_2
   (e) 0.219 g of La gave a 0.256 g residue with O_2

9. From a specific heat measurement, the approximate atomic weight of a metal is found to be 136. A 0.3167 g sample of this metal is heated to constant weight in air to convert it to the oxide, yielding a residue that weighs 0.3890 g. Find the true atomic weight of the metal.

10. The following values of isotopic atomic weights and abundances are obtained with a mass spectrometer. Compute the chemical atomic weights for the elements involved. The isotopic weights are in parentheses.
    (a) For neon: 90.51% (19.99872), 0.28% (20.99963), and 9.21% (21.99844).
    (b) For sulfur: 95.06% (31.98085), 0.74% (32.98000), 4.18% (33.97710), and 0.02% (35.97800).

11. It is found experimentally that when a metal M is heated in chlorine gas, 0.540 g of M gives 2.67 g of metal chloride. The formula of the chloride is not known.

(a) Compute possible values of the atomic weight of M, for each of the following formulas: MCl, MCl$_2$, MCl$_3$, MCl$_4$.

(b) It is found by other methods that the atomic weight of M is about 27. Which of the above formulas is the correct one?

12. A metal forms two different chlorides. Analysis shows one to be 54.7% Cl and the other to be 64.4% Cl by weight. What are the possible values of the atomic weight of the metal?

13. An organic compound containing C, H, O, and S is subjected to two analytical procedures. When a 9.33 mg sample is burned, it gives 19.50 mg of CO$_2$ and 3.99 mg of H$_2$O. A separate 11.05 mg sample is fused with Na$_2$O$_2$, and the resulting sulfate is precipitated as BaSO$_4$, which (when washed and dried) weighs 20.4 mg. The amount of oxygen in the original sample is obtained by difference. Determine the empirical formula of this compound.

PROBLEMS B

14. Find the percentage composition of (the percentage by weight of each element in) each of the following compounds.

(a) NH$_3$ (g) (NH$_4$)$_2$CrO$_4$
(b) N$_2$H$_4$ (h) CaCN$_2$
(c) HN$_3$ (i) PtP$_2$O$_7$
(d) Zn(NO$_2$)$_2$ (j) BiONO$_3 \cdot$ H$_2$O
(e) Zn(NO$_3$)$_2$ (k) Streptomycin, C$_{21}$H$_{29}$O$_{12}$N$_7$
(f) Zn(NO$_3$)$_2 \cdot$ 6H$_2$O

15. What is the weight of 1.00 mole of each compound in Problem 14?
16. How many moles are in 1.00 lb of each compound in Problem 14?
17. Determine the number of molecules in
(a) 50.0 g of mercury
(b) 0.500 lb of chloroform, CHCl$_3$
(c) 1.00 nanogram of HCl
(d) 25.0 ml of benzene, C$_6$H$_6$, whose density is 0.879 g/ml
(e) a copper bar 1″ × 2″ × 24″. The density of copper is 8.92 g/ml.

18. From the following analytical results (percentage by weight), determine the empirical formulas for the compounds analyzed.

(a) 42.9% C, 57.1% O   (f) 32.4% Na, 22.6% S, 45.0% O
(b) 27.3% C, 72.7% O   (g) 79.3% Tl, 9.9% V, 10.8% O
(c) 53.0% C, 47.0% O   (h) 25.8% P, 26.7% S, 47.5% F
(d) 19.3% Na, 26.8% S, 53.9% O   (i) 19.2% P, 2.5% H, 78.3% I
(e) 29.1% Na, 40.5% S, 30.4% O   (j) 14.2% Ni, 61.3% I, 20.2% N,
                                             4.3% H

19. Weighed samples of the following hydrates are heated to drive off the water, and then the cooled residues are weighed. From the data given, find the formulas of the hydrates.

(a) 0.520 g of $NiSO_4 \cdot x\,H_2O$ gave a residue of 0.306 g

(b) 0.895 g of $MnI_2 \cdot x\,H_2O$ gave a residue of 0.726 g

(c) 0.654 g of $MgSO_4 \cdot x\,H_2O$ gave a residue of 0.320 g

(d) 1.216 g of $CdSO_4 \cdot x\,H_2O$ gave a residue of 0.988 g

(e) 0.783 g of $KAl\,(SO_4)_2 \cdot x\,H_2O$ gave a residue of 0.426 g

20. Weighed samples of the following metals are completely converted to compounds by heating them in the presence of the specified elements, and then are reweighed to find the increase in weight. The excess of the nonmetal is easily removed in each case. From the data given find the formulas of the compounds formed.

(a) 0.753 g of Ca gave a 0.792 g residue with H_2

(b) 0.631 g of Al gave a 1.750 g residue with S

(c) 0.137 g of Pb gave a 0.243 g residue with Br_2

(d) 0.211 g of U gave a 0.249 g residue with O_2

(e) 0.367 g of Co gave a 0.463 g residue with P

21. The following values of isotopic weights and abundances are obtained with a mass spectrometer. Compute the chemical atomic weights for the elements involved. The isotopic weights are in parentheses.

(a) For magnesium: 78.70% (23.98504), 10.13% (24.98584), and 11.17% (25.98259).

(b) For titanium: 7.95% (45.9661), 7.75% (46.9647), 73.45% (47.9631), 5.51% (48.9646), and 5.34% (49.9621).

22. A metal forms two different chlorides. Analysis shows one to be 40.3% metal and the other to be 47.4% metal by weight. What are the possible values of the atomic weight of the metal?

23. A sample of an organic compound containing C, H, and O, which weighs 12.13 mg, gives 30.6 mg of CO_2 and 5.36 mg of H_2O in combustion. The amount of oxygen in the original sample is obtained by difference. Determine the empirical formula of this compound.

24. A 5.135 g sample of impure limestone ($CaCO_3$) yields 2.050 g of CO_2 (which was absorbed in a soda-lime tube) when treated with an excess of acid. Assuming that limestone is the only component that would yield CO_2, calculate the percentage purity of the limestone sample.

11

Gases

In order for chemists to prepare and handle gases under a variety of conditions, they must understand the relationships between the weight, volume, temperature, and pressure of a gas sample. Measurement of the first three of these quantities is relatively straightforward. Because the measurement of pressure can present some complications, we discuss it before we consider the interrelationship of all four variables.

MEASUREMENT OF GAS PRESSURE

The most common laboratory instrument used to measure gas pressure is a manometer, a glass U-tube partially filled with a liquid (Figure 11-1). Mercury is the most commonly used liquid, because it is fairly nonvolatile, chemically inactive, and dense, and it does not dissolve gases or wet (adhere to) glass. Because mercury does not wet glass, its meniscus will curve upward instead of down, and its position is recorded as that of the horizontal plane tangent to the top of the meniscus.

Figure 11-1 illustrates the measurement of gas pressure with a manometer. In part (a), both sides of the mercury column are at the same pressure, and the mercury level is the same in both tubes. In part (b), the righthand tube has been evacuated and, as a result, the mercury has risen until the weight of the mer-

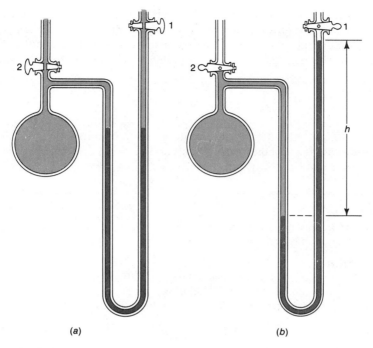

FIGURE 11-1
Mercury manometer.   **(a)** Atmospheric pressure in both tubes. **(b)** Gas
pressure in flask, and vacuum in righthand tube.

cury column of height h exactly balances the gas pressure in the lefthand tube.
When one side of the manometer is open to the atmosphere and the other side is
evacuated, the instrument functions as a barometer; this could be the arrangement in Figure 11-1(b), with stopcock 2 opened to the atmosphere. At sea level
and a temperature of 0°C, the average barometric pressure corresponds to a
760.0 mm column of mercury. Such a pressure is designated as one atmosphere
(1 atm). It is necessary to specify the temperature when using a mercury manometer, because mercury expands with an increase in temperature. Because of
the mercury expansion, a longer column of mercury will be needed at a higher
temperature in order to provide the same mass of mercury required to balance a
given gas pressure.

Various units are used to express gas pressures. Because pressure is defined
as force per unit area, one should employ units of force in measurements and
calculations. In practice, however, it is convenient to use units more directly
related to the measurements (such as the height of the mercury manometer
column), or to use mass units per unit area (such as g/cm² or lb/in²). In scientific
work the commonly used unit is the *torr*, defined as the pressure that will

support a column of mercury exactly 1 mm in height at a temperature of 0°C. In order to correct mercury manometer readings, H (in mm) taken at normal laboratory temperatures (t in °C) to values P (in torr) that would have been observed had the mercury temperature been 0°C, one can use the simple formula

$$P = H[1 - \alpha t] \qquad (11\text{-}1)$$

where $\alpha = 1.63 \times 10^{-4}$ deg^{-1} if the scale is made of brass, or $\alpha = 1.72 \times 10^{-4}$ deg^{-1} if the scale is etched on glass. Conversions from one set of pressure units to others are illustrated in the following problems.

PROBLEM:
Express a pressure of 1 atm in terms of the following units: (a) grams per cm²; (b) pounds per in²; (c) dynes per cm².

SOLUTION:
(a) We assume that the mercury column of the barometer has a diameter of 0.60 cm. At a temperature of 0°C, the density of mercury is 13.60 g/ml, and the column height corresponding to 1 atm is 76.00 cm.

$$\text{Cross-sectional area of column} = (\pi)(0.30 \text{ cm})^2$$

$$\text{Volume of mercury} = (\pi)(0.30 \text{ cm})^2(76.00 \text{ cm})$$

$$\text{Mass of mercury} = (\pi)(0.30 \text{ cm})^2(76.00 \text{ cm})(13.60 \text{ g/cm}^3)$$

$$1 \text{ atm} = \frac{\text{mass}}{\text{area}} = \frac{(\pi)(0.30 \text{ cm})^2(76.00 \text{ cm})(13.60 \text{ g/cm}^3)}{(\pi)(0.30 \text{ cm})^2}$$

$$= 1033 \text{ g/cm}^2$$

Note that the result does *not* depend on the cross-sectional area we assumed for the mercury column, because this value cancels in the computation.
(b) To convert 1 atm to pounds per in², we use the approximate factors of 454 g/lb and 2.54 cm/in.

$$\text{Pounds per in}^2 = \left(1033 \; \frac{\text{g}}{\text{cm}^2}\right)\left(2.54 \; \frac{\text{cm}}{\text{in}}\right)^2\left(\frac{1 \text{ lb}}{454 \text{ g}}\right)$$

$$1 \text{ atm} = 14.7 \text{ lb/in}^2$$

(c) To convert 1 atm to dynes/cm², multiply the mass in grams by the acceleration due to gravity.

$$\text{Dynes per cm}^2 = \left(1033 \; \frac{\text{g}}{\text{cm}^2}\right)\left(980.7 \; \frac{\text{dynes}}{\text{g}}\right)$$

$$1 \text{ atm} = 1.013 \times 10^6 \text{ dynes/cm}^2$$

THE IDEAL GAS LAW

Every chemistry textbook describes the basic experiments that relate the volume of a gas sample to its pressure and Kelvin temperature. For a given quantity of gas it may be shown, by combining Boyle's law ($PV = k''$, at constant T) and Charles' law ($V = k'T$, at constant P), that

$$\frac{PV}{T} = k \text{ (a constant)} \tag{11-2}$$

This equation predicts the behavior of gases very well except at relatively low temperatures and at relatively high pressures. Just how low the temperature must be or how high the pressure must be before serious deviations from this equation are observed will vary from one gas to another. Under extreme conditions, these equations cannot be used without correction for the volume occupied by the molecules themselves or for the attractive forces between neighboring molecules. For every gas, conditions exist under which the molecules condense to a liquid that occupies a fairly incompressible volume. A hypothetical gas, called an *ideal gas,* would obey Equation 11-2 under all conditions, and would possess zero volume at a temperature of 0 K. Most gases obey this equation at normal temperatures and pressures.

Experiments have shown that one mole of *any* gas (behaving ideally) at the *standard conditions* of 760.0 torr (1 atm) and 273.2 K occupies a volume of 22.42 liters. These values make it possible to evaluate k for one mole of gas in Equation 11-2:

$$k = \frac{PV}{T} = \frac{(760.0 \text{ torr})\left(22.4 \frac{\text{liters}}{\text{mole}}\right)}{273.2 \text{ K}} = 62.37 \frac{\text{torr liter}}{\text{mole K}}$$

This constant, known as the "ideal gas constant," is given the special symbol R. Equation 11-2 would be written as $PV = RT$ for one mole of gas; for the general case of n moles, it becomes the important ideal gas equation,

$$PV = nRT \tag{11-3}$$

When pressures are measured in atm, it is convenient to have the value of R in these units too. This may be computed as before, but substituting 1 atm for P in Equation 11-2.

$$R = \frac{PV}{T} = \frac{(1 \text{ atm})\left(22.42 \frac{\text{liters}}{\text{mole}}\right)}{273.2 \text{ K}} = 0.08206 \frac{\text{liter atm}}{\text{mole K}}$$

Some other frequently used values of R are

$$R = 1.987 \; \frac{\text{cal}}{\text{mole K}}$$

$$= 8.314 \times 10^7 \; \frac{\text{ergs}}{\text{mole K}}$$

The following problems illustrate a variety of applications of Equation 11-3. In each case the *important* thing to see is how the given situation is related to the ideal gas law. *Don't* try to think of each problem as needing a special equation; learn to reason from the one equation.

PROBLEM:

What is the volume of one mole of *any* gas at room conditions (740 torr and 27°C)?

SOLUTION:

$$V = \frac{nRT}{P}$$

$$= \frac{(1 \text{ mole})\left(62.4 \; \frac{\text{torr liter}}{\text{mole K}}\right)(300 \text{ K})}{740 \text{ torr}}$$

$$= 25.3 \text{ liters}$$

PROBLEM:

What will be the pressure exerted by *any* gas in a sealed vessel if its temperature is raised to 200°C from 20°C, where its pressure is 600 torr?

SOLUTION:

To avoid making a big production out of this problem, you need to realize that the vessel is sealed. Therefore, V and n are constant during this change, and Equation 11-3 can be written as

$$\frac{P_1}{T_1} = \frac{P_2}{T_2} = \frac{nR}{V} = \text{constant}$$

By substituting the given conditions and solving for P_2, we get

$$P_2 = \frac{P_1 T_2}{T_1} = \frac{(600 \text{ torr})(473 \text{ K})}{(293 \text{ K})} = 969 \text{ torr}$$

PROBLEM:

What is the density of NH_3 gas at 67°C and 800 torr?

SOLUTION:

$$\text{Density} = D = \frac{w \text{ g}}{V \text{ liter}}$$

$$\text{Weight} = w = (n \text{ moles}) \left(M \frac{\text{g}}{\text{mole}} \right)$$

Therefore,

$$D = \left(\frac{n}{V} \right) M$$

Equation 11-3 can be rearranged to give

$$\frac{n}{V} = \frac{P}{RT}$$

and thus density of *any* gas will be given by

$$D = \frac{PM}{RT}$$

Taking the specific gas NH_3 ($M = 17.0$ g/mole) at the given conditions, we have

$$D = \frac{(800 \text{ torr}) \left(17.0 \frac{\text{g}}{\text{mole}} \right)}{\left(62.4 \frac{\text{torr liter}}{\text{mole K}} \right)(340 \text{ K})} = 0.641 \frac{\text{g}}{\text{liter}}$$

PROBLEM:

What is the molecular weight of a gas if a 0.0866 g sample in a 60.0 ml bulb has a pressure of 400 torr at 20°C?

SOLUTION:

Because the number of moles of a substance can always be calculated from its weight (w) and mole weight (M) by

$$n = \frac{w \text{ g}}{M \text{ g/mole}}$$

Equation 11-3 can be written as

$$PV = \frac{wRT}{M}$$

Rearranging and solving for M gives

$$M = \frac{wRT}{PV} = \frac{(0.0866 \text{ g}) \left(62.4 \frac{\text{torr liter}}{\text{mole K}} \right)(293 \text{ K})}{(400 \text{ torr})(0.060 \text{ liter})} = 66.0 \frac{\text{g}}{\text{mole}}$$

If the empirical formula of this compound had been found to be $(CH_2F)_x$, this mole weight determination could be used to find the *true* formula. The mole weight of the CH_2F unit is 33.0. The true mole weight must be some integral multiple (x) of 33.0:

$$x = \frac{66.0}{33.0} = 2$$

The true formula is $C_2H_4F_2$.

PROBLEM:

What is the apparent molecular weight of air, assuming that it contains 78% nitrogen, 21% oxygen, and 1% argon by volume?

SOLUTION:

The apparent molecular weight of a gas mixture will be the total weight of the mixture divided by the total number of moles in the mixture (that is, the average weight of one mole of molecules). The ideal gas law, Equation 11-3, can't distinguish between gases; it works equally well for a pure gas or a gas mixture. For a given temperature and pressure the volume of a single gas (A) is given by

$$V_A = n_A \left(\frac{RT}{P}\right)$$

and for a mixture of gases (A, B, and C) by

$$V = (n_A + n_B + n_C) \left(\frac{RT}{P}\right)$$

If you divide the first equation by the second, you get

$$\frac{V_A}{V} = \frac{n_A}{n_A + n_B + n_C} = \frac{n_A}{n} = \text{mole fraction}$$

and conclude that, at a given temperature and pressure, the fraction (or percentage) of the volume occupied by each component is just equal to its fraction (or percentage) of the total moles present.

If we arbitrarily take 100 moles of air, the percentages by volume indicate that we have 78 moles of N_2, 21 moles of O_2 and 1 mole of Ar. We know the molecular weights of the individual components, so we calculate the total weight as follows:

$$78 \text{ moles of } N_2 \text{ weigh} \quad (78 \text{ moles}) \left(28.0 \frac{g}{\text{mole}}\right) = 2184 \text{ g}$$

$$21 \text{ moles of } O_2 \text{ weigh} \quad (21 \text{ moles}) \left(32.0 \frac{g}{\text{mole}}\right) = 672 \text{ g}$$

$$1 \text{ mole of Ar weighs} \quad (1 \text{ mole}) \left(40.0 \frac{g}{\text{mole}}\right) = 40 \text{ g}$$

$$\text{Total weight of 100 moles} = 2896 \text{ g}$$

$$\text{Average weight of one mole} = \frac{2896 \text{ g}}{100 \text{ moles}} = 29.0 \frac{\text{g}}{\text{mole}}$$

$$= \text{apparent molecular weight of air}$$

PROBLEM:
How many molecules are there in 3.00 liters of a gas at a temperature of 500°C and a pressure of 50.0 torr?

SOLUTION:
If we find the number of moles of gas using Equation 11-3, we can convert that to molecules by using Avogadro's number.

$$n = \frac{PV}{RT} = \frac{(50.0 \text{ torr})(3.00 \text{ liters})}{\left(62.4 \frac{\text{torr liter}}{\text{mole K}}\right)(773 \text{ K})} = 3.11 \times 10^{-3} \text{ moles}$$

$$\text{No. of molecules} = (3.11 \times 10^{-3} \text{ moles}) \left(6.02 \times 10^{23} \frac{\text{molecules}}{\text{mole}}\right)$$

$$= 1.87 \times 10^{21} \text{ molecules}$$

DALTON'S LAW OF PARTIAL PRESSURES

The pressure of a gas is due to the impacts of the molecules on the walls of the container. The greater the number of molecules, the higher the pressure. In a gas mixture, the pressure that each gas would exert if it occupied the same volume by itself at the same temperature is called the *partial pressure*. The ideal gas law, Equation 11-3, can't distinguish between gases; it works equally well for a pure gas or a gas mixture. For a given temperature and volume the pressure of a single gas (A) is given by

$$P_A = n_A \left(\frac{RT}{V}\right)$$

and for a mixture of gases (A, B, and C) by

$$P = (n_A + n_B + n_C) \left(\frac{RT}{V}\right)$$

If you divide the first equation by the second, you get

$$\frac{P_A}{P} = \frac{n_A}{n_A + n_B + n_C} = \frac{n_A}{n} = \text{mole fraction}$$

and conclude that, at a given volume and temperature, the fraction (or percentage) of the total pressure exerted by each component is just equal to its fraction (or percentage) of the total moles present. Thus, to draw on a previous problem, if we know that 21.0% of the moles of air is O_2, we know that 21.0% of the air pressure is due to O_2. If the barometric pressure is 740 torr, then the partial pressure of O_2 is

$$P_{O_2} = (0.210)(740 \text{ torr}) = 155 \text{ torr}$$

A corollary of this observation is *Dalton's law of partial pressures,* that the total pressure (P) of a gas mixture is equal to the sum of the partial pressures of the components, i.e.,

$$P = P_A + P_B + P_C + \cdots \tag{11-4}$$

In the laboratory work of general chemistry, we have important applications of partial pressures. Gases (such as oxygen) that are not very soluble in water are collected in bottles by displacement of water. As the gas bubbles rise through the water, they become saturated with vapor, and the collected gas is a mixture of water vapor and the original gas. When the bottle is filled, it is at atmospheric pressure, or

$$P_{\text{gas}} + P_{H_2O} = \text{barometric pressure} = P_B$$

To obtain the partial pressure of the gas (P_{gas}), we must subtract the water-vapor pressure from the barometric pressure

$$P_{\text{gas}} = P_B - P_{H_2O}$$

Fortunately, water-vapor pressures are known accurately over the entire liquid range of water (see Table 11-1) and do not have to be determined experimentally each time.

When gases are collected over mercury, no correction is needed for the vapor pressure of Hg because it is so small (about 2×10^{-3} torr at room temperature). Collection over Hg has the further advantage that gases are insoluble in it. Disadvantages are its high cost and the toxicity of its vapor.

PROBLEM:
A 250 ml flask is filled with oxygen, collected over water at a barometric pressure of 730 torr and a temperature of 25°C. What will be the volume of the oxygen sample, dry, at standard conditions?

SOLUTION:
Changing the temperature and pressure of a gas sample changes only the volume, *not* the number of moles (that is, n remains constant). As a consequence, the ideal

TABLE 11-1
Vapor Pressures of Water at Various Temperatures

t (°C)	p (torr)	t (°C)	p (torr)	t (°C)	p (torr)
0	4.58	15	12.8	29	30.0
1	4.93	16	13.6	30	31.8
2	5.29	17	14.5	31	33.7
3	5.69	18	15.5	32	35.7
4	6.10	19	16.5	33	37.7
5	6.54	20	17.5	34	39.9
6	7.01	21	18.7	35	42.2
7	7.51	22	19.8	40	55.3
8	8.05	23	21.1	50	92.5
9	8.61	24	22.4	60	149.4
10	9.21	25	23.8	70	233.7
11	9.84	26	25.2	80	355.1
12	10.5	27	26.7	90	525.8
13	11.2	28	28.3	100	760.0
14	12.0				

gas equation (Equation 11-3) can be rearranged to give

$$\frac{P_1 V_1}{T_1} = \frac{P_2 V_2}{T_2} = nR = \text{constant}$$

and

$$V_2 = \frac{P_1 V_1 T_2}{P_2 T_1}$$

In this problem, however, we must use the actual partial pressure of the O_2 in the flask, not the total pressure. The partial pressure of O_2 is easily obtained by subtracting the vapor pressure of H_2O at 25°C (24 torr, from Table 11-1) from the total pressure (730 torr) to give 706 torr. Therefore

$$V_2 = \frac{(706 \text{ torr})(250 \text{ ml})(273 \text{ K})}{(760 \text{ torr})(298 \text{ K})}$$

$$= 213 \text{ ml at standard conditions}$$

PROBLEM:
What is the volume of one mole of N_2 (or *any* gas) measured over water at 730 torr and 30°C?

SOLUTION:
From Table 11-1, we find that the partial pressure of water vapor is 32 torr at 30°C. Thus, the partial pressure of the N_2 is 730 torr − 32 torr = 698 torr. Applying the ideal gas equation, we obtain

$$V = \frac{(1 \text{ mole}) \left(62.4 \dfrac{\text{torr liter}}{\text{mole K}} \right) (303 \text{ K})}{698 \text{ torr}}$$

$$= 27.1 \text{ liter}$$

GRAHAM'S LAW OF DIFFUSION AND EFFUSION

When we put two gases together, the molecules diffuse throughout the container, so that within a short time the mixture is homogeneous, or of uniform concentration throughout. Not all gases diffuse at the same rate, however: the lighter the molecule, the more rapid the diffusion process.

If different gases are put into a container at the same temperature and pressure and then allowed to effuse (leak out) through a pinhole in the container, you can compare their rates (r) of effusion (measured in ml/min). The simplest way to do this is to determine the times (t) required for equal volumes to effuse through the pinhole. The rates are just inversely proportional to the times; the shorter the time, the faster the rate. Such a comparison of any two gases shows that these rates of effusion (and diffusion) are related to the molecular weights of the gases according to the equation

$$\frac{r_1}{r_2} = \frac{t_2}{t_1} = \sqrt{\frac{M_2}{M_1}} \tag{11-5}$$

which is known as *Graham's law of diffusion and effusion*. This same relationship can be derived theoretically from the kinetic theory of gases. This equation offers a simple way to determine the molecular weights of gases.

PROBLEM:
The molecular weight of an unknown gas is found by measuring the time required for a known volume of the gas to effuse through a small pinhole, under constant pressure. The apparatus is calibrated by measuring the time needed for the same volume of O_2 (mol wt = 32) to effuse through the same pinhole, under the same conditions. The time found for O_2 is 60 sec, and that for the unknown gas is 120 sec. Compute the molecular weight of the unknown gas.

SOLUTION:
If we use Graham's law, and let gas 1 be O_2 and gas 2 be the unknown gas, then

$$\frac{t_2}{t_{O_2}} = \sqrt{\frac{M_2}{M_{O_2}}}$$

$$\frac{120 \text{ sec}}{60 \text{ sec}} = 2 = \left(\frac{M_2 \text{ g/mole}}{32 \text{ g/mole}} \right)^{\frac{1}{2}}$$

By squaring both sides we get

$$4 = \frac{M_2}{32}$$

$$M_2 = 128 \frac{g}{mole}$$

PROBLEMS A

1. A graduated tube, sealed at the upper end, has a mercury-filled leveling bulb connected to the lower end. The gas volume is 25.0 ml when the mercury level is the same in both tubes. The barometric pressure is 732 torr. What is the volume when the level on the open side is 10.0 cm above the level of the closed side?

2. If a barometer were filled with a liquid of density 1.60 g/ml, what would be the reading when the mercury barometer read 730 torr? The density of mercury is 13.56 g/ml.

3. A mercury barometer reading of 728.3 mm is obtained at 23°C with a brass scale. What is the barometer reading "corrected to 0°C"—that is, in torr?

4. What is the atmospheric pressure in lbs/in^2 when the barometer reading is 720 torr?

5. A student collects 265 ml of a gas over Hg at 25°C and 750 torr. What is the volume at standard conditions?

6. What is the volume at room conditions (740 torr and 25°C) of 750 ml of a gas at standard conditions?

7. A sealed vessel containing methane, CH_4, at 730 torr and 27°C is put into a box cooled with "dry ice" (−78°C). What pressure will the CH_4 exert under these conditions?

8. Liquid nitrogen (boiling point, −195.8°C) is commonly used as a cooling agent. A vessel containing helium at 10 lb/in^2 at the temperature of boiling water is sealed off and then cooled with boiling liquid nitrogen. What will be its pressure expressed in torr?

9. Two liters of N_2 at 1.0 atm, 5.0 liters of H_2 at 5.0 atm, and 3.0 liters of CH_4 at 2.0 atm are mixed and transferred to a 10.0 liter vessel. What is the resulting pressure?

10. A low pressure easily achieved with a diffusion pump and a mechanical vacuum pump is 1.00×10^{-6} torr. Calculate the number of molecules still present in 1.00 ml of gas at this pressure and at 0°C.

11. A sample of nitrous oxide is collected over water at 24°C and 735 torr. The volume is 235 ml. What is the volume at standard conditions?

12. (a) What volume will 0.500 g of O_2 occupy at 750 torr and 26°C over water?
    (b) What volume will it occupy if collected over mercury at the same conditions?

13. A sample of $NaNO_2$ is tested for purity by heating it with an excess of NH_4Cl and collecting the evolved N_2 over water. The volume collected is 567.3 ml at a barometric pressure of 741 torr. The temperature is 22°C. What volume would the nitrogen, dry, occupy at standard conditions?

14. What is the weight of 250 ml of N_2 measured at 740 torr and 25°C?

15. What volume will be occupied by 1.00 g of O_2 measured over water at 27°C and 730 torr?

16. What is the molecular weight of a gas if 250 ml measured over water at 735 torr and 28°C weighs 1.25 g?

17. What is the density of chlorine gas (Cl_2) at 83°F and 723 torr?

18. Calculate the density of N_2O (a) at standard conditions; (b) at 730 torr and 25°C, dry.

19. It is found experimentally that 0.563 g of a vapor at 100°C and 725 torr has a volume of 265 ml. Find the molecular weight.

20. A compound has the formula C_8H_{18}. What volume will 1.00 g of this material have at 735 torr and 99°C?

21. What is the apparent mole weight of a gas mixture composed of 20.0% H_2, 70.0% CO_2, and 10.0% NO? (Composition given is percentage by volume.)

22. An oxygen-containing gas mixture at 1.00 atm is subjected to the action of yellow phosphorus, which removes the oxygen. In this way it is found that oxygen makes up 35.0% by volume of the mixture. What is the partial pressure of O_2 in the mixture?

23. A mixture of gases contained in a vessel at 0.500 atm is found to comprise 15.0% N_2, 50.0% N_2O, and 35.0% CO_2 by volume.
    (a) What is the partial pressure of each gas?
    (b) A bit of solid KOH is added to remove the CO_2. Calculate the resulting total pressure, and the partial pressures of the remaining gases.

24. A vessel whose volume is 235.0 ml, and whose weight evacuated is 13.5217 g + a tare vessel, is filled with an unknown gas at a pressure of 725 torr and a temperature of 19°C. It is then closed, wiped with a damp cloth, and hung in the balance case to come to equilibrium with the tare vessel. The tare vessel has about the same surface area and is needed to minimize surface moisture effects. This second weighing is 13.6109 g + the tare vessel. What is the mole weight of the gas?

25. Two or three milliliters of a liquid that boils at about 50°C are put into an Erlenmeyer flask. The flask is closed with a polystyrene stopper that has a fine glass capillary running through it. The gas-containing part of the flask is then completely immersed in a bath of boiling water, which (at the elevation of the experiment) boils at 99.2°C. After a short time the air has been completely

swept out through the capillary, and the excess liquid has boiled away, leaving the flask filled only with the vapor of the liquid. At this point the flask is removed from the boiling water and cooled. The vapor condenses to liquid, and air rushes in to fill the flask again. Whereas the flask, when empty and dry, weighed 45.3201 g, after the experiment it weighs 46.0513 g. The barometric pressure during the experiment is 735 torr. The volume of the flask is determined by filling the flask with water, inserting the stopper to its previous position, and squeezing out the excess water through the capillary. The volume of water so held is 263.2 ml. What is the mole weight of the liquid?

26. The liquid used in Problem 25 is analyzed and found to be 54.5% C, 9.10% H, and 36.4% O. What is the true molecular formula of this liquid?

27. An automobile tire has a gauge pressure of 32 lb/in^2 at 20°C when the prevailing atmospheric pressure is 14.7 lb/in^2. What is the gauge pressure if the temperature rises to 50°C?

28. It takes 1 min and 37 sec for a given volume of chlorine (Cl_2) to effuse through a pinhole under given conditions of temperature and pressure. How long will it take for the same volume of water vapor to effuse through the same hole under the same conditions?

29. Argon effuses through a hole (under prescribed conditions of temperature and pressure) at the rate of 3.0 ml/min. At what velocity will xenon effuse through the same hole under the same conditions?

30. A rubber balloon weighing 5.0 g is 12 inches in diameter when filled with hydrogen at 730 torr and 25°C. How much will the balloon lift in addition to its own weight? (Assume the density of air to be 1.2 g/liter under these conditions.)

31. (a) If the balloon of Problem 30 were filled with ammonia gas (NH_3) under the same conditions, would it rise?
    (b) If so, how much weight would it lift?

32. What must be the composition of a mixture of H_2 and O_2 if it inflates a balloon to a diameter of 15 inches and yet the balloon *just barely* rises from a table top? (The balloon weighs 6.0 g, and the pressure and temperature are 740 torr and 30°C.)

33. Oxygen is commonly sold in 6.0 ft^3 steel cylinders at a pressure of 2000 lb/in^2 (at 70°F). What weight of oxygen does such a cylinder contain? (Assume oxygen to be an ideal gas under these conditions.)

34. The average breath that an 18-year-old takes when not exercising is about 300 ml at 20°C and 750 torr. His respiratory rate is about 20 breaths/min.
    (a) What volume of air, corrected to standard conditions, does an average 18-year-old breathe each day?
    (b) What weight of air does he breathe each day? (Assume that air is 21% oxygen and 79% nitrogen by volume.)

35. The percentage of CO_2 in normal air is 0.035% by volume, and that in the exhaled air of the average 18-year-old is about 4.0%.

(a) What volume of CO_2, at standard conditions, does the average 18-year-old make each day?

(b) What weight of CO_2 does he make each day?

36. An organic compound containing C, H, O, and N is analyzed. When a sample weighing 0.01230 g is burned, it produces 18.62 mg of CO_2 (absorbed in a soda-lime tube) and 7.62 mg of H_2O [absorbed in a tube containing $Mg(ClO_4)_2$]. When another sample, weighing 0.00510 g, is burned, the CO_2 and H_2O are absorbed, and the N_2 formed is collected in a measuring tube. At 730 torr and 22°C, the N_2 gas displaces an equal volume of mercury, which is weighed and found to weigh 15.000 g. The density of mercury is 13.56 g/ml. Calculate the empirical formula of the compound.

PROBLEMS B

37. A graduated gas tube, sealed at the upper end, has a mercury-filled leveling bulb connected to the lower end. The gas volume is 17.2 ml when the leveling bulb is 8 cm above the other mercury level. What will be the gas volume when the leveling bulb is 8 cm below the other mercury level? The barometric pressure is 738 torr.

38. If a barometer were filled with a silicone fluid whose vapor pressure is very low but whose density is 1.15 g/ml, what would be the barometer reading when the atmospheric pressure is 710 torr? (The density of mercury is 13.56 g/ml.)

39. A gas pressure is measured as 826.4 mm with a mercury manometer in a lab whose temperature is 24°C. A brass scale is used. Express this gas pressure in torr.

40. A gas pressure is 690 torr. What is its pressure in dynes/cm²?

41. What is the volume at 730 torr and 27°C of 350 ml of H_2S at standard conditions?

42. A 50.0 ml quartz vessel is filled with O_2 at 300 torr and at 35°C. It is then heated to 1400°C in an electric furnace. What will be the oxygen pressure at the higher temperature?

43. A sample of NH_3 gas collected over mercury measures 595 ml at 19°C and 755 torr. What will be its volume at standard conditions?

44. What will be the final gas pressure when 3.0 liters of CO at 2.0 atm, 6.0 liters of Ar at 4.0 atm, and 2.0 liters of C_2H_6 at 5.0 atm are mixed and transferred to a 8.0 liter vessel?

45. A 375 ml sample of hydrogen is collected over water at 18°C and 720 torr. What is its volume at standard conditions?

46. What will be the difference in volume occupied by 0.100 g of hydrogen at 740 torr and 19°C if it is collected over water instead of mercury?

47. A mixture of copper and zinc is analyzed for the percentage of zinc by adding an excess of HCl and collecting the evolved H_2 over water. (Copper will not react with HCl.) The volume of H_2 collected is 229.5 ml at a barometric pressure of 732 torr. The temperature is 29°C. What volume would the H_2 occupy at standard conditions, dry?

48. In some recent work at the Bell Telephone Research Laboratory, a low pressure of 10^{-11} torr of mercury was used. This is an unusually low pressure, one not easily obtained. Calculate the number of molecules still remaining in 1 ml of gas at this pressure at 27°C.

49. A 1 g sample of helium occupies 5.6 liters at standard conditions. What will be its weight when expanded to a pressure of 0.10 atm?

50. What volume will 5.00 g of methyl alcohol, CH_3OH, occupy at 720 torr and 98°C?

51. What is the weight of 420 ml of NH_3 measured at 735 torr and 27°C?

52. What volume will be occupied by 2.50 g of CO measured over water at 27°C and 725 torr?

53. What is the molecular weight of a gas if 365 ml of it measured over water at 727 torr and 30°C weighs 1.42 g?

54. What is the density of NH_3 gas at 78°F and 741 torr?

55. Calculate the density of C_2H_6 (a) at standard conditions;   (b) at 725 torr and 27°C, dry.

56. What is the apparent mole weight of a gas mixture whose composition by volume is 60.0% NH_3, 25.0% NO, and 15.0% N_2?

57. If 0.670 g of a vapor at 100°C and 735 torr has a volume of 249 ml, what must its mole weight be?

58. A mixture of gases contained in a vessel at 1.30 atm is found to be 60.0% NH_3, 25.0% NO, and 15.0% N_2 by volume.
    (a) What is the partial pressure of each gas?
    (b) A bit of solid P_4O_{10} is added to remove the NH_3. Calculate the resulting total pressure, and the partial pressures of the remaining gases.

59. A gas mixture containing CO_2 is subjected, at 1.00 atm, to the action of KOH, which removes the CO_2. In this way, the CO_2 is found to be 27.0% by volume of the mixture. What is the partial pressure of CO_2 in the mixture?

60. A vessel, whose volume is 205.3 ml and whose weight evacuated is 5.3095 g + a tare flask, is filled with an unknown gas to a pressure of 750 torr at a temperature of 27°C. It is then cleaned, wiped with a damp cloth, and hung in the balance case to come to equilibrium with the tare vessel. The tare vessel has about the same surface area and is needed to minimize effects of surface moisture. The second weighing is 5.6107 g + the tare flask. What is the mole weight of the gas?

61. Two or three milliliters of a liquid that boils at about 60°C are put into an Erlenmeyer flask. The flask is closed with a polystyrene stopper that has a fine glass capillary running through it. The gas-containing part of the flask is then completely immersed in a bath of boiling water, which (at the elevation of the experiment) boils at 98.8°C. After a short time the air has been completely swept out through the capillary, and the excess liquid has boiled away, leaving the flask filled only with the vapor of the liquid. At this point the flask is removed from the boiling water and cooled. The vapor condenses to liquid, and air rushes in to fill the flask again. The flask, when dry and empty, had weighed 39.5762 g; after the experiment it weighs 40.3183 g. The barometric pressure during the experiment is 730 torr. The volume of the flask is determined by filling the flask with water, inserting the stopper to its previous position, and squeezing out the excess water through the capillary. The volume of water so held is 239.6 ml. What is the mole weight of the liquid?

62. The liquid used in Problem 61 is analyzed and found to be 24.2% C, 4.05% H, and 71.7% Cl. What is the true molecular formula of this liquid?

63. An automobile tire has a gauge pressure of 35.0 lb/in² when the atmospheric pressure is 14.7 lb/in² and the temperature is 40°F. What will be its gauge pressure if its temperature rises to 120°F?

64. Hydrogen chloride effuses through a hole (under prescribed conditions of temperature and pressure) at the rate of 2.70 ml/min. At what velocity will helium effuse through the same hole under the same conditions?

65. It takes 45 sec for a given volume of CO to effuse through a pinhole under given conditions of temperature and pressure. How long will it take for the same volume of Br_2 vapor to effuse through the same hole under the same conditions?

66. Under certain prescribed conditions, O_2 effuses through a pinhole at the rate of 3.65 ml/min. A mixture of CO and CO_2 effuses through the same pinhole under the same conditions at the rate of 3.21 ml/min. Calculate the percentage of CO in the gas mixture.

67. A rubber balloon weighing 10 g is 15 inches in diameter when inflated with helium at 735 torr and 75°F. How much weight will the balloon lift in addition to its own weight? (Assume the density of air to be 1.20 g/liter under these conditions.)

68. If the balloon of Problem 67 were filled with methane, CH_4, at 735 torr and 75°F, would it rise? If so, how much additional weight would it lift?

69. A mixture of N_2 and H_2 has a density of 0.267 g/liter at 700 torr and 30°C. For this mixture, calculate (a) the apparent molecular weight, (b) the percentage composition by volume, and (c) the number of molecules in one ml.

70. Show the mathematical steps needed in order to combine Boyle's and Charles' laws to give Equation 11-2 on p 159.

71. Use the methods of calculus to combine Boyle's and Charles' laws to give Equation 11-2 on p 159.

Stoichiometry II: Calculations Based on Chemical Equations

A chemical equation is a statement of experimental fact. It gives on the left side the reactants and on the right side the products of the reaction. Because no atoms are produced or destroyed in a nonnuclear chemical reaction, the equation must be so balanced that every atom originally present in the reactants is accounted for in the products. This means that the combined weight of the reaction products is exactly equal to the combined weight of the original reactants.

All of the important stoichiometric calculations that relate the weights and volumes of starting materials to the weights and volumes of products typically involve just three simple steps.

1. Find how many moles correspond to the given quantity of some substance in the reaction.

2. Use the balanced chemical equation to find, from the number of moles of the given substance, the number of moles of the substance sought in the calculation.

3. Convert the number of moles of the substance sought to the units requested in the statement of the problem.

These three steps are illustrated in the following problems.

PROBLEM:

Oxygen is prepared by heating $KClO_3$.

(a) What weight of O_2 is obtained from 3.00 g $KClO_3$?

(b) What is the volume of O_2, measured at standard conditions?

(c) What volume does the O_2 occupy if collected over water at 730 torr and 25°C?

SOLUTION:

The first step is to write the balanced equation for the reaction (if it is not given). This step requires knowledge of the experimental facts. We note in the text that, when $KClO_3$ is decomposed by heating, the products are KCl and O_2, so we start with the unbalanced equation

$$KClO_3 \xrightarrow{\Delta} KCl + O_2$$

(The Δ symbol indicates that heating is necessary.) To account for the three moles of oxygen atoms in $KClO_3$ we need $\frac{3}{2}$ moles of O_2 in the products:

$$KClO_3 \xrightarrow{\Delta} KCl + \tfrac{3}{2}O_2$$

This is now a balanced equation, but we prefer to eliminate fractional numbers of moles, so we multiply all terms by 2, getting the final equation

$$2KClO_3 \xrightarrow{\Delta} 2KCl + 3O_2$$

We now examine the problem, asking two questions: (a) what is given, and (b) what is sought? We see (a) that the weight of $KClO_3$ used is given, and (b) that we seek the amount of O_2 produced. We then proceed with the three steps listed earlier.

1. From the weight of $KClO_3$ given, compute the number of moles that are given.

$$\text{Moles of } KClO_3 \text{ given} = \frac{3.00 \text{ g } KClO_3}{122.6 \dfrac{\text{g } KClO_3}{\text{mole } KClO_3}} = 0.0245 \text{ moles } KClO_3$$

2. From the moles of $KClO_3$ given, compute the number of moles of O_2 produced. The chemical equation shows that 3 moles of O_2 are produced from 2 moles of $KClO_3$. Therefore,

$$\text{Moles of } O_2 \text{ produced} = \left(\frac{3 \text{ moles } O_2}{2 \text{ moles } KClO_3}\right)(0.0245 \text{ moles } KClO_3)$$

$$= 0.0368 \text{ moles } O_2$$

3. From the moles of O_2 produced, express the amount of O_2 in the units specified in the statement of the problem, as follows.

(a) Weight of $O_2 = (0.0368 \text{ moles } O_2)\left(32.0 \dfrac{\text{g } O_2}{\text{mole } O_2}\right)$

$$= 1.18 \text{ g } O_2$$

(b) The volume of O_2 will be given by the ideal gas law (see p 159). For standard conditions, $T = 273.2$ K and $P = 760.0$ torr, so

$$V = \frac{nRT}{P} = \frac{(0.0368 \text{ moles}) \left(62.4 \frac{\text{torr liter}}{\text{mole K}} \right) (273 \text{ K})}{760 \text{ torr}}$$

$$= 0.825 \text{ liters } O_2 \text{ at standard conditions}$$

(c) The volume of O_2 will again be given by the ideal gas law, but the partial pressure of O_2 must be used (not the combined pressures of O_2 and water vapor). The partial pressure of O_2 is calculated from Dalton's law, using the vapor pressure of H_2O (24 torr) from Table 11-1.

$$P_{O_2} = 730 \text{ torr} - 24 \text{ torr} = 706 \text{ torr}$$

$$V = \frac{nRT}{P} = \frac{(0.0368 \text{ moles}) \left(62.4 \frac{\text{torr liter}}{\text{mole K}} \right) (298 \text{ K})}{706 \text{ torr}}$$

$$= 0.969 \text{ liters } O_2$$

PROBLEM:

Chlorine is prepared by the reaction

$$2NaMnO_4 + 10NaCl + 8H_2SO_4 \rightarrow 2MnSO_4 + 6Na_2SO_4 + 5Cl_2 + 8H_2O$$

or

$$2MnO_4^- + 10Cl^- + 16H^+ \rightarrow 2Mn^{2+} + 5Cl_2 + 8H_2O$$

What weights of (a) pure NaCl and (b) 90.0% pure $NaMnO_4$ are needed to prepare 500 ml of Cl_2 gas measured dry at 25°C and 730 torr?

SOLUTION:

We follow the three simple steps.

1. From the volume of Cl_2 that we are given (to prepare), compute the moles of Cl_2 that are given, using the ideal gas equation.

$$n = \frac{PV}{RT} = \frac{(730 \text{ torr})(0.500 \text{ liter})}{\left(62.4 \frac{\text{torr liter}}{\text{mole K}} \right) (298 \text{ K})}$$

$$= 0.0196 \text{ moles } Cl_2 \text{ given}$$

2. From the moles of Cl_2 given, compute the number of moles of NaCl and $NaMnO_4$ required. The chemical equation shows that 10 moles of NaCl and 2 moles of $NaMnO_4$ are required for 5 moles of Cl_2, therefore the needed

$$\text{moles of NaCl} = \left(\frac{10 \text{ moles NaCl}}{5 \text{ moles } Cl_2} \right) (0.0196 \text{ moles } Cl_2) = 0.0392 \text{ moles}$$

$$\text{moles of NaMnO}_4 = \left(\frac{2 \text{ moles NaMnO}_4}{5 \text{ moles } Cl_2} \right) (0.0196 \text{ moles } Cl_2) = 0.00784 \text{ moles}$$

3. From the moles of NaCl and $NaMnO_4$ required, express these quantities in the units specified in the statement of the problem.

(a) Weight of pure NaCl = (0.0392 moles NaCl) $\left(58.5 \dfrac{\text{g NaCl}}{\text{mole NaCl}}\right)$

$= 2.29$ g NaCl

(b) If x = the required grams of 90.0% pure NaMnO$_4$, you can see that 0.90x g must contain 0.00784 moles of NaMnO$_4$. That is,

$$0.90x \text{ g} = (.00784 \text{ moles NaMnO}_4)\left(141.9 \dfrac{\text{g NaMnO}_4}{\text{mole NaMnO}_4}\right)$$

$= 1.11$ g pure NaMnO$_4$

$$x = \dfrac{1.11}{0.90} = 1.23 \text{ g impure NaMnO}_4$$

PROBLEM:
(a) What volume of oxygen at 20°C and 750 torr is needed to burn 3.00 liters of propane, C$_3$H$_8$, also at 20°C and 750 torr?
(b) What volume of air (21.0% O$_2$ by volume) would be required under the same conditions? The products of combustion are solely CO$_2$ and H$_2$O.

SOLUTION:
First, you must have a balanced chemical equation on which to base your calcula-tion. Because the three C atoms of C$_3$H$_8$ are converted to 3CO$_2$, and the 8H atoms are converted to 4H$_2$O, you can readily see that the 10 oxygen atoms needed in this much CO$_2$ and H$_2$O must come from 5O$_2$. Therefore,

$$C_3H_8 + 5O_2 \xrightarrow{\Delta} 3CO_2 + 4H_2O$$

Second, you must realize that, when the two substances you are interested in are both *gases,* you can make a much simpler calculation than that involved in the "three simple steps." You recall (see p 160) that equal volumes of gases under the same conditions of temperature and pressure contain the same number of moles (or molecules). The chemical equation shows that you need 5 *moles* of O$_2$ per mole of C$_3$H$_8$; therefore, you will need 5 times the *volume* of O$_2$ as the volume of C$_3$H$_8$ under the same conditions. Therefore,

(a) Volume of O$_2$ = $\left(\dfrac{5 \text{ moles O}_2}{1 \text{ mole C}_3\text{H}_8}\right)(3.00 \text{ liters C}_3\text{H}_8)$

$= 15.0$ liters of O$_2$

(b) If V = the required volume of air (also a gas) that is 21.0% O$_2$, you can see that 0.210V liters of air must provide 15.0 liters of O$_2$. That is,

$$0.210V = 15.0 \text{ liters O}_2$$

$$V = \dfrac{15.0}{0.210} = 71.4 \text{ liters of air}$$

PROBLEM:
Sulfur dioxide is prepared by heating iron pyrites, FeS_2, in the presence of air. The reaction is

$$4FeS_2 + 11O_2 \xrightarrow{\Delta} 2Fe_2O_3 + 8SO_2$$

(a) How many tons of SO_2 can be obtained from 20.0 tons of FeS_2?
(b) What volume of air, in cubic feet at 0.90 atm and 77°F, is required for the treatment of 20.0 tons of FeS_2? Assume the air to be 21.0% O_2 by volume.

SOLUTION:
First, you need to know that it is not necessary to convert tons to grams to moles, and then at the end reconvert moles to grams to tons. You recall that the atomic weight scale (inside back cover) is a *relative* atomic weight scale habitually used with gram as the mass unit. For problems like this where tons (or lb, or oz, or whatever) are involved, it is easier to use ton as the mass unit and to use ton molecular weights (ton-moles) instead of gram molecular weights (moles). We still use the three-step approach.

1. From the weight of FeS_2 given, compute the number of ton-moles that are given.

$$\text{Ton-moles of } FeS_2 \text{ given} = \frac{20.0 \text{ tons } FeS_2}{120.0 \dfrac{\text{tons } FeS_2}{\text{ton-mole } FeS_2}} = 0.167 \text{ ton-moles } FeS_2$$

2. From the ton-moles of FeS_2 given, compute the number of ton-moles of SO_2 produced and the ton-moles of O_2 required. The chemical equation shows that 8 ton-moles of SO_2 are produced and 11 ton-moles of O_2 are required for every 4 ton-moles of FeS_2. Therefore,

$$\text{ton-moles of } SO_2 \text{ produced} = \left(\frac{8 \text{ ton-moles } SO_2}{4 \text{ ton-moles } FeS_2}\right)(0.167 \text{ ton-moles } FeS_2)$$

$$= 0.333 \text{ ton-moles } SO_2$$

$$\text{ton-moles of } O_2 \text{ required} = \left(\frac{11 \text{ ton-moles } O_2}{4 \text{ ton-moles } FeS_2}\right)(0.167 \text{ ton-moles } FeS_2)$$

$$= 0.458 \text{ ton-moles } O_2$$

3. From the ton-moles of SO_2 and O_2 produced, express the amounts of SO_2 and O_2 in the units specified in the statement of the problem.

(a) Weight of $SO_2 = (0.333 \text{ ton-moles } SO_2)\left(64.1 \dfrac{\text{tons } SO_2}{\text{ton-mole } SO_2}\right)$

$$= 21.3 \text{ tons } SO_2 \text{ produced}$$

(b) The *volume* of O_2 (or air) must be calculated from the ideal gas equation, where the gas constant R is always on a *per mole* basis (that is, a *gram-molecular-weight* basis). The simplest approach in this case is to first convert 0.458 ton-moles of O_2 to gram moles (just plain moles), and then use the ideal gas equation.

$$\text{Moles of } O_2 = (0.458 \text{ ton-moles } O_2)\left(2000 \; \frac{\text{lb-moles } O_2}{\text{ton-mole } O_2}\right)\left(454 \; \frac{\text{g-moles } O_2}{\text{lb-mole } O_2}\right)$$

$$= 4.16 \times 10^5 \text{ moles } O_2$$

The ideal gas equation will give the *liters* of O_2 at the specified conditions (77°F = 25°C, and 0.9 atm = 0.9 × 760 torr/atm = 684 torr).

$$V = \frac{nRT}{P} = \frac{(4.16 \times 10^5 \text{ moles})\left(62.4 \; \dfrac{\text{torr liter}}{\text{mole K}}\right)(298 \text{ K})}{684 \text{ torr}}$$

$$= 1.13 \times 10^7 \text{ liters of pure } O_2$$

If X = volume of air (21.0% O_2 by volume) that is required, then $0.210X$ must provide 1.13×10^7 liters of pure O_2.

$$0.21X = (1.13 \times 10^7 \text{ liters } O_2)\left(10^3 \; \frac{\text{cm}^3}{\text{liter}}\right)\left(\frac{1 \text{ in}}{2.54 \text{ cm}}\right)^3\left(\frac{1 \text{ ft}}{12 \text{ in}}\right)^3$$

$$= 3.99 \times 10^5 \text{ ft}^3 \text{ of pure } O_2$$

$$X = \frac{3.99 \times 10^5}{0.21} = 1.90 \times 10^6 \text{ ft}^3 \text{ of air}$$

PROBLEM:
A 0.2052 g mixture of copper and aluminum is analyzed for the percentage of aluminum by adding an excess of H_2SO_4 and collecting the evolved H_2 over water. (Copper will not react with H_2SO_4.) The volume of H_2 collected is 229.5 ml at a barometric pressure of 732 torr. The temperature is 29°C. Calculate the percentage of aluminum in the original sample. The chemical reaction is

$$2Al + 3H_2SO_4 \rightarrow Al_2SO_4 + 3H_2$$

or
$$2Al + 6H^+ \rightarrow 2Al^{3+} + 3H_2$$

SOLUTION:
We follow the standard "three simple steps" to find, from the given amount of H_2, how much Al must be present. At the end, this amount of Al is stated in terms of how much sample was used (that is, the % purity of the sample).

1. From the volume of H_2 given (produced), compute the number of moles of H_2 that are produced, by using the ideal gas equation. The pressure must be corrected for the vapor pressure of water (30 torr at 29°C, from Table 11-1).

$$n = \frac{PV}{RT} = \frac{(732 \text{ torr} - 30 \text{ torr})(0.2295 \text{ liter})}{\left(62.4 \; \dfrac{\text{torr liter}}{\text{mole K}}\right)(302 \text{ K})}$$

$$= 8.55 \times 10^{-3} \text{ moles } H_2$$

2. From the moles of H_2 produced, calculate the number of moles of Al that must have been present. The chemical equation shows that 2 moles of Al are required for 3 moles of H_2. Therefore,

$$\text{moles of Al} = \left(\frac{2 \text{ moles Al}}{3 \text{ moles H}_2}\right)(8.55 \times 10^{-3} \text{ moles H}_2)$$

$$= 5.70 \times 10^{-3} \text{ moles of Al in original sample}$$

3. From the moles of Al present, calculate the weight of Al present. From the weight of Al present, then calculate the percentage in the original sample.

$$\text{Weight of Al present} = (5.70 \times 10^{-3} \text{ moles Al}) \left(27.0 \frac{\text{g Al}}{\text{mole Al}}\right)$$

$$= 0.1539 \text{ g Al}$$

$$\% \text{ Al present} = \frac{(0.1539 \text{ g Al})}{(0.2052 \text{ g sample})} \times 100 = 75.0\% \text{ Al}$$

All of these illustrative problems have been worked in the three distinct steps, in order to emphasize the reasoning involved. With a little practice, you can combine two or three of these steps into one operation (or set-up), greatly increasing the efficiency in using your calculator.

PROBLEMS A

1. Balance the following equations, which show the starting materials and the reaction products. It is not necessary to supply any additional reactants or products. A Δ sign indicates that heating is necessary.

(a) $KNO_3 \xrightarrow{\Delta} KNO_2 + O_2$

(b) $Pb(NO_3)_2 \xrightarrow{\Delta} PbO + NO_2 + O_2$

(c) $Na + H_2O \rightarrow NaOH + H_2$

(d) $Fe + H_2O \xrightarrow{\Delta} Fe_3O_4 + H_2$

(e) $C_2H_5OH + O_2 \xrightarrow{\Delta} CO_2 + H_2O$

(f) $Fe_3O_4 + H_2 \xrightarrow{\Delta} Fe + H_2O$

(g) $CO_2 + NaOH \rightarrow NaHCO_3$

(h) $MnO_2 + HCl \rightarrow H_2O + MnCl_2 + Cl_2$

(i) $Zn + KOH \rightarrow K_2ZnO_2 + H_2$

(j) $Cu + H_2SO_4 \xrightarrow{\Delta} H_2O + SO_2 + CuSO_4$

(k) $Al(NO_3)_3 + NH_3 + H_2O \rightarrow Al(OH)_3 + NH_4NO_3$

(l) $Al(NO_3)_3 + NaOH \rightarrow NaAlO_2 + NaNO_3 + H_2O$

2. Some common gases may be prepared in the laboratory using reactions represented by the following balanced equations. A Δ sign indicates that heating is necessary. For purity, air must be swept out of the apparatus before the gas is collected, and some gases must be dried with a suitable desiccant.

$$FeS + 2HCl \rightarrow H_2S + FeCl_2$$

$$CaCO_3 + 2HCl \rightarrow CO_2 + CaCl_2 + H_2O$$

$$2NH_4Cl + CaO \xrightarrow{\Delta} 2NH_3 + CaCl_2 + H_2O$$

$$NaCl + H_2SO_4 \xrightarrow{\Delta} HCl + NaHSO_4$$

$$NH_4Cl + NaNO_3 \xrightarrow{\Delta} N_2O + NaCl + 2H_2O$$

$$2Al + 3H_2SO_4 \rightarrow Al_2(SO_4)_3 + 3H_2$$

$$CaC_2 + 2H_2O \rightarrow C_2H_2 + Ca(OH)_2$$

(a) What weight of FeS is needed to prepare (*i*) 4.50 moles of H_2S? (*ii*) 1.00 lb of H_2S?

(b) How many tons of limestone ($CaCO_3$) are needed to prepare 5.00 tons of "dry ice" (CO_2), assuming that 30% of the CO_2 produced is wasted in converting it to the solid?

(c) How many grams of NH_4Cl and CaO are needed to make 0.100 mole of NH_3?

(d) How many grams of 95% pure NaCl are needed to produce 2.00 lb of HCl?

(e) What volume of commercial HCl (36% HCl by weight, density = 1.18 g/ml) and what weight of limestone (90% pure) are needed to produce 2.00 kg of CO_2?

(f) Commercial sulfuric acid that has a density of 1.84 g/ml and is 95% H_2SO_4 by weight is used for the production of HCl. (*i*) What weight of commercial acid is needed for the production of 365 g of HCl? (*ii*) What volume of acid is needed for the production of 365 g of HCl?

(g) Commercial sulfuric acid that has a density of 1.45 g/ml and is 55.1% H_2SO_4 by weight is used for the production of H_2. What (*i*) weight and (*ii*) volume of this commercial acid are needed for the production of 50.0 g of H_2 gas?

(h) A manufacturer supplies 1 lb cans of calcium carbide whose purity is labeled as 85%. How many grams of acetylene can be prepared from 1.00 lb of this product if the label is correct?

(i) A whipped cream manufacturer wishes to produce 500 lb of N_2O for her chain of soda fountains. What is the cost of the necessary NH_4Cl and $NaNO_3$ if they cost $840 and $1230 per ton, respectively?

(j) A 0.795 g sample of impure limestone is tested for purity by adding H_2SO_4 (instead of HCl as shown in the second equation). After the generated gas is passed over $Mg(ClO_4)_2$ to dry it, it is passed over soda lime (a mixture of sodium and calcium hydroxides), which absorbs the CO_2. The soda-lime tube increases in weight by 0.301 g. What is the percentage of $CaCO_3$ in the original sample?

(k) The purity of a 0.617 g sample of impure FeS is tested by passing the H_2S produced by the HCl (as in the first equation) into a dilute solution of $AgNO_3$. The precipitate of Ag_2S is filtered off, washed, and gently dried.

The weight of the Ag_2S produced is 1.322 g. How pure was the original sample of FeS?

3. How many grams of zinc are needed to prepare 3.00 liters of hydrogen at standard conditions? The reaction is

$$Zn + H_2SO_4 \rightarrow ZnSO_4 + H_2$$

4. (a) How many grams of zinc are needed to prepare 3.00 liters of H_2 collected over water at 750 torr and 26°C?
   (b) How many moles of H_2SO_4 are used?

5. What volume of O_2 at 730 torr and 25°C will react with 3.00 liters of H_2 at the same conditions?

6. What volume of steam at 1000°C and 1 atm is needed to produce 10^6 ft^3 of H_2, under the same conditions, by the reaction

$$4H_2O + 3Fe \xrightarrow{\Delta} Fe_3O_4 + 4H_2$$

7. What volume of Cl_2 at 730 torr and 27°C is needed to react with 7.00 g of sodium metal by the reaction

$$2Na + Cl_2 \rightarrow 2NaCl$$

8. (a) How many grams of MnO_2 are needed to prepare 5.00 liters of Cl_2 at 750 torr and 27°C?
   (b) How many moles of HCl are needed for the reaction? The reaction is

$$MnO_2 + 4HCl \rightarrow MnCl_2 + Cl_2 + 2H_2O$$

9. How much H_2S gas at 725 torr and 25°C is needed to react with the copper in 1.5 g of $CuSO_4$? The reaction is

$$CuSO_4 + H_2S \rightarrow CuS + H_2SO_4$$

10. (a) What volume of O_2 at 730 torr and 60°F is needed to burn 500 g of octane, C_8H_{18}?
    (b) What volume of air (21% O_2 by volume) is needed to provide this amount of O_2? Balance the equation before working the problem.

$$C_8H_{18} + O_2 \xrightarrow{\Delta} CO_2 + H_2O$$

11. Chlorine is prepared by the reaction

$$2KMnO_4 + 16HCl \rightarrow 2KCl + 2MnCl_2 + 5Cl_2 + 8H_2O$$

   (a) What weight of $KMnO_4$ is needed to prepare 2.50 liters of Cl_2 at standard conditions?
   (b) How many moles of HCl are used?
   (c) What volume of solution is needed if there are 12.0 moles of HCl per liter?
   (d) What weight of $MnCl_2$ is obtained from the reaction?

12. Nitric oxide is prepared by the reaction

$$3Cu + 8HNO_3 \rightarrow 3Cu(NO_3)_2 + 2NO + 4H_2O$$

(a) What weight of copper and (b) how many moles of HNO_3 are needed to prepare 500 ml NO, measured over water at 730 torr and 25°C?

(c) If the nitric acid solution contains 10.0 moles/liter, what volume of the solution is used?

13. Arsenic compounds may be detected easily by the Marsh test. In this test, some metallic zinc is added to an acid solution of the material to be tested, and the mixture is heated. The arsenic is liberated as arsine, AsH_3, which may be decomposed by heat to give an "arsenic mirror." The reaction is

$$4Zn + H_3AsO_4 + 8HCl \xrightarrow{\Delta} 4ZnCl_2 + AsH_3 + 4H_2O$$

What volume of AsH_3 at 720 torr and 25°C is evolved by 7.00×10^{-7} g of arsenic, the smallest amount of arsenic that can be detected with certainty by this method?

14. H_2S gas will cause immediate unconsciousness at a concentration of 1 part per 1000 by volume. What weight of FeS is needed to fill a room 20 ft × 15 ft × 9 ft with H_2S at this concentration? Barometric pressure is 740 torr, and the temperature is 80°F. The reaction is

$$FeS + 2HCl \rightarrow FeCl_2 + H_2S$$

15. The Mond process separates nickel from other metals by passing CO over the hot metal mixture. The nickel reacts to form a volatile compound (called nickel carbonyl), which is then swept away by the gas stream. The reaction is

$$Ni + 4CO \xrightarrow{\Delta} Ni(CO)_4$$

How many cubic feet of CO at 3.00 atm and 65°F are needed to react with 1.00 ton of nickel?

16. A cement company produces 100 tons of cement per day. Its product contains 62.0% CaO, which is prepared by calcining limestone by the reaction

$$CaCO_3 \xrightarrow{\Delta} CaO + CO_2$$

What volume of CO_2 at 735 torr and 68°F is sent into the air around the plant each day as a result of this calcination?

17. The Ostwald process of making HNO_3 involves the air oxidation of NH_3 over a platinum catalyst. The first two steps in this process are

$$4NH_3 + 5O_2 \xrightarrow{\Delta} 6H_2O + 4NO$$

$$2NO + O_2 \rightarrow 2NO_2$$

How many cubic feet of air (21% O_2 by volume) at 27°C and 1.00 atm are needed for the conversion of 50.0 tons of NH_3 to NO_2 by this process?

18. How many cubic feet of air (21% O_2 by volume) are needed for the production of 10^5 ft^3 NO_2 at the same conditions as those used to measure the air? (See Problem 17 for the equations involved.)

19. The du Pont company has developed a nitrometer, an apparatus for the rapid routine analysis of nitrates, which measures the volume of NO liberated by

the reaction of concentrated H_2SO_4 with nitrates in the presence of metallic mercury, by the reaction

$$2KNO_3 + 4H_2SO_4 + 3Hg \rightarrow K_2SO_4 + 3HgSO_4 + 4H_2O + 2NO$$

In a simple form of this apparatus, the NO is collected over water in a graduated tube and its volume, temperature, and pressure are measured. A 1.000 g sample containing a mixture of KNO_3 and K_2SO_4 is treated in this manner, and 37.50 ml of NO is collected over water at a temperature of 23°C and a pressure of 732.0 torr. Calculate the percentage of KNO_3 in the original sample.

20. A commercial laboratory wished to speed up its routine analysis for HNO_3 in a mixture of acids, using the nitrometer mentioned in Problem 19. To do this, it collects the NO over mercury, uses enough concentrated H_2SO_4 to make correction for water vapor unnecessary, thermostats its graduated tube at 25.0°C, and takes all pressure measurements at 730.0 torr. The tube is graduated to 100.0 ml. What weight of original acid sample should always be taken so that the buret reading under these conditions also indicates directly the percentage of HNO_3 in the original sample?

21. In the Dumas method for measuring the total nitrogen in an organic compound, the compound is mixed with CuO and heated in a stream of pure CO_2. All the gaseous products are passed through a heated tube of Cu turnings, to reduce any oxides of nitrogen to N_2, and then through a 50% solution of KOH to remove the CO_2 and water. The N_2 is not absorbed, and its volume is measured by weighing the mercury (density $= 13.56$ g/ml) that the N_2 displaces from the apparatus.
    (a) A 20.1 mg sample of a mixture of glycine, $CH_2(NH_2)COOH$, and benzoic acid, $C_7H_6O_2$, yields N_2 at 730 torr and 21°C. This N_2 displaces 5.235 g of mercury. Calculate the percentage of glycine in the original mixture.
    (b) A 4.71 mg sample of a compound containing C, H, O, and N is subjected to a Dumas nitrogen determination. The N_2, at 735 torr and 27°C, displaces 10.5532 g of mercury. A carbon–hydrogen analysis shows that this compound contains 3.90% H and 46.78% C. Determine the empirical formula of this compound.

PROBLEMS B

22. Balance the following equations, which show the starting materials and the reaction products. It is not necessary to supply any additional reactants or products. A Δ sign indicates that heating is necessary.
    (a) $H_3BO_3 \xrightarrow{\Delta} H_4B_6O_{11} + H_2O$
    (b) $C_6H_{12}O_6 \rightarrow C_2H_5OH + CO_2$
    (c) $CaC_2 + N_2 \xrightarrow{\Delta} CaCN_2 + C$
    (d) $CaCN_2 + H_2O \rightarrow CaCO_3 + NH_3$
    (e) $BaO + C + N_2 \xrightarrow{\Delta} Ba(CN)_2 + CO$
    (f) $C_6H_6 + O_2 \xrightarrow{\Delta} CO_2 + H_2O$

(g) $C_7H_{16} + O_2 \xrightarrow{\Delta} CO_2 + H_2O$

(h) $H_3PO_3 \xrightarrow{\Delta} H_3PO_4 + PH_3$

(i) $MnO_2 + KOH + O_2 \xrightarrow{\Delta} K_2MnO_4 + H_2O$

(j) $KMnO_4 + H_2SO_4 \rightarrow K_2SO_4 + Mn_2O_7 + H_2O$

(k) $CO + Fe_3O_4 \xrightarrow{\Delta} FeO + CO_2$

(l) $ZnS + O_2 \xrightarrow{\Delta} ZnO + SO_2$

23. Some common gases may be prepared in the laboratory by reactions represented by the following equations. A Δ sign indicates that heating is necessary. For purity, air must be swept out of the apparatus before the gas is collected, and some gases must be dried with a suitable desiccant.

$$2NaHSO_3 + H_2SO_4 \rightarrow 2SO_2 + 2H_2O + Na_2SO_4$$

$$MnO_2 + 4HCl \xrightarrow{\Delta} Cl_2 + MnCl_2 + 2H_2O$$

$$2Na_2O_2 + 2H_2O \rightarrow 4NaOH + O_2$$

$$SiO_2 + 2H_2F_2 \rightarrow SiF_4 + 2H_2O$$

$$2HCOONa + H_2SO_4 \xrightarrow{\Delta} 2CO + 2H_2O + Na_2SO_4$$

$$NH_4Cl + NaNO_2 \xrightarrow{\Delta} N_2 + NaCl + 2H_2O$$

$$2Al + 2NaOH + 2H_2O \rightarrow 2NaAlO_2 + 3H_2$$

$$Al_4C_3 + 12H_2O \rightarrow 3CH_4 + 4Al(OH)_3$$

(a) What weight of $NaHSO_3$ is needed to prepare (*i*) 1.30 moles of SO_2?   (*ii*) 2.00 lb of SO_2?

(b) How many moles of pyrolusite, MnO_2, are needed to prepare (*i*) 100 g of Cl_2?   (*ii*) 2.60 moles of Cl_2?

(c) How many pounds of sand, SiO_2, are needed to prepare 10.0 lb of SiF_4, assuming that 25% of the sand is inert material and does not produce SiF_4?

(d) How many grams of sodium formate, $HCOONa$, are needed to make 0.250 mole of CO?

(e) How many pounds of aluminum hydroxide are produced along with 12.0 moles of CH_4 (methane)?

(f) How many moles of NH_4Cl are needed to prepare 1.33 moles of N_2 by the sixth reaction?

(g) How many grams of 90% pure Na_2O_2 (sodium peroxide) are needed to prepare 2.50 lb of O_2?

(h) Commercial sulfuric acid that has a density of 1.84 g/ml and is 95% H_2SO_4 by weight is used for the production of CO by the fifth reaction.   (*i*) What weight of commercial acid is needed for the production of 560 g of CO?   (*ii*) What volume of acid is needed for the production of 560 g of CO?

(i) What volume of commercial HCl (36% HCl by weight, density = 1.18 g/ml) and weight of pyrolusite (85% MnO_2) are needed to produce 5.00 kg of Cl_2 by the second reaction?

(j) Commercial sulfuric acid that has a density of 1.52 g/ml and is 62% H_2SO_4 by weight is used for the production of SO_2 by the first reaction. What (*i*) weight and (*ii*) volume of this commercial acid are needed for the production of 720 g of SO_2 gas?

(k) Assume that an excess of metallic aluminum is added to a solution containing 100 g NaOH, and that as a result all the NaOH is used up. When the reaction is complete, the excess aluminum metal is filtered off and the excess water evaporated. How many grams of $NaAlO_2$ are obtained?

24. How many grams of aluminum are needed for the preparation of 5.50 liters of H_2 at standard conditions? The reaction is

$$2Al + 3H_2SO_4 \rightarrow Al_2(SO_4)_3 + 3H_2$$

25. What weight of $(NH_4)_2SO_4$ is needed for the preparation of 5.00 moles of NH_3 gas? The reaction is

$$(NH_4)_2SO_4 + Ca(OH)_2 \rightarrow CaSO_4 + 2NH_3 + 2H_2O$$

26. What volume of O_2 at 0.90 atm and 75°F is needed to burn 21.0 liters of propane gas, C_3H_8, under the same conditions? The reaction is

$$C_3H_8 + 5O_2 \xrightarrow{\Delta} 3CO_2 + 4H_2O$$

27. What volume of NO can react with 100 liters of air (21.0% O_2 by volume) at the same conditions of temperature and pressure. The reaction is

$$2NO + O_2 \rightarrow 2NO_2$$

28. An interesting lecture demonstration is the "Vesuvius" experiment, in which a small mound of $(NH_4)_2Cr_2O_7$ is heated to commence decomposition. It then continues its decomposition unaided, gives off heat, light, and sparks, and leaves a "mountain" of Cr_2O_3. The reaction is

$$(NH_4)_2Cr_2O_7 \xrightarrow{\Delta} N_2 + 4H_2O + Cr_2O_3$$

What volume of N_2 at 730 torr and 31°C is produced from 1.60 moles of $(NH_4)_2Cr_2O_7$?

29. What volume of H_2S at 720 torr and 85°F is needed for the precipitation of the bismuth in 50.0 g of $BiCl_3$? The reaction is

$$2BiCl_3 + 3H_2S \rightarrow Bi_2S_3 + 6HCl$$

30. What volume of H_2S at 740 torr and 20°C is needed to precipitate the nickel from 50.0 g of $Ni_3(PO_4)_2 \cdot 7H_2O$ as NiS?

31. HCN gas is fatal at a concentration of 1 part per 500 by volume and is very dangerous within one hour at a concentration of 1 part per 10,000 by volume. What weight of NaCN is needed to fill a classroom 20 ft × 15 ft × 9 ft with HCN at a concentration of 1 part per 10,000? (The barometric pressure is 740 torr, and the temperature is 80°F.) The reaction is

$$2NaCN + H_2SO_4 \rightarrow Na_2SO_4 + 2HCN$$

32. (a) How many grams of aluminum are needed for the production of 5.50 liters of H_2 over H_2O at 730 torr and 18°C?

(b) How many moles of H_2SO_4 are needed? (The reaction is given in Problem 24.)

33. What volume of O_2, collected over water at 735 torr and 20°C, can be produced from 4.00 lb of sodium peroxide, Na_2O_2, according to the reaction

$$2Na_2O_2 + 2H_2O \rightarrow 4NaOH + O_2$$

34. (a) What weight of $NaNO_3$ is needed to produce 5.00 liters of N_2O collected over water at 737 torr and 27°C? The reaction is

$$NH_4Cl + NaNO_3 \overset{\Delta}{\rightarrow} NaCl + N_2O + 2H_2O$$

(b) How many moles of NH_4Cl are needed?

35. (a) What volume of NH_3 at 700 torr and 50°C is needed to make 5.00 lb of $(NH_4)_3PO_4$? The reaction is

$$3NH_3 + H_3PO_4 \rightarrow (NH_4)_3PO_4$$

(b) What volume of H_3PO_4 (density = 1.69 g/ml, 85% H_3PO_4 by weight) is needed for this preparation?

36. (a) What volume of H_2SO_4 (density = 1.84 g/ml, 95% H_2SO_4 by weight) is needed to produce 8.30 liters of H_2 collected over water at 740 torr and 18°C? The reaction is

$$Mg + H_2SO_4 \rightarrow MgSO_4 + H_2$$

(b) How many moles of Mg are used?

37. What volume of air (21% O_2 by volume) at 720 torr and 68°F is needed to burn 1.00 lb of butane gas, C_4H_{10}? Balance the equation before solving the problem.

$$C_4H_{10} + O_2 \overset{\Delta}{\rightarrow} CO_2 + H_2O$$

38. In the process of photosynthesis, plants use CO_2 and water to produce sugars according to the overall reaction

$$11H_2O + 12CO_2 \rightarrow C_{12}H_{22}O_{11} + 12O_2$$

(a) What volume of CO_2 at 30°C and 730 torr is used by a plant in making 1.00 lb of sucrose, $C_{12}H_{22}O_{11}$?
(b) What volume of air is deprived of its normal amount of CO_2 in this process? (Air contains 0.035% CO_2 by volume. Assume a barometric pressure of 750 torr and a temperature of 70°F.)

39. A big national industry produces ethyl alcohol, C_2H_5OH, by the enzymatic action of yeast on sugar, by the reaction

$$C_6H_{12}O_6 \xrightarrow{\text{zymase}} 2C_2H_5OH + 2CO_2$$

What volume of CO_2 at 720 torr and 30°C is formed during the production of 100 gal of alcohol that is 95% C_2H_5OH by weight? (The density of 95% C_2H_5OH is 0.800 g/ml.)

40. The adsorbing capacity of charcoal may be greatly increased by "steam activation." The principal action of the steam passing over the very hot charcoal is to widen the pores as the result of the reaction

$$H_2O + C \xrightarrow{\Delta} H_2 + CO$$

The mixture of CO and H_2, known as water gas, can be used as a fuel for other manufacturing concerns, as well as to heat up the coke for this process.

(a) How many cubic feet of water gas at 730 torr and 68°F are obtained from the activation of 50.0 tons of charcoal if it is activated to a 60% weight loss?

(b) How many gallons of water (density = 1 g/ml) will this take?

41. What volume of water gas (CO_2 + H_2) can be prepared from 1500 liters of steam at 900°C and 1 atm? (The steam and water gas are measured under the same conditions. See Problem 40 for the equation.)

42. In the contact process for making H_2SO_4, S is burned with air to SO_2, and the SO_2–air mixture is then passed over a V_2O_5 catalyst, which converts it to SO_3:

$$S + O_2 \xrightarrow{\Delta} SO_2$$
$$2SO_2 + O_2 \xrightarrow{\Delta} 2SO_3$$

(a) What volume of air (21% O_2) at 27°C and 1.00 atm is needed for the conversion of 1.00 ton of sulfur to SO_3? (Use a 15% excess of oxygen.)

(b) How many gallons of H_2SO_4 (95% H_2SO_4 by weight, density = 1.84 g/ml) can be produced from 1.00 ton of sulfur?

43. A common biological determination is for amino-acid nitrogen. This determination is made by the van Slyke method, in which the amino groups ($-NH_2$) in protein material react with HNO_2 to produce N_2 gas, the volume of which is measured. A 0.530 g sample of a biological material containing glycine, $CH_2(NH_2)COOH$, yields 37.2 ml of N_2 gas collected over water at a pressure of 737 torr and 27°C. What is the percentage of glycine in the original sample? The reaction is

$$CH_2(NH_2)COOH + HNO_2 \rightarrow CH_2(OH)COOH + H_2O + N_2$$

13

Stoichiometry III: Calculations Based on Concentrations of Solutions

When two or more substances are mixed together in a manner that is homogeneous and uniform at the molecular level, the mixture is called a *solution*. The component (usually a liquid) that is present in much larger quantity than the others is called the *solvent;* the other components are the *solutes*. The *concentration* of a solution describes the amount of solute present in a given amount of solution. When a solution is involved in a reaction, the stoichiometric calculations must take into account two quantities not previously discussed: the concentration of the solution, and its volume.

PREPARATION OF SOLUTIONS

The ways in which chemists most frequently express concentrations involve the mole as the concentration unit (rather than the gram), because reactions are between molecules as the basic entities.

Molarity

A solution that contains one mole of solute per liter of solution is known as a one *molar* solution; it is abbreviated 1.00 M. In general,

$$\text{molarity of solution} = M = \frac{\text{moles of solute}}{\text{liter of solution}}$$

It is simple to prepare solutions of known molarity from solids and non-volatile liquids that can be weighed on an analytical balance, and then dissolved and diluted *to* a known volume in a volumetric flask. When a reagent-grade sample is accurately weighed and diluted with care, as in the following problem, the resulting solution is said to be a *standard solution*.

PROBLEM:
Prepare 250.0 ml of a 0.1250 M $AgNO_3$ solution, using solid $AgNO_3$.

SOLUTION:
From the given volume and concentration you can calculate how many grams of $AgNO_3$ to weigh out:

$$\text{wt. of } AgNO_3 \text{ needed} = (0.2500 \text{ liter})\left(0.1250 \ \frac{\text{moles } AgNO_3}{\text{liter}}\right)\left(169.9 \ \frac{\text{g } AgNO_3}{\text{mole } AgNO_3}\right)$$

$$= 5.309 \text{ g } AgNO_3$$

Transfer the 5.309 g $AgNO_3$ to a 250.0 ml volumetric flask, dissolve it in some distilled water, then dilute *to* the mark (see p 86). Shake vigorously to get a uniform solution. *Don't* add 250.0 ml of water to the weighed sample, because the resulting solution may actually be larger or smaller than 250.0 ml due to interaction of solute and solvent.

Many crystals contain "water of crystallization," which must be included in the weight of the material weighed out for preparing a solution. Allowance is made for this by using the molecular weight of the hydrate in your calculations, *not* the mole weight of the anhydrous form.

PROBLEM:
Prepare 100.0 ml of 0.2000 M $CuSO_4$, starting with solid $CuSO_4 \cdot 5H_2O$.

SOLUTION:
From the given volume and concentration of $CuSO_4$, you can calculate the moles of $CuSO_4$ required. Furthermore, the formula shows that 1 mole of $CuSO_4 \cdot 5H_2O$ is required per mole of $CuSO_4$. Thus the weight (W) of $CuSO_4 \cdot 5H_2O$ needed is

$$W = (0.1000 \text{ liter})\left(0.2000 \ \frac{\text{moles } CuSO_4}{\text{liter}}\right)\left(1 \ \frac{\text{mole } CuSO_4 \cdot 5H_2O}{\text{mole } CuSO_4}\right)\left(249.6 \ \frac{\text{g } CuSO_4 \cdot 5H_2O}{\text{mole } CuSO_4 \cdot 5H_2O}\right)$$

$$= 4.992 \text{ g } CuSO_4 \cdot 5H_2O \text{ needed}$$

Transfer the 4.992 g to a 100 ml volumetric flask, dissolve it in some distilled water, then dilute *to* the mark. The fact that some of the water in the solution comes from the weighed sample is irrelevant. The source of the water is never a matter of concern.

Many times it is convenient to obtain a substance from its solution. What volume of solution should you use to get the quantity of solute you want?

PROBLEM:

What volume of 0.250 M Na_2CrO_4 will be needed in order to obtain 8.10 g of Na_2CrO_4?

SOLUTION:

If V = liters of solution needed, it must supply the number of moles contained in 8.10 g of Na_2CrO_4.

$$\text{Moles of } Na_2CrO_4 \text{ needed} = \frac{8.10 \text{ g } Na_2CrO_4}{162.0 \dfrac{\text{g } Na_2CrO_4}{\text{mole } Na_2CrO_4}} = 0.0500 \text{ moles}$$

$$\text{Moles of } Na_2CrO_4 \text{ in } V \text{ liters} = \left(0.250 \frac{\text{moles } Na_2CrO_4}{\text{liter}}\right)(V \text{ liters})$$

$$= 0.250V \text{ moles}$$

$$0.250V = 0.0500$$

$$V = 0.200 \text{ liter} = 200 \text{ ml needed}$$

This problem illustrates the two most common ways of calculating moles of a compound: (a) weight divided by mole weight, and (b) molarity times volume in liters.

Molality

When discussing the colligative properties of a solution (Chapter 21), it is more important to relate the moles of solute to a constant amount of solvent rather than to the volume of the solution, as in the case of molarity. In practice this is accomplished by using a *kilogram of solvent* instead of a liter of solution as the reference. A solution that contains one mole of solute per kilogram of solvent is known as a one *molal* solution; it is abbreviated 1.00 *m*. In general,

$$\text{molality of solution} = m = \frac{\text{moles of solute}}{\text{kg of solvent}}$$

Because the density of water is approximately 1 g/ml, the molarities and molalities of water solutions will have about the same value. This will not be true for most other solvents.

It is simple to prepare solutions of known molality. The volume of the final solution doesn't enter into it at all.

PROBLEM:

Prepare a 2.00 m naphthalene ($C_{10}H_8$) solution using 50.0 g CCl_4 as the solvent.

SOLUTION:

You are given the molality of the solution and the weight of the solvent, from which you can find x, the number of moles of $C_{10}H_8$ needed.

$$\text{Molality} = m = \frac{x \text{ moles } C_{10}H_8}{0.0500 \text{ kg } CCl_4} = 2.00$$

$$x = 0.100 \text{ moles } C_{10}H_8 \text{ needed}$$

$$\text{Weight of } C_{10}H_8 \text{ needed} = (0.100 \text{ moles } C_{10}H_8)\left(128 \ \frac{\text{g } C_{10}H_8}{\text{mole } C_{10}H_8}\right)$$

$$= 12.8 \text{ g } C_{10}H_8$$

To prepare the solution, dissolve 12.8 g $C_{10}H_8$ in 50.0 g CCl_4. If you knew that the density of CCl_4 is 1.59 g/ml, you could measure out

$$\frac{50.0 \text{ g}}{1.59 \ \frac{\text{g}}{\text{ml}}} = 31.4 \text{ ml } CCl_4$$

Mole Fraction

Another way of expressing concentrations that is used commonly with gases (see p 162) and colligative properties (see p 328) is *mole fraction,* which is defined (for a given component) as being the moles of component in question divided by the total moles of all components in solution. For a solution that has three components (A, B, and C), the mole fraction of A is given by

$$\text{mole fraction of A} = X_A = \frac{\text{moles of A}}{\text{moles of A} + \text{moles of B} + \text{moles of C}}$$

If you think about it, it's also easy to make a solution of a given mole fraction.

PROBLEM:

Prepare a 0.0348 mole fraction solution of sucrose ($C_{12}H_{22}O_{11}$, mole weight = 342 g/mole), using 100 g (that is, 100 ml) of water.

SOLUTION:

You are given the mole fraction of sucrose and the moles of water

$$\frac{100 \text{ g}}{18 \text{ } \dfrac{\text{g}}{\text{mole}}} = 5.55 \text{ moles}$$

What you don't know is x, the moles of sucrose needed. By definition,

$$\text{mole fraction} = 0.0348 = \frac{x \text{ moles sucrose}}{(5.55 \text{ moles } H_2O) + (x \text{ mole sucrose})}$$

Therefore,

$$x = \frac{(5.55)(0.0348)}{1 - 0.0348} = 0.200 \text{ mole sucrose}$$

$$\text{Weight of sucrose needed} = (0.200 \text{ mole sucrose})\left(342 \text{ } \frac{\text{g sucrose}}{\text{mole sucrose}}\right)$$

$$= 68.4 \text{ g sucrose}$$

Prepare the solution by dissolving 68.4 g sucrose in 100 g water.

Commercial Concentrated Solutions

Many solutions can't be made accurately, or at all, by weighing out the solute and dissolving it in the proper amount of solvent. For example, pure substances such as HCl and NH_3 are gases; H_2SO_4 and HNO_3 are fuming hygroscopic corrosive liquids; NaOH and KOH avidly absorb water and CO_2 from the air. In such circumstances, the customary procedure is to purchase the chemicals in the form of extremely concentrated solutions, then dilute them to the desired strength. The following problem is typical.

PROBLEM:
Commercial concentrated sulfuric acid is labeled as having a density of 1.84 g/ml and being 96.0% H_2SO_4 by weight. Calculate the molarity of this solution.

SOLUTION:
This is a typical "conversion" problem in which we want to go from grams of solution per liter to moles of H_2SO_4 per liter.

$$\left(1840 \text{ } \frac{\text{g solution}}{\text{liter}}\right)\left(0.960 \text{ } \frac{\text{g } H_2SO_4}{\text{g solution}}\right)\left(\frac{1 \text{ mole } H_2SO_4}{98.1 \text{ g } H_2SO_4}\right) = 18.0 \text{ } \frac{\text{moles } H_2SO_4}{\text{liter}}$$

$$= 18.0 \text{ M}$$

PROBLEM:
What are (a) the molality and (b) the mole fraction of the commercial H_2SO_4 solution in the previous problem?

SOLUTION:

(a) To find the molality we need to know, for a given amount of solution, the moles of H_2SO_4 and the kg of H_2O. If we take a liter of solution, we shall have 1840 g of solution, of which 4.0% is water, so $(0.040)(1840 \text{ g}) = 74 \text{ g } H_2O$. Because there are 18.0 moles of H_2SO_4 in this liter, we have

$$\text{molality} = m = \frac{18.0 \text{ moles } H_2SO_4}{0.074 \text{ kg } H_2O} = 243 \text{ molal } H_2SO_4$$

(b) To find the mole fraction, we need to know (for a given amount of solution) the moles of H_2SO_4 and the moles of H_2O. If we take a liter of solution, we shall have 18.0 moles of H_2SO_4 and

$$\frac{74 \text{ g } H_2O}{18 \dfrac{\text{g } H_2O}{\text{mole } H_2O}} = 4.1 \text{ moles of } H_2O$$

Therefore the mole fraction is

$$X_{H_2SO_4} = \frac{18.0 \text{ moles } H_2SO_4}{(18.0 \text{ moles } H_2SO_4) + (4.1 \text{ moles } H_2O)} = 0.814$$

Note that it is not possible to convert from molarity to molality or mole fraction unless some information about the density or weight composition of the solution is given.

Dilution

One of the most common ways to prepare a solution is to dilute a concentrated solution that has already been prepared. There is a fundamental principle that underlies all dilutions: the number of moles of solute is the same after dilution as before. It is only the moles of solvent that have been changed (increased). This principle makes dilution calculations simple. If M_1 and M_2 are the molarities before and after dilution, and V_1 and V_2 are the initial and final volumes of solution, then

moles of solute before dilution = moles of solute after dilution

$$\left(M_1 \frac{\text{moles}}{\text{liter}}\right)(V_1 \text{ liters}) = \left(M_2 \frac{\text{moles}}{\text{liter}}\right)(V_2 \text{ liters})$$

$$M_1 V_1 = M_2 V_2$$

The following problem illustrates the use of this equation.

PROBLEM:

What volume of 18.0 M H_2SO_4 is needed for the preparation of 2.00 liters of 3.00 M H_2SO_4?

SOLUTION:

We are given the initial and final concentrations of the two solutions, along with the final volume. Therefore,

$$\left(18.0 \ \frac{moles}{liter}\right)(V \text{ liters}) = (2.00 \text{ liters})\left(3.00 \ \frac{moles}{liter}\right)$$

$$V = \frac{(2.00)(3.00)}{(18.0)} = 0.333 \text{ liter}$$

To prepare the solution, measure out 333 ml of 18.0 M H_2SO_4 and dilute it *to* 2.00 liters in a volumetric flask. Shake well for uniformity. Because the interaction of concentrated H_2SO_4 with H_2O evolves *much* heat and can cause hazardous splattering, it is better to do a partial dilution with about one liter of water first in a beaker. Then, after cooling, transfer the contents to the volumetric flask, and complete the dilution. The final dilution to the mark must be made with the solution at room temperature.

PREPARATIVE REACTIONS

Solutions often are involved in reactions used to prepare compounds or to perform chemical analyses. In either case, you will be faced with stoichiometrical calculations similar to those in Chapters 10 and 12, but you must also consider the concentrations and volumes of the solutions that are used. We shall use the "three simple steps" as outlined on p 173.

PROBLEM:

Cupric nitrate is prepared by dissolving a weighed amount of copper metal in a nitric acid solution:

$$3Cu + 8HNO_3 \rightarrow 3Cu(NO_3)_2 + 2NO + 4H_2O$$
$$3Cu + 2NO_3^- + 8H^+ \rightarrow 3Cu^{2+} + 2NO + 4H_2O$$

What volume of 6.00 M HNO_3 should be used to prepare 10.0 g of $Cu(NO_3)_2$?

SOLUTION:

1. From the given weight of $Cu(NO_3)_2$, calculate the moles of $Cu(NO_3)_2$ that are given (to prepare):

$$\text{moles of } Cu(NO_3)_2 \text{ given} = \frac{10.0 \text{ g } Cu(NO_3)_2}{187.5 \ \dfrac{\text{g } Cu(NO_3)_2}{\text{mole } Cu(NO_3)_2}} = 0.0533 \text{ mole } Cu(NO_3)_2$$

2. From the given number of moles of $Cu(NO_3)_2$, calculate the moles of HNO_3 required. The chemical equation shows that 8 moles of HNO_3 are required for 3 moles of $Cu(NO_3)_2$. Therefore,

$$\text{moles of HNO}_3 \text{ required} = \left(\frac{8 \text{ moles HNO}_3}{3 \text{ moles Cu(NO}_3)_2} \right) [0.0533 \text{ moles Cu(NO}_3)_2]$$

$$= 0.142 \text{ moles HNO}_3$$

3. From the number of moles of HNO_3 required, find the volume of HNO_3 solution as specified in the statement of the problem. Because the HNO_3 solution is 6.00 moles/liter, we must use a volume that will provide 0.142 moles. That is,

$$V = \frac{0.142 \text{ moles HNO}_3}{6.00 \dfrac{\text{moles HNO}_3}{\text{liter}}} = 0.0237 \text{ liter}$$

You will have to add at least 23.7 ml of 6.00 M HNO_3 in order to prepare 10.0 g of $Cu(NO_3)_2$.

TITRATION REACTIONS

Standardization

The concentrations of solutions that are made from concentrated commercial solutions (such as the H_2SO_4 solution on p 192) are usually not known very accurately because the label information may be only approximate, or the concentrations may change with the opening and closing of the bottles. For example, concentrated HCl and NH_3 tend to lose their solutes (the odors are terrible!); NaOH and KOH tend to react with CO_2 from the air; and H_2SO_4 and H_3PO_4 are diluted as they pick up H_2O from the air. As a result, standard solutions of these compounds can't be made from concentrated commercial solutions. However, it *is* possible to make a dilute solution of approximately the desired concentration (see p 194), and then *standardize* it by reaction with a known weight of a pure compound or with a known volume of another standard solution.

The amount of a solution needed to react with a given quantity of another substance often is determined by a process known as *titration*. The reagent solution, the titrant, is added from a buret (a graduated tube) to the sample solution until chemically equivalent amounts of the two are present. The endpoint of the titration corresponds to the use of chemically equivalent amounts; it often is signaled by a color change in the solution due to the presence of a suitable indicator. The solution in the buret may be a standard solution that is being used for analysis of the material in the titration flask, or it may be a solution that is being standardized by reaction with a known amount of material in the flask. In any case, the volume of solution added from the buret to reach the endpoint is always recorded. The two following problems illustrate standardization of solutions by titration.*

---

* In Chapter 20 we present an alternative approach to solving all of the problems in the balance of this chapter.

PROBLEM:
Titration of a 0.7865 g sample of pure potassium hydrogen phthalate requires 35.73 ml of a NaOH solution. Calculate the molarity of the NaOH solution. This acid–base reaction is

$$NaOH + KHC_8H_4O_4 \rightarrow NaKC_8H_4O_4 + H_2O$$
$$OH^- + HC_8H_4O_4^- \rightarrow C_8H_4O_4^{2-} + H_2O$$

SOLUTION:
1. From the given grams of $KHC_8H_4O_4$, calculate the moles:

$$\text{moles of } KHC_8H_4O_4 \text{ given} = \frac{0.7865 \text{ g } KHC_8H_4O_4}{204.1 \dfrac{\text{g } KHC_8H_4O_4}{\text{mole } KHC_8H_4O_4}} = 3.854 \times 10^{-3} \text{ moles}$$

2. The chemical equation shows 1 mole of NaOH needed per mole of $KHC_8H_4O_4$ at the endpoint, so

$$\text{moles of NaOH used} = \left(\frac{1 \text{ mole NaOH}}{1 \text{ mole } KHC_8H_4O_4}\right)(3.854 \times 10^{-3} \text{ moles } KHC_8H_4O_4)$$
$$= 3.854 \times 10^{-3} \text{ moles NaOH}$$

3. The moles of NaOH used were dissolved in 35.73 ml of solution. Therefore, the molarity of the solution is

$$\text{molarity} = \frac{3.854 \times 10^{-3} \text{ moles NaOH}}{0.03573 \text{ liters}} = 0.1079 \frac{\text{moles NaOH}}{\text{liter}}$$

You have "standardized" the solution and found it to be 0.1079 M NaOH.

PROBLEM:
A solution of H_2SO_4 is prepared by diluting commercial concentrated acid. A volume of 42.67 ml of this solution is required to titrate exactly 50.00 ml of the NaOH solution standardized in the preceding problem. Calculate the molarity of the H_2SO_4 solution. The titration reaction is

$$2NaOH + H_2SO_4 \rightarrow Na_2SO_4 + 2H_2O$$
$$OH^- + H^+ \rightarrow H_2O$$

SOLUTION:
1. From the given volume and concentration of NaOH solution, calculate the moles of NaOH given (used):

$$\text{moles of NaOH used} = (0.05000 \text{ liters})\left(0.1079 \frac{\text{moles NaOH}}{\text{liter}}\right)$$
$$= 5.395 \times 10^{-3} \text{ moles NaOH}$$

2. The chemical equation shows that 1 mole of H_2SO_4 is used per 2 moles of NaOH at the endpoint, so the number of moles of H_2SO_4 used is

$$\text{moles of H}_2\text{SO}_4 = \left(\frac{1 \text{ mole H}_2\text{SO}_4}{2 \text{ moles NaOH}}\right)(5.395 \times 10^{-3} \text{ moles NaOH})$$

$$= 2.698 \times 10^{-3} \text{ moles H}_2\text{SO}_4$$

3. The moles of H_2SO_4 used were dissolved in 42.67 ml of solution. Therefore, the molarity is

$$\text{molarity} = \frac{2.698 \times 10^{-3} \text{ moles H}_2\text{SO}_4}{0.04267 \text{ liters}} = 0.06322 \frac{\text{moles H}_2\text{SO}_4}{\text{liter}}$$

You have "standardized" the H_2SO_4 solution and found it to be 0.06322 M H_2SO_4.

Analysis

Sometimes it is not possible to perform a "direct" titration as illustrated in the last two problems. In an "indirect" titration, a known (but excess) volume of standard solution is used to insure a complete reaction with whatever is being analyzed for; then, at the end of the analysis, the unused quantity of the standard solution is determined by what is called a *back-titration*. For example, such a procedure is useful for the determination of a gas that can be bubbled through the standard reaction solution. Another example is illustrated by the next problem.

PROBLEM:
A 0.3312 g sample of impure Na_2CO_3 is dissolved in exactly 50.00 ml of the H_2SO_4 solution standardized in the preceding problem. There is CO_2 gas liberated in this reaction:

$$Na_2CO_3 + H_2SO_4 \rightarrow Na_2SO_4 + CO_2\uparrow + H_2O$$

$$CO_3^{2-} + 2H^+ \rightarrow CO_2\uparrow + H_2O$$

The CO_2 gas is totally driven off by gentle heating, so that it will not interfere with the endpoint determination. The unused portion of the H_2SO_4 is then titrated. The back-titration reaction is

$$2NaOH + H_2SO_4 \rightarrow Na_2SO_4 + 2H_2O$$

$$OH^- + H^+ \rightarrow H_2O$$

The back-titration requires 9.36 ml of 0.1079 M NaOH. Calculate the percentage of Na_2CO_3 in the original sample.

SOLUTION:
At the final endpoint, the H_2SO_4 has been exactly used up by two reactions: one with NaOH, and the other with Na_2CO_3. We know the moles of H_2SO_4 and NaOH involved because we know the volumes and concentrations of their solutions. If we subtract from the total moles of H_2SO_4 the number of moles used by the

NaOH, we will know the number of moles of H_2SO_4 used by the Na_2CO_3. This will tell us how much Na_2CO_3 must have been present originally.

$$\text{Total moles } H_2SO_4 \text{ at outset} = \left(0.06322 \; \frac{\text{moles } H_2SO_4}{\text{liter}}\right) (0.05000 \text{ liters})$$

$$= 0.003161 \text{ moles } H_2SO_4$$

moles of H_2SO_4 used by NaOH

$$= \left(0.1079 \; \frac{\text{moles NaOH}}{\text{liter}}\right) (0.00936 \text{ liters}) \left(\frac{1 \text{ mole } H_2SO_4}{2 \text{ moles NaOH}}\right)$$

$$= 0.000505 \text{ moles } H_2SO_4$$

Moles of H_2SO_4 *not* used by NaOH $= 0.003161 - 0.000505$

$$= 0.002656 \text{ moles } H_2SO_4 \text{ used by } Na_2CO_3$$

The chemical equation shows that 1 mole of Na_2CO_3 is used per mole of H_2SO_4, so

weight of Na_2CO_3 present

$$= (0.002656 \text{ moles } H_2SO_4) \left(\frac{1 \text{ mole } Na_2CO_3}{1 \text{ mole } H_2SO_4}\right) \left(106.0 \; \frac{\text{g } Na_2CO_3}{\text{mole } Na_2CO_3}\right)$$

$$= 0.2815 \text{ g } Na_2CO_3 \text{ in sample}$$

$$\% \; Na_2CO_3 = \frac{0.2815 \text{ g } Na_2CO_3}{0.3312 \text{ g sample}} \times 100 = 85.00\% \; Na_2CO_3$$

PROBLEM:
An iron ore is a mixture of Fe_2O_3 and inert impurities. One method of iron analysis is to dissolve the ore sample in HCl, convert all of the iron to the Fe^{2+} state with metallic zinc, then titrate the solution with a standard $KMnO_4$ solution. The reaction is

$$KMnO_4 + 5FeCl_2 + 8HCl \rightarrow KCl + MnCl_2 + 5FeCl_3 + 4H_2O$$
$$MnO_4^- + 5Fe^{2+} + 8H^+ \rightarrow Mn^{2+} + 5Fe^{3+} + 4H_2O$$

At the endpoint, there is no more $FeCl_2$ for reaction, so the further addition of purple $KMnO_4$ will make the whole solution purple because it can no longer be converted to colorless $MnCl_2$. If a 0.3778 g ore sample requires 38.60 ml of 0.02105 M $KMnO_4$ for titration, calculate the percentage of iron in the original sample.

SOLUTION:
1. From the given amount of $KMnO_4$, calculate the moles of $KMnO_4$ used:

$$\text{moles } KMnO_4 \text{ used} = \left(0.02105 \; \frac{\text{moles } KMnO_4}{\text{liter}}\right) (0.03860 \text{ liter})$$

$$= 8.125 \times 10^{-4} \text{ moles } KMnO_4$$

2. From the given moles of $KMnO_4$, calculate the moles of Fe used up. The chemical equation shows that 5 moles of Fe (as $FeCl_2$) are used per mole of $KMnO_4$, so

$$\text{moles of Fe used up} = (8.125 \times 10^{-4} \text{ moles } KMnO_4)\left(5\ \frac{\text{moles Fe}}{\text{mole } KMnO_4}\right)$$

$$= 4.063 \times 10^{-3} \text{ moles Fe present}$$

3. From the number of moles of Fe present, calculate the weight of Fe, and from that the percentage of Fe in the original sample.

$$\text{Weight of Fe present} = (4.063 \times 10^{-3} \text{ moles Fe})\left(55.8\ \frac{\text{g Fe}}{\text{mole Fe}}\right)$$

$$= 0.2267 \text{ g Fe}$$

$$\% \text{ Fe present} = \frac{0.2267 \text{ g Fe}}{0.3778 \text{ g sample}} \times 100 = 60.00\% \text{ Fe}$$

PROBLEMS A

1. Tell how you would prepare each of the following solutions.
   (a) 3.00 liters of 0.750 M NaCl from solid NaCl
   (b) 55.0 ml of 2.00 M $ZnSO_4$ from solid $ZnSO_4 \cdot 7H_2O$
   (c) 180 ml of 0.100 M $Ba(NO_3)_2$ from solid $Ba(NO_3)_2$
   (d) 12 liters of 6.0 M KOH from solid KOH
   (e) 730 ml of 0.0700 M $Fe(NO_3)_3$ from solid $Fe(NO_3)_3 \cdot 9H_2O$

2. Tell how you would prepare each of the following solutions.
   (a) 15.0 ml of 0.200 M H_2SO_4 from 6.00 M H_2SO_4
   (b) 280 ml of 0.600 M $CoCl_2$ from 3.00 M $CoCl_2$
   (c) 5.70 liters of 0.0300 M $ZnSO_4$ from 2.50 M $ZnSO_4$
   (d) 60.0 ml of 0.00350 M $K_3Fe(CN)_6$ from 0.800 M $K_3Fe(CN)_6$
   (e) 25.0 ml of 2.70 M $UO_2(NO_3)_2$ from 8.30 M $UO_2(NO_3)_2$

3. Find molarities and molalities of each of the following solutions.

Solution	Density (g/ml)	Weight percentage
(a) KOH	1.344	35.0
(b) HNO_3	1.334	54.0
(c) H_2SO_4	1.834	95.0
(d) $MgCl_2$	1.119	29.0
(e) $Na_2Cr_2O_7$	1.140	20.0
(f) $Na_2S_2O_3$	1.100	12.0
(g) Na_3AsO_4	1.113	10.0
(h) $Al_2(SO_4)_3$	1.253	22.0

4. Give the weight of metal ion in each of the following solutions.
   (a) 250 ml of 0.10 M $CuSO_4$ (e) 75 ml of 0.10 M $AlCl_3$
   (b) 125 ml of 0.050 M $CdCl_2$ (f) 1.5 liters of 3.0 M $AgNO_3$
   (c) 50 ml of 0.15 M $MgSO_4$ (g) 2.0 liters of 0.333 M $Fe_2(SO_4)_3$
   (d) 50 ml of 0.075 M Na_2SO_4

5. The density of a 7.00 M HCl solution is 1.113 g/ml. Find the percentage of HCl by weight.

6. (a) What is the percentage of HNO_3 by weight in a 21.2 M solution of HNO_3 whose density is 1.483 g/ml?
   (b) What is the molality of this solution?

7. If 3.0 liters of 6.0 M HCl are added to 2.0 liters of 1.5 M HCl, what is the resulting concentration? Assume the final volume to be exactly 5.0 liters.

8. What volume of 15.0 M HNO_3 should be added to 1250 ml of 2.00 M HNO_3 to prepare 14.0 liters of 1.00 M HNO_3? Water is added to make the final volume exactly 14.0 liters.

9. The density of a 2.04 M $Cd(NO_3)_2$ solution is 1.382 g/ml. If 500 ml of water is added to 750 ml of this solution, (a) what will be the percentage by weight of $Cd(NO_3)_2$ in the new solution?   (b) What will be its molality?

10. If 40.00 ml of an HCl solution is titrated by 45.00 ml of 0.1500 M NaOH, what is the molarity of the HCl?

11. We need 35.45 ml of an NaOH solution to titrate a 2.0813 g sample of pure benzoic acid, $HC_7H_5O_2$. What is the molarity of the NaOH solution?

12. What is the molarity of a $K_2C_2O_4$ solution if 35.00 ml of it is needed for the titration of 47.65 ml of 0.06320 M $KMnO_4$ solution? The reaction is

$$2MnO_4^- + 5C_2O_4^{2-} + 16H^+ \rightarrow 2Mn^{2+} + 10CO_2 + 8H_2O$$

13. How many milliliters of a solution containing 31.52 g $KMnO_4$ per liter will react with 3.814 g $FeSO_4 \cdot 7H_2O$? The reaction is

$$MnO_4^- + 5Fe^{2+} + 8H^+ \rightarrow Mn^{2+} + 5Fe^{3+} + 4H_2O$$

14. It takes 35.00 ml of 0.1500 M KOH to react with 40.00 ml of H_3PO_4 solution. The titration reaction is

$$H_3PO_4 + 2OH^- \rightarrow HPO_4^{2-} + 2H_2O$$

What is the molarity of the H_3PO_4 solution?

15. A 19.75 ml sample of vinegar of density 1.061 g/ml requires 43.24 ml of 0.3982 M NaOH for titration. What is the percentage by weight of acetic acid, $HC_2H_3O_2$, in the vinegar?

16. If we add 39.20 ml of 0.1333 M H_2SO_4 to 0.4550 g of a sample of soda ash that is 59.95% Na_2CO_3, what volume of 0.1053 M NaOH is required for back-titration?

17. A 0.500 g sample of impure $CaCO_3$ is dissolved in 50.0 ml of 0.100 M HCl, and the residual acid titrated by 5.00 ml of 0.120 M NaOH. Find the percentage of $CaCO_3$ in the sample. The reaction is

$$CaCO_3 + 2HCl \rightarrow CaCl_2 + H_2O + CO_2$$

18. It takes 45.00 ml of a given HCl solution to react with 0.2435 g calcite, $CaCO_3$. This acid is used to determine the percent purity of a $Ba(OH)_2$ sample, as follows.

Wt of impure $Ba(OH)_2$ sample = 0.4367 g

Vol of HCl used = 35.27 ml

Vol of NaOH used for back-titration = 1.78 ml

1.200 ml of HCl titrates 1.312 ml of NaOH

What is the percentage of $Ba(OH)_2$ in the sample?

19. (a) What weight of AgCl can be obtained by precipitating all the Ag^+ from 50 ml of 0.12 M $AgNO_3$?
    (b) What weight of NaCl is required to precipitate the AgCl?
    (c) What volume of 0.24 M HCl would be needed to precipitate the AgCl?

20. What volume of 10.0 M HCl is needed to prepare 6.40 liters H_2S at 750 torr and 27°C? The reaction is

$$FeS + 2HCl \rightarrow FeCl_2 + H_2S$$

21. What volume of 12.5 M NaOH is needed to prepare 25.0 liters of H_2 at 735 torr and 18°C by the reaction

$$2Al + 2NaOH + 2H_2O \rightarrow 2NaAlO_2 + 3H_2$$

22. (a) What weight of silver and what volume of 6.00 M HNO_3 are needed for the preparation of 500 ml of 3.00 M $AgNO_3$?
    (b) What volume of NO, collected over water at 725 torr and 27°C, will be formed? The reaction is

$$3Ag + 4HNO_3 \rightarrow 3AgNO_3 + NO + 2H_2O$$

23. Fuming sulfuric acid is a mixture of H_2SO_4 and SO_3. A 2.500 g sample of fuming sulfuric acid requires 47.53 ml of 1.1513 M NaOH for titration. What is the percentage of SO_3 in the sample?

PROMBLEMS B

24. Tell how you would prepare each of the following solutions.
    (a) 125 ml of 0.62 M NH_4Cl from solid NH_4Cl
    (b) 2.75 liters of 1.72 M $Ni(NO_3)_2$ from solid $Ni(NO_3)_2 \cdot 6H_2O$
    (c) 65.0 ml of 0.25 M $Al(NO_3)_3$ from solid $Al(NO_3)_3 \cdot 9H_2O$

   (d) 230 ml of 0.460 M LiOH from solid LiOH

   (e) 7.57 liters of 1.10 M $KCr(SO_4)_2$ from solid $KCr(SO_4)_2 \cdot 12H_2O$

25. Tell how you would prepare each of the following solutions.
   (a) 750 ml of 0.55 M H_3PO_4 from 3.60 M H_3PO_4
   (b) 12 liters of 3.0 M NH_3 from 15 M NH_3
   (c) 25 ml of 0.020 M $Pr_2(SO_4)_3$ from 0.50 M $Pr_2(SO_4)_3$
   (d) 365 ml of 0.0750 M $K_4Fe(CN)_6$ from 0.950 M $K_4Fe(CN)_6$

26. Tell how you would prepare each of the following solutions (weight percentage given).
   (a) 650 ml of 0.350 M $AlCl_3$ from a 16.0% solution whose density is 1.149 g/ml
   (b) 1.35 liters of 4.35 M NH_4NO_3 from a 62.0% solution whose density is 1.294 g/ml
   (c) 465 ml of 3.70 M H_3PO_4 from an 85.0% solution whose density is 1.689 g/ml
   (d) 75.0 ml of 1.25 M $CuCl_2$ from a 36.0% solution whose density is 1.462 g/ml
   (e) 8.32 liters of 1.50 M $ZnCl_2$ from a 60.3% solution whose density is 1.747 g/ml

27. Find the molalities of the original solutions used in Problem 26.

28. The density of a 3.68 M sodium thiosulfate solution is 1.269 g/ml.
   (a) Find the percentage of $Na_2S_2O_3$ by weight.
   (b) What is the molality of this solution?

29. The solubility of Hg_2Cl_2 is 7.00×10^{-4} g per 100 g of water at 30°C.
   (a) Assuming that the density of water (1 g/ml) is not appreciably affected by the presence of this amount of Hg_2Cl_2, determine the molarity of the saturated solution of Hg_2Cl_2.
   (b) What is the molality of this solution?

30. A pharmaceutical house wishes to prepare a nonirritating nose-drop preparation. To do this, it will put the active agent in a "normal saline" solution, which is merely 0.90% NaCl by weight. What quantities of material will be needed if it is desired to make 3000 gal of nose-drop solution that is normal saline and contains 0.10% active agent? (The density of 0.90% NaCl solution is 1.005 g/ml.)

31. If 500 ml of 3.00 M H_2SO_4 is added to 1.50 liters of 0.500 M H_2SO_4, what is the resulting concentration?

32. What volume of 15.0 M NH_3 should be added to 3.50 liters of 3.00 M NH_3 in order to give 6.00 liters of 5.00 M NH_3 on dilution with water?

33. The density of a 1.660 M $Na_2Cr_2O_7$ solution is 1.244 g/ml.
   (a) Find the percentage of $Na_2Cr_2O_7$ by weight.
   (b) If 1.50 liters of water are added to 1.00 liter of this solution, what is the percentage by weight of $Na_2Cr_2O_7$ in the new solution?

34. If 35.0 ml of 0.750 M $Al(NO_3)_3$ are added to 100 ml of 0.150 M $Al(NO_3)_3$, what will be the resulting concentration?

35. What volume of 4.00 M NaOH should be added to 5.00 liters of 0.500 M NaOH in order to get 15.0 liters of 1.00 M NaOH on dilution with water?

36. How would you make a standard solution that contains 10.0γ of Hg^{2+} per liter of solution? (Start with $HgCl_2$, and use volumetric flasks, pipets, and a balance sensitive to 0.2 mg. A γ is a millionth of a gram.)

37. It takes 31.00 ml of 0.2500 M H_2SO_4 to react with 48.00 ml NH_3 solution. What is the molarity of the NH_3?

38. How many milligrams of Na_2CO_3 will react with 45.00 ml of 0.2500 M HCl?

39. What is the molarity of a $KMnO_4$ solution if 30.00 ml of it is needed for the titration of 45.00 ml of 0.1550 M $Na_2C_2O_4$ solution? (See Problem 12 for the equation.)

40. What is the molarity of a ceric sulfate solution if 46.35 ml is required for the titration of a 0.2351 g sample of $Na_2C_2O_4$ that is 99.60% pure? The reaction is

$$2Ce^{4+} + C_2O_4^{2-} \rightarrow 2Ce^{3+} + 2CO_2$$

41. A 0.2120 g sample of pure iron wire is dissolved, reduced to Fe^{2+}, and titrated by 40.00 ml $KMnO_4$ solution. The reaction is

$$MnO_4^- + 5Fe^{2+} + 8H^+ \rightarrow Mn^{2+} + 5Fe^{3+} + 4H_2O$$

Find the molarity of the $KMnO_4$ solution.

42. A 0.220 g sample of H_2SO_4 is diluted with water and titrated by 40.0 ml of 0.100 M NaOH. Find the percentage by weight of H_2SO_4 in the sample.

43. A sample of vinegar weighs 14.36 g and requires 42.45 ml of 0.2080 M NaOH for titration. Find the percentage of acetic acid, $HC_2H_3O_2$, in the vinegar.

44. From the following data, compute the molarity of (a) the H_2SO_4 solution, and (b) the KOH solution:

Wt of sulfamic acid, $H(NH_2)SO_3$ = 0.2966 g

Vol of KOH to neutralize the $H(NH_2)SO_3$ = 34.85 ml

31.08 ml of H_2SO_4 titrates 33.64 ml of KOH

45. A crystal of calcite, $CaCO_3$, is dissolved in excess HCl, and the solution boiled to remove the CO_2. The unneutralized acid is then titrated by a base solution that has previously been compared with the acid solution. Calculate the molarity of (a) the acid, and (b) the base solutions.

Wt of calcite = 1.9802 g

Vol of HCl added to calcite = 45.00 ml

Vol of NaOH used in back-titration = 14.43 ml

30.26 ml of acid titrates 21.56 ml of base

46. A 2.500 g sample of an ammonium salt of technical grade is treated with concentrated NaOH, and the NH_3 that is liberated is distilled and collected in

50.00 ml of 1.2000 M HCl; 3.65 ml of 0.5316 M NaOH is required for back-titration. Calculate the percentage of NH_3 in the sample.

47. A 0.5000 g sample of impure CaO is added to 50.00 ml of 0.1000 M HCl. The excess HCl is titrated by 5.00 ml of 0.1250 M NaOH. Find the percentage of CaO in the sample. The reaction is

$$CaO + 2HCl \rightarrow CaCl_2 + H_2O$$

48. Silver nitrate solution is prepared by dissolving 85.20 g of pure $AgNO_3$ and diluting to 500.0 ml.
    (a) What is its molarity?
    (b) A $CaCl_2$ sample is titrated by 40.00 ml of this $AgNO_3$ solution. What is the weight of $CaCl_2$ in the sample?

49. (a) What weight of MnO_2 and (b) what volume of 12.0 M HCl are needed for the preparation of 750 ml of 2.00 M $MnCl_2$? (c) What volume of Cl_2 at 745 torr and 23°C will be formed? The reaction is

$$MnO_2 + 4HCl \rightarrow MnCl_2 + Cl_2 + 2H_2O$$

50. What volume of 12.0 M HCl is needed to prepare 3.00 liters of Cl_2 at 730 torr and 25°C by the reaction

$$2KMnO_4 + 16HCl \rightarrow 2MnCl_2 + 5Cl_2 + 8H_2O + 2KCl$$

51. (a) What volume of 6.00 M HNO_3 and what weight of copper are needed for the production of 1.50 liters of a 0.500 M $Cu(NO_3)_2$ solution?
    (b) What volume of NO, collected over water at 745 torr and 18°C, will be produced at the same time? The reaction is

$$3Cu + 8HNO_3 \rightarrow 3Cu(NO_3)_2 + 2NO + 4H_2O$$

52. What volume of 10.0 M HCl is needed to prepare 12.7 liters of CO_2 at 735 torr and 35°C? The reaction is

$$CaCO_3 + 2HCl \rightarrow CaCl_2 + CO_2 + H_2O$$

53. A 20.00 ml sample of a solution containing $NaNO_2$ and $NaNO_3$ is acidified with H_2SO_4 and then treated with an excess of NaN_3 (sodium azide). The hydrazoic acid so formed reacts with and completely removes the HNO_2. It does not react with the nitrate. The reaction is

$$HN_3 + HNO_2 \rightarrow H_2O + N_2 + N_2O$$

The volume of N_2 and N_2O is measured over water and found to be 36.50 ml at 740.0 torr and 27°C. What is the molar concentration of $NaNO_2$ in this solution?

54. You have a 0.5000 g mixture of oxalic acid, $H_2C_2O_4 \cdot 2H_2O$, and benzoic acid, $HC_7H_5O_2$. This sample requires 47.53 ml of 0.1151 M KOH for titration. What is the percentage composition of this mixture?

Thermochemistry

Thermochemistry is the study of the thermal (heat) changes that are associated with physical and chemical changes. Thermodynamics is much broader in scope because it includes the study of all forms of energy, including work. Some of these aspects are considered in later chapters.

CHANGES IN TEMPERATURE

If heat is applied to a substance, the temperature is raised; if heat is withdrawn, the temperature is lowered. The unit of heat is the *calorie* (cal), which is defined as the quantity of heat required to raise the temperature of 1 g of water 1 degree Celsius. We shall not deal with problems so accurately as to be concerned about the very small difference between a "15° calorie" (the heat needed to raise 1 g of water from 14.5°C to 15.5°C), or the "mean calorie" (1 one-hundredth of the heat needed to raise 1 g of water from 0°C to 100°C).

The number of calories required to raise the temperature of an object 1°C is called the *heat capacity* of the object. The *molar heat capacity* of a substance is the number of calories needed to raise the temperature of a mole of the substance 1°C.

The *specific heat* of any substance is the number of calories required to raise one gram of it 1°C. From our definition of the calorie, it follows that the specific heat of water is 1 cal/g °C.

PROBLEM:

The specific heat of Fe_2O_3 is 0.151 cal/g °C. How much heat is needed to raise the temperature of 200 g of Fe_2O_3 from 20.0°C to 30.0°C? What is the molar heat capacity of Fe_2O_3?

SOLUTION:

Because 0.151 cal raises 1 g of Fe_2O_3 1°C,

$$\text{total calories} = (200 \text{ g})\left(0.151 \frac{\text{cal}}{\text{g °C}}\right)(10.0°C)$$

$$= 302 \text{ cal needed}$$

The molar heat capacity is the product of specific heat and mole weight.

$$\text{Molar heat capacity} = \left(0.151 \frac{\text{cal}}{\text{g °C}}\right)\left(159.6 \frac{\text{g}}{\text{mole}}\right)$$

$$= 24.1 \frac{\text{cal}}{\text{mole °C}}$$

CALORIMETRY

The amount of heat that is absorbed or liberated in a physical or chemical change can be measured in a well-insulated vessel called a calorimeter (Figure 14-1). Calorimetry is based on the principle that the observed temperature change resulting from a chemical reaction can be simulated with an electrical heater. The electrical measurements of current (I), heater resistance (R), and duration (t) of heating make it possible to calculate how much heat is equivalent to the amount of heat produced by the chemical change, using the formula $I^2Rt/4.184$. By using weighed quantities of reactants, one can calculate the heat change per gram or per mole.

The formula is derived as follows. Electrical current, measured in amperes, is the *rate* of flow of electrical charge (coulombs); by definition, it is

$$\text{amperes} = \frac{\text{coulombs}}{\text{seconds}}$$

The common relationship between volts (E), amperes (I), and resistance (R) is known as Ohm's law:

$$E = IR$$

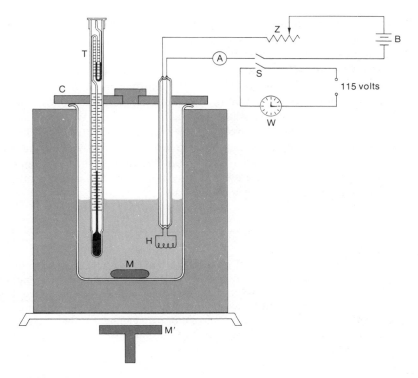

FIGURE 14-1
One type of calorimeter.

Electrical energy is given by

$$\text{energy} = (\text{volts})(\text{amperes})(\text{seconds}) = (\text{volts})\left(\frac{\text{coulombs}}{\text{seconds}}\right)(\text{seconds})$$

$$= EIt = I^2Rt \text{ volt coulombs (or joules)}$$

By definition,

1 joule = 1 volt coulomb (the SI-approved unit of energy)

1 calorie = 4.184 joules (the so-called mechanical equivalent of heat)

By combining these laws and definitions, we get

$$\text{electrical energy} = \frac{I^2Rt}{4.184} \text{ calories}$$

The insulation surrounding the calorimeter minimizes heat loss or gain through the walls. The cover (C) supports the electrical heater (H); it has two holes, one to permit the insertion of a sensitive differential thermometer (T), and one through which reactants can be added. The solution is stirred by a Teflon-covered magnet (M) that is rotated by the motor-driven magnet (M'). The electrical heating is controlled and measured as follows. The double-pole switch (S) controls two things simultaneously: the timer (W), and the battery (B), which supplies a constant current to the immersion heater (H). The current (I) that flows through the heater is read from the ammeter (A), and the duration (t) of time that the current flows is read from the timer. The resistance (R) of the heater is known from a separate measurement. The following problems are based on the calorimeter shown in Figure 14-1.

PROBLEM:

A calorimeter contains 200 ml of 0.100 M NaCl solution at room temperature (25.038°C according to the thermometer). With the magnetic stirrer going, exactly 10.00 ml of 1.000 M AgNO$_3$ solution (also at room temperature) is added dropwise through the porthole of the calorimeter. The temperature, as indicated by the thermometer, rises to 25.662°C. A current of 0.700 ampere is then passed through the electrical heater (whose resistance is 6.50 ohms) for a period of 4 min 5 sec, and the temperature rises by 0.742°C. Calculate the number of calories produced when one mole of AgCl precipitates from aqueous solution.

SOLUTION:

The electrical heater and thermometer tell you how many calories it takes to raise the reaction mixture by 1°C:

$$\frac{\text{energy}}{°C} = \frac{\left[\dfrac{(0.700\ \text{amp})^2(6.50\ \text{ohm})(245\ \text{sec})}{4.184\ \dfrac{\text{joule}}{\text{cal}}}\right]}{0.742°C} = \frac{186.5\ \text{cal}}{0.742°C}$$

$$= 251\ \frac{\text{cal}}{°C}$$

The chemical reaction caused the temperature of the same mixture to rise by 0.624°C, corresponding to an energy release of

$$\text{energy from reaction} = (0.624°C)\left(251\ \frac{\text{cal}}{°C}\right) = 157\ \text{cal}$$

You will note that there was added (0.200 liter) (0.100 mole/liter) = 0.0200 mole NaCl and (0.0100 liter) (1.00 mole/liter) = 0.0100 mole AgNO$_3$. Only half of the NaCl is used, and just 0.0100 mole of AgCl is formed. It is the formation of this 0.0100 mole of AgCl that produces the 157 cal, so

$$\text{energy per mole} = \frac{157 \text{ cal}}{0.0100 \text{ mole AgCl}} = 15,700 \frac{\text{cal}}{\text{mole AgCl}}$$

If you want to measure the specific heat of a liquid, you need know only the electrical energy needed to heat a known weight of the liquid and the measured temperature change, but you must use a *calibrated* calorimeter, so that a correction can be made for the amount of electrical energy that was absorbed by the calorimeter walls rather than by the liquid.

PROBLEM:
A calorimeter requires a current of 0.800 amp for 4 min 15 sec to raise the temperature of 200.0 ml of H_2O by 1.100°C. The same calorimeter requires 0.800 amp for 3 min 5 sec to raise the temperature of 200.0 ml of another liquid (whose density is 0.900 g/ml) by 0.950°C. The heater resistance is 6.50 ohms. Calculate the specific heat of the liquid.

SOLUTION:
The total heat energy produced by the electrical heater in water was used to raise the temperature of the water and the calorimeter walls by 1.100°C; it is

$$\text{total energy} = \frac{(0.800 \text{ amp})^2(6.50 \text{ ohms})(255 \text{ sec})}{4.184} = 254 \text{ cal}$$

The energy required to raise just the water by 1.100°C (assuming the density and specific heat of water are both 1.000) is

$$\text{energy for water} = (200.0 \text{ g})\left(1.00 \frac{\text{cal}}{\text{g °C}}\right)(1.100°C) = 220 \text{ cal}$$

The energy required to raise the temperature of the calorimeter walls is the difference between the total energy and that required for the water—that is, $254 - 220 = 34$ cal. The heat capacity of the calorimeter (calories required to raise that part of its walls in contact with the liquid by 1.0°C) is

$$\text{heat capacity of calorimeter} = \frac{34 \text{ cal}}{1.10°C} = 31 \frac{\text{cal}}{°C}$$

The total electrical energy produced when the heater was in the liquid is

$$\frac{(0.800 \text{ amp})^2(6.50 \text{ ohms})(185 \text{ sec})}{4.184} = 184 \text{ cal}$$

Of this total amount, the part required to raise the calorimeter walls by 0.950°C is

$$\text{energy for calorimeter} = (31 \text{ cal/°C})(0.950°C) = 29 \text{ cal}$$

The difference between 184 cal and 29 cal is 155 cal; this is the amount required to raise the temperature of the liquid by 0.950°C. The amount of heat required to raise 1 g of the liquid by 1°C (its *specific heat*) is

$$\text{specific heat of liquid} = \frac{155 \text{ cal}}{(200.0 \text{ ml})(0.900 \text{ g/ml})(0.950°C)}$$

$$= 0.910 \frac{\text{cal}}{\text{g °C}}$$

An alternative method of calorimetry that gives less accurate results, but is simpler in concept, uses only a single insulated container and a thermometer. Temperature changes in the calorimeter are brought about by adding hot (or cold) objects of known weight and temperature. Calculations are based on the principle that the heat lost by the added hot object is equal to that gained by the water in the calorimeter and the calorimeter walls. This simple approach is illustrated in the next two problems.

PROBLEM:

The temperature in a calorimeter containing 100 g of water is 22.7°C. Fifty grams of water are heated to boiling (99.1°C at this location) and quickly poured into the calorimeter. The final temperature is 44.8°C. From these data, calculate the heat capacity of the calorimeter.

SOLUTION:

The heat loss from the hot water is equal to the heat gain by the calorimeter and the water initially in it.

$$\text{Heat lost by hot water} = (\text{wt of } H_2O)(\text{sp ht of } H_2O)(\text{temp change})$$

$$= (50 \text{ g})\left(1.00 \frac{\text{cal}}{\text{g °C}}\right)(99.1°C - 44.8°C)$$

$$= 2715 \text{ cal}$$

$$\text{Heat gained by calorimeter water} = (\text{wt of } H_2O)(\text{sp ht of } H_2O)(\text{temp change})$$

$$= (100 \text{ g})\left(1.00 \frac{\text{cal}}{\text{g °C}}\right)(44.8°C - 22.7°C)$$

$$= 2210 \text{ cal}$$

$$\text{Heat gained by calorimeter} = (\text{ht capacity of calorimeter})(\text{temp change})$$

$$= \left(x \frac{\text{cal}}{°C}\right)(44.8°C - 22.7°C)$$

$$= 22.1x \text{ cal}$$

We equate the heat lost to the heat gained and solve for x, the heat capacity of the calorimeter:

$$2715 \text{ cal} = 2210 \text{ cal} + 22.1x \text{ cal}$$

$$x = \frac{2715 - 2210}{22.1} = 22.8 \frac{\text{cal}}{°\text{C}}$$

PROBLEM:

The calorimeter of the preceding problem is used to measure the specific heat of a metal sample. A 100 g sample of water is put into the calorimeter at a temperature of 24.1°C. A 45.32 g sample of metal filings is put into a dry test tube that is immersed in a bath of boiling water until the metal is at the temperature of the latter, 99.1°C. The hot metal is then quickly poured into the calorimeter and the water stirred by a thermometer that is read at frequent intervals until the temperature reaches a maximum of 27.6°C. Compute the specific heat of the metal.

SOLUTION:

The heat lost by the metal sample is equal to the heat gained by the calorimeter.

Heat gained by water = (wt of H_2O)(sp ht of H_2O)(temp change)

$$= (100 \text{ g}) \left(1.00 \frac{\text{cal}}{\text{g } °\text{C}} \right) (27.6°\text{C} - 24.1°\text{C})$$

$$= 350 \text{ cal}$$

Heat gained by calorimeter = (ht capacity of calorimeter)(temp change)

$$= \left(22.8 \frac{\text{cal}}{\text{g } °\text{C}} \right) (27.6°\text{C} - 24.1°\text{C})$$

$$= 80 \text{ cal}$$

Total heat gained = 350 cal + 80 cal

$$= 430 \text{ cal}$$

Heat lost by metal = (wt of metal)(sp ht of metal)(temp change)

$$430 \text{ cal} = (45.32 \text{ g}) \left(x \frac{\text{cal}}{\text{g } °\text{C}} \right) (99.1°\text{C} - 27.6°\text{C})$$

$$x = \frac{430 \text{ cal}}{(45.32 \text{ g})(71.5°\text{C})}$$

$$\text{Specific heat} = 0.133 \frac{\text{cal}}{\text{g } °\text{C}}$$

THE RULE OF DULONG AND PETIT

Many years ago, Pierre Dulong and Alexis Petit observed that the molar heat capacity for most solid elements is approximately 6.2 cal/mole °C. That is, the number of calories required to raise one mole of a solid element by 1.0°C is given by

$$\left(\frac{g}{mole}\right)\left(\frac{cal}{g\ °C}\right) \cong 6.2\ \frac{cal}{mole\ °C}$$

This rule of Dulong and Petit provides a simple way to find the approximate values of the atomic weights of solid elements. For example, if you had an unknown solid element in a finely divided state, you could put a weighed sample of it into your calibrated calorimeter and quickly find its specific heat; then, using the rule of Dulong and Petit, you could find its approximate atomic weight.

PROBLEM:

A 50.0 g sample of a finely divided metal, insoluble and unreactive to water, is put into 200.0 ml of water in the calorimeter that was calibrated in the preceding problems. A current of 0.800 amp is passed through the heater for 15 min 50 sec in order to raise the temperature by 4.00°C. What is the specific heat of the metal?

SOLUTION:

The total electrical energy required to raise the water, the calorimeter, and the metal by 4.00°C is

$$\frac{(0.800\ amp)^2(6.50\ ohms)(950\ sec)}{4.184} = 945\ cal$$

Together, the water and the calorimeter require 231 cal/°C, so the total energy required by them for 4.00°C is

$$\left(231\ \frac{cal}{°C}\right)(4.00°C) = 924\ cal$$

The difference, $945 - 924 = 21$ cal, is required to raise the 50.0 g of metal by 4.00°C. Therefore, the specific heat of the metal is

$$\frac{21.0\ cal}{(50.0\ g)(4.00°C)} = 0.105\ \frac{cal}{g\ °C}$$

$$\text{Approximate atomic weight of the metal} = \frac{6.2}{0.105} = 59\ \frac{g}{mole}$$

ENTHALPY OF TRANSITION

When substances melt or vaporize, they absorb energy but do not change temperature. Instead, this energy is used to overcome the mutual attraction of the molecules or ions and permit them to move more independently than they could in their former state; the new state with its added energy always has less molecular order. For example, liquid water at 0°C is a less-ordered state than crystalline water at 0°C, and water vapor at 100°C is chaotic in its molecular organization compared to liquid water at 100°C.

These statements are made more precise and quantitative in the following way. It is said that, at a given temperature and pressure, the molecules of each substance in "state 1" (say, liquid) have a heat content of H_1, whereas in "state 2" (say, vapor) they have a heat content of H_2. The "heat of transition" (in this case, vaporization) is simply "the change in heat content" (ΔH_T):

$$\Delta H_T = H_2 - H_1$$

To avoid the use of the ambiguous term "heat" in connection with "heat content," it is customary to use the term *enthalpy*. At a given temperature and pressure, every substance possesses a characteristic amount of enthalpy (H), and the heat changes associated with chemical and physical changes at constant pressure are called changes in enthalpy (ΔH); ΔH_T is the enthalpy of transition. Two common enthalpies of transition are $\Delta H_f = 1435$ cal/mole for the enthalpy of fusion (melting) of ice at 0°C, and $\Delta H_v = 9713$ cal/mole for the enthalpy of vaporization of water at 100°C.

Energy also is involved in transitions from one allotropic form to another, or from one crystal form to another. To change a mole of red phosphorus to yellow phosphorus, we must supply 4.22 kilocalories ($\Delta H_T = +4.22$ kcal/mole), and when 1 mole of yellow silicon disulfide changes to white silicon disulfide, 3.11 kcal is liberated ($\Delta H_T = -3.11$ kcal/mole).

In the following problem we apply the principles involved in both specific heat and heats of transition.

PROBLEM:

What is the resulting temperature if 36.0 grams of ice at 0°C are put into 200 g of H_2O at 25.0°C, contained in the calibrated calorimeter used in the preceding problems?

SOLUTION:

The energy required to melt the ice is supplied by the water and the calorimeter walls which, as a result, are cooled. Let T be the final temperature.

$$\text{Calories needed to melt ice at 0°C} = \frac{(36.0 \text{ g})(1435 \text{ cal/mole})}{(18.0 \text{ g/mole})} = 2870 \text{ cal}$$

$$\text{Calories needed to raise } H_2O \text{ from 0° to } T = (36.0 \text{ g})\left(1 \frac{\text{cal}}{\text{g °C}}\right)(T\text{°C}) = 36.0T \text{ cal}$$

$$\text{Total calories needed} = 2870 + 36.0T$$

Calories lost by water going from $T°$ to $0°C = (200 \text{ g})\left(1 \dfrac{\text{cal}}{\text{g }°C}\right)(25°C - T°C)$

Calories lost by calorimeter walls from $T°$ to $0°C = \left(31 \dfrac{\text{cal}}{°C}\right)(25°C - T°C)$

$$\text{Total calories lost} = (231)(25.0 - T)$$

Going on the principle that

$$\text{total calories needed} = \text{total calories lost}$$

we have

$$2870 + 36.0T = (231)(25.0 - T)$$
$$267T = 2905$$
$$T = 10.9°C, \text{ the resultant temperature}$$

ENTROPY OF TRANSITION

The enthalpy of transition, divided by the absolute temperature at which it occurs, is a common measure of the change in molecular order that occurs during the transition. We refer to this change in molecular order as the "change in entropy" (ΔS_T), or the "entropy of transition." If ΔS_T is positive, the change results in an increase in molecular *disorder*. Changes in state offer some of the simplest examples from which one can obtain a feeling for the relationship between changes in entropy and changes in molecular order. Crystals have a very high degree of order; in them, the movement of atoms, ions, or molecules is restricted primarily to vibration about their locations in the crystalline lattice. When crystals melt, the component atoms, ions, or molecules can move fairly independently of each other in the liquid, slowly changing their neighbors by diffusion; the molecular order represented by the lattice disappears. When liquids vaporize, the component atoms or molecules, now in the gaseous phase, move about independently in a chaotic, random manner. Each stage, melting and vaporization, represents an increase in molecular chaos, and is described in terms of an increase in entropy.

Just as ΔH_T represents the difference in enthalpies between "state 2" and "state 1," so does ΔS_T represent the difference between the entropies in "state 2" and "state 1":

$$\Delta S_T = S_2 - S_1$$

$$\Delta S_T = \frac{\Delta H_T}{T} \quad \frac{\text{cal}}{\text{mole }°C}$$

For water,

$$\Delta S_f = \frac{1435 \dfrac{cal}{mole}}{273°C} = 5.26 \frac{cal}{mole \ °C} = \text{entropy of fusion at } 0°C$$

$$\Delta S_v = \frac{9713 \dfrac{cal}{mole}}{373°C} = 26.0 \frac{cal}{mole \ °C} = \text{entropy of vaporization at } 100°C$$

For very many liquids, the entropy of vaporization at the normal boiling point is approximately 21 cal/mole °C; water is not typical. The units for changes in entropy are the same as those for molar heat capacity, and care must be used to avoid confusion. When referring to an entropy change, a cal/mole °C is often called an *entropy unit,* abbreviated e.u. In order to avoid later misunderstanding, note now that this method of calculating ΔS from $\Delta H/T$ is valid only under equilibrium conditions. For transitions, for example, this method can be used only at temperatures where the two phases in question can coexist in equilibrium with each other.

ENTHALPY OF REACTION

Most reactions either liberate or absorb heat. To say that heat is liberated means that the atoms, in the molecular arrangement they have as products, must possess less energy than they did in their arrangement as reactants, and that this difference in energy is evolved as heat; the reaction is exothermic. When it is important to show this heat change, one way to do so is to include it as part of the chemical equation, as illustrated by the burning of methane gas:

$$CH_{4(g)} + 2O_{2(g)} \rightarrow CO_{2(g)} + 2H_2O_{(l)} + 212,800 \text{ cal}$$

Another, more useful, way is to say that the enthalpy of the reactants ("state 1") is H_1, that the enthalpy of the products ("state 2") is H_2, and that the "heat of reaction" is simply the "change in enthalpy" (ΔH):

$$\Delta H = H_2 - H_1 = \text{heat of reaction} = \text{enthalpy of reaction}$$

The actual amount of heat we measure experimentally for a given reaction depends somewhat on (a) the temperature of the experiment and (b) whether the experiment is run at constant volume or constant pressure. The basic reasons for this are that (a) each reactant and product has a characteristic specific heat that varies individualistically with temperature, and (b) at constant pressure, some of the heat of reaction may expand or compress gases if they are not confined to a fixed volume. We shall avoid these complications here by saying that $\Delta H°$ refers only to changes occurring at 25°C and at a *constant* pressure of 1 atm; that is, the reactants and the products are in their *standard states*.

Also, ΔH will always be negative for an exothermic reaction, because the products collectively have a smaller enthalpy than the reactants. For the burning of CH_4,

$$CH_{4(g)} + 2O_{2(g)} \rightarrow CO_{2(g)} + 2H_2O_{(l)} \qquad \Delta H° = -212,800 \text{ cal}$$

For an endothermic reaction, ΔH is positive. For example,

$$2HCl_{(g)} \rightarrow H_{2(g)} + Cl_{2(g)} \qquad \Delta H° = +44,120 \text{ cal}$$

If we had written the previous equation in the reverse order, the sign of ΔH would have been negative. The positive sign means that it takes energy to decompose HCl, and the negative sign means that energy is liberated when HCl is formed from the elements, H_2 and Cl_2:

$$\tfrac{1}{2}H_{2(g)} + \tfrac{1}{2}Cl_{2(g)} \rightarrow HCl_{(g)} \qquad \Delta H° = -22,060 \text{ cal}$$

The proper way to interpret the calorimeter experiment on p 208 is to say that the enthalpy of reaction between Ag^+ and Cl^- is $-15,700$ cal/mole AgCl:

$$Ag^+_{(aq)} + Cl^-_{(aq)} \rightarrow AgCl_{(s)} \downarrow \qquad \Delta H = -15,700 \text{ cal}$$

The energy changes associated with chemical reactions are determined *solely* by the state of the reactants and the state of the products, and are totally independent of the path or method of preparation. As a result, if a reaction can be considered to be the sum of two or more other reactions, ΔH for that reaction must be the sum of the ΔH values for the other reactions; this is known as *Hess's law*. For example, CO_2 may be made directly from the elements, or indirectly by first making CO which is subsequently burned to CO_2:

$$C_{(s)} + \tfrac{1}{2}O_{2(g)} \rightarrow CO_{(g)} \qquad \Delta H° = -26.42 \text{ kcal}$$
$$CO_{(g)} + \tfrac{1}{2}O_{2(g)} \rightarrow CO_{2(g)} \qquad \Delta H° = -67.63 \text{ kcal}$$
$$C_{(s)} + O_{2(g)} \rightarrow CO_{2(g)} \qquad \Delta H° = -94.05 \text{ kcal}$$

You can see that the sum of the first two reactions gives the third, just as the sum of the $\Delta H°$ values for the first two gives the $\Delta H°$ for the third.

When a reaction produces a compound from *elements* in their common physical state at 25.0°C and 1 atm, the value of $\Delta H°_f$ is called the standard *enthalpy of*

TABLE 14-1
Standard Enthalpies of Formation at 25°C and 1 atm (ΔH_f° in kcal/mole)

Substance	State	ΔH_f°	Substance	State	ΔH_f°	Substance	State	ΔH_f°
Any element	Normal	0.00	CO	g	−26.42	N (atom)	g	85.56
Ag^+	aq	25.31	CO_2	g	−94.05	Na (atom)	g	25.98
AgCl	s	−30.36	Cu^{2+}	aq	15.49	Na^+	aq	−57.30
Br (atom)	g	26.71	F (atom)	g	18.30	NH_3	g	−11.04
Br^-	aq	−28.90	F^-	aq	−78.66	NH_3	aq	−19.32
BrCl	g	3.51	FeO	s	−63.70	NH_4^+	aq	−31.74
C (atom)	g	171.70	Fe_2O_3	s	−196.50	NO	g	21.60
C (diamond)	s	0.45	H (atom)	g	52.09	NO_2	g	8.09
Ca^{2+}	aq	−129.77	H^+	aq	0.00	NO_3^-	aq	−49.37
Cd^{2+}	aq	−17.30	HBr	g	−8.66	N_2O	g	19.49
CH_3OH	l	−57.02	HCl	g	−22.06	O (atom)	g	59.16
CH_3OH	g	−48.08	HI	g	6.20	OH^-	aq	−54.96
C_2H_5OH	l	−66.36	H_2O	l	−68.32	P (atom)	g	75.18
C_2H_5OH	g	−56.24	H_2O	g	−57.80	PCl_3	g	−73.22
C_4H_{10}	g	−29.81	H_2S	g	−4.82	PCl_5	g	−95.35
C_6H_6	l	11.72	I (atom)	g	25.48	S (atom)	g	53.25
C_6H_6	g	19.82	I^-	aq	−13.37	S^{2-}	aq	10.00
Cl (atom)	g	29.01	ICl	g	4.20	SO_2	g	−70.96
Cl^-	aq	−40.02	K^+	aq	−60.04	SO_4^{2-}	aq	−216.90
ClO_4^-	aq	−31.41	Li^+	aq	−66.54	Zn^{2+}	aq	−36.43

NOTE: The state aq represents a very dilute aqueous solution.

formation. The standard enthalpy of formation of 1 mole of HCl is $\Delta H_f^\circ = -22.06$ kcal/mole. The standard enthalpies of formation of hundreds of compounds have been determined and are listed in tables in chemistry handbooks. A few are listed here in Table 14-1. The standard enthalpy of formation of all *elements,* in their common form at 25.0°C and 1 atm pressure, is assumed to be zero.

One real value of tables of standard enthalpies of formation is that they permit the *calculation* of the standard enthalpy of any reaction for which all the reactants and products are listed; it is not necessary to do an experimental measurement. Based on Hess's law, the basic premise of the use of tables is that the enthalpy of reaction is the difference between the sum of the enthalpies of the formation of the products and the sum of the enthalpies of formation of the reactants. That is,

$$\Delta H° = \Sigma \text{ (standard enthalpies of formation of products)}$$
$$- \Sigma \text{ (standard enthalpies of formation of reactants)}$$
$$= \Sigma(\Delta H_f^\circ)_{\text{products}} - \Sigma(\Delta H_f^\circ)_{\text{reactants}}$$

In the following problems, we first write the chemical equation. Then, below each substance, we write its standard enthalpy of formation, multiplied by the number of moles of the substance used in the balanced equation. The standard enthalpy of reaction is the difference between the sum of the enthalpies of formation of the products and the sum of the enthalpies of formation of the reactants.

PROBLEM:

Compute the standard enthalpy of reaction for the gaseous dissociation of PCl_5 into PCl_3 and Cl_2.

SOLUTION:

Write the balanced chemical equation and take the needed values of ΔH_f° from Table 14-1.

$$PCl_{5(g)} \rightleftarrows PCl_{3(g)} + Cl_{2(g)}$$

$$(1 \text{ mole})\left(-95.35 \ \frac{\text{kcal}}{\text{mole}}\right) \quad (1 \text{ mole})\left(-73.22 \ \frac{\text{kcal}}{\text{mole}}\right) \quad (1 \text{ mole})\left(0.00 \ \frac{\text{kcal}}{\text{mole}}\right)$$

$$\Sigma(\Delta H_f^{\circ})_{\text{products}} = (1)(-73.22) + (1)(0.00) = -73.22 \text{ kcal}$$

$$\Sigma(\Delta H_f^{\circ})_{\text{reactants}} = (1)(-95.35) = -95.35 \text{ kcal}$$

$$(\Delta H^{\circ})_{\text{reaction}} = (-73.22 \text{ kcal}) - (-95.35 \text{ kcal}) = +22.13 \text{ kcal}$$

The positive sign of the answer indicates that the reaction is endothermic and that the dissociation at 25.0°C requires 22,130 cal/mole.

PROBLEM:

What is the standard enthalpy of combustion of ethyl alcohol, C_2H_5OH?

SOLUTION:

Write the balanced chemical equation and take the needed values of ΔH_f° from Table 14-1.

$$C_2H_5OH_{(l)} + 3O_{2(g)} \rightarrow 2CO_{2(g)} + 3H_2O_{(l)}$$

$$(1 \text{ mole})\left(-66.36 \ \frac{\text{kcal}}{\text{mole}}\right) \quad (3 \text{ moles})\left(0.00 \ \frac{\text{kcal}}{\text{mole}}\right) \quad (2 \text{ moles})\left(-94.05 \ \frac{\text{kcal}}{\text{mole}}\right) \quad (3 \text{ moles})\left(-68.32 \ \frac{\text{kcal}}{\text{mole}}\right)$$

$$\Sigma(\Delta H_f^{\circ})_{\text{products}} = (2)(-94.05) + (3)(-68.32) = -393.06 \text{ kcal}$$

$$\Sigma(\Delta H_f^{\circ})_{\text{reactants}} = (1)(-66.36) + (3)(0.00) = -66.36 \text{ kcal}$$

$$(\Delta H^{\circ})_{\text{reaction}} = (-393.06 \text{ kcal}) - (-66.36 \text{ kcal}) = -326.70 \text{ kcal}$$

The reaction is exothermic.

Enthalpies of combustion are relatively simple to determine, and they often are used to find other energy values that are very difficult or impossible to determine directly. For example, the enthalpy of formation of methyl alcohol corresponds to the reaction

$$2H_{2(g)} + C_{(s)} + \tfrac{1}{2}O_{2(g)} \rightarrow CH_3OH_{(l)}$$

yet it is impossible to make CH_3OH directly from the elements. However, we can burn CH_3OH in an excess of O_2 in a bomb calorimeter and measure the heat produced ($\Delta H° = -173.67$ kcal/mole) in the reaction

$$CH_3OH_{(l)} \quad + \quad \tfrac{3}{2}O_{2(g)} \quad \rightarrow \quad CO_{2(g)} \quad + \quad 2H_2O_{(l)}$$

$$\text{(1 mole)} \left(\Delta H_f° \; \frac{\text{kcal}}{\text{mole}}\right) \quad \text{(1.5 moles)} \left(0.00 \; \frac{\text{kcal}}{\text{mole}}\right) \quad \text{(1 mole)} \left(-94.05 \; \frac{\text{kcal}}{\text{mole}}\right) \quad \text{(2 moles)} \left(-68.32 \; \frac{\text{kcal}}{\text{mole}}\right)$$

$$\Sigma(\Delta H_f°)_{\text{products}} = (1)(-94.05) + (2)(-68.32) = -230.69 \text{ kcal}$$

$$\Sigma(\Delta H_f°)_{\text{reactants}} = (1)(\Delta H_f°) + (1.5)(0.00) = \Delta H_f°$$

$$(\Delta H°)_{\text{reaction}} = (-230.69 \text{ kcal}) - \Delta H_f° = -173.67 \text{ kcal}$$

$$\Delta H_f° = (-230.69 \text{ kcal}) + (173.67 \text{ kcal}) = -57.02 \text{ kcal}$$

$$= \text{standard enthalpy of formation of } CH_3OH_{(l)}$$

You will note that Table 14-1 also contains standard enthalpies of formation for ions in aqueous solution. It is worth noting here that calorimetry was a strong argument favoring the view that all strong acids and bases exist in dilute water solution only as ions and not as molecules. No matter which combination of strong acid and base is used in a neutralization reaction, the heat value obtained is always very close to the value of $\Delta H° = -13.36$ kcal per mole of water formed. The implication of course is that, although the *chemicals are different* in each case, the *reaction is the same;* it must be

$$H_{(aq)}^+ + OH_{(aq)}^- \rightarrow H_2O_{(l)} \qquad \Delta H° = -13.36 \text{ kcal}$$

If weak acids or bases are used, the observed heat values are always less and quite variable. In essence, the reaction is also the same for the weak ones, except that some of the enthalpy of the reaction (some of the -13.36 kcal) must be used to remove the H^+ or OH^- from the weak acids or bases. The subscript aq refers to the fact that the substance in question is in dilute aqueous solution; $\Delta H_f°$ for $H_{(aq)}^+$ in Table 14-1 has also been set arbitrarily equal to zero, just as were the elements in their standard states.

BOND ENERGIES (ENTHALPIES)

The term *bond energy* is defined as the ΔH required to break a bond between two atoms in an isolated gaseous molecule, producing the dissociated fragments in the isolated gaseous state. At first, you might think that the value of $\Delta H° =$ $+22.06$ kcal/mole would be the H–Cl bond energy, but it is not. The value of $+22.06$ represents the difference between the energy required to dissociate the HCl molecule and the energy liberated when the H atoms and Cl atoms combine to form H_2 and Cl_2 molecules. The bond energy, however, corresponds to the reaction

$$HCl_{(g)} \rightarrow H_{(g)} + Cl_{(g)}$$

We could calculate ΔH for this if we knew (for the elements) the enthalpy of formation of molecules from their atoms. Some crystalline elements (especially metals) vaporize as monatomic gases, and it is not too difficult to determine their heats of sublimation. Some elements—such as H_2, O_2, and Br_2—are diatomic gases that dissociate into atoms at high temperature; these dissociation energies may also be determined. Table 14-1 also includes the standard enthalpies of formation of a number of atoms; these are based on the normal physical form of the element at 25.0°C. For HCl we find

$$\Sigma(\Delta H^°_f)_{products} = (+52.09 \text{ kcal}) + (+29.01 \text{ kcal}) = 81.10 \text{ kcal}$$

$$\Sigma(\Delta H^°_f)_{reactants} = -22.06 \text{ kcal}$$

$$(\Delta H°)_{reaction} = (81.10 \text{ kcal}) - (-22.06 \text{ kcal}) = +103.16 \text{ kcal}$$

The bond energy is 103.16 kcal. There is additional discussion of bond energies on pp 113–115.

CHANGES IN INTERNAL ENERGY

In the definition of enthalpy change (p 215) and in all of the examples of heat changes and transfers we have discussed, there has been the limitation of *constant pressure*.

Most experiments are performed at constant atmospheric pressure in vessels and flasks open to the air. In the illustrative examples involving the combustion of C_2H_5OH and CH_3OH, however, the measurements had to be carried out in a heavy-walled "bomb" calorimeter at *constant volume*.

In comparing the heat effects associated with these two different limitations, we must look at three different constant-pressure situations.

1. If there are more moles of gaseous products than gaseous reactants in the balanced chemical equation, then the *extra* gaseous moles will expand against the atmospheric pressure and the work energy required for this will come at the expense of some of the heat that is liberated. A smaller amount of heat will be liberated than if the reaction had occurred at constant volume.

2. If there are fewer moles of gaseous products than gaseous reactants in the balanced chemical equation, then the atmospheric pressure will do work on the reaction mixture as it contracts due to the *diminished* number of gaseous moles, and this work energy will be added to the heat that is liberated. A larger amount of heat will be liberated than if the reaction had occurred at constant volume.

3. If there are the same number of gaseous moles of products as reactants, there will be no contraction or expansion of the reaction mixture. The heat liberated at constant pressure will be the same as at constant volume.

We next offer a simple way to calculate the heat effect at constant pressure from that observed at constant volume, or vice versa. First, note that the product of P and V always has the units of energy. A simple bit of evidence for this observation comes from the ideal gas law, using the value of 1.987 cal/mole K for R:

$$PV = (n \text{ moles})\left(1.987 \ \frac{\text{cal}}{\text{mole K}}\right)(T \text{ K}) = 1.987(nT) \text{ calories}$$

It is said that every substance has an *internal energy* (designated as E), and that the heat effect associated with a change at a *constant volume* and temperature is ΔE. As the molecules go from "state 1" to "state 2," $\Delta E = E_2 - E_1$. This effect is exactly analogous to the heat effect that is associated with a change at *constant pressure* and temperature: $\Delta H = H_2 - H_1$. The variables H and E are related by the potential of the system to expand or contract—that is, to the potential to be affected by PV work—by the explicit function

$$H = E + PV$$

For a *change,*

$$\Delta H = \Delta E + \Delta(PV)$$

If the change is a chemical reaction at constant temperature and pressure, $\Delta(PV)$ becomes $P\Delta V$ because the pressure remains constant, and it comes as a result of the difference in the number of gaseous moles of products and reactants in the balanced chemical equation:

$$\Delta(PV) = P\Delta V = (\Delta n)RT$$

where $\Delta n = n_2$ (gaseous moles of products) $- n_1$ (gaseous moles of reactants). For chemical reactions, then,

$$\Delta H = \Delta E + (\Delta n)RT$$

This equation makes it simple to calculate enthalpies of reaction from values of ΔE obtained at constant volume, and vice versa.

PROBLEM:
The combustion of $C_2H_5OH_{(l)}$ in a bomb calorimeter at constant volume gives a value of $\Delta E^\circ = -326.1$ kcal/mole. Calculate the value of ΔH°.

SOLUTION:
The balanced chemical equation

$$C_2H_5OH_{(l)} + 3O_{2(g)} \rightarrow 2CO_{2(g)} + 3H_2O_{(l)}$$

shows that $\Delta n = 2 - 3 = -1$. Therefore, at standard conditions (25.0°C),

$\Delta H^\circ = \Delta E^\circ + (\Delta n)RT$

$= (-326.1 \text{ kcal}) + (-1 \text{ mole})\left(1.987 \dfrac{\text{cal}}{\text{mole K}}\right)\left(10^{-3} \dfrac{\text{kcal}}{\text{cal}}\right)(298 \text{ K})$

$= -326.7 \text{ kcal (per mole of } C_2H_5OH)$

THE FIRST LAW OF THERMODYNAMICS

A concise summary of the principles of the last section is given by the *first law of thermodynamics,* which states that any change in internal energy (ΔE) of a system is just equal to the difference between the heat (Q) it absorbs and the work (W) it performs:

$$\Delta E = Q - W$$

In other words, energy cannot be created or destroyed; all energy must be accounted for.

A "system" is any carefully defined object or collection of materials that is under discussion or study. For example, it may be the substances in a chemical reaction mixture, the contents of a calorimeter, a solid of prescribed dimensions or amount, or a gas at a given temperature, pressure, and volume. Everything in the lab or the universe that exchanges heat or work with the system is called "the surroundings."

For a chemical reaction at constant volume, $W = 0$, so $\Delta E = Q$. If the reaction absorbs heat from the surroundings, Q is positive, and the reaction is endothermic (ΔE is positive).

For a chemical reaction at constant pressure, $Q = \Delta H$, and $W = (\Delta n)RT$, so $\Delta H = \Delta E + (\Delta n)RT$. If the reaction absorbs heat from the surroundings, ΔH is positive, and the reaction is endothermic; ΔE may be larger or smaller than ΔH depending on the sign of Δn.

PROBLEMS A

1. Calculate the approximate specific heat of each of the following elements: (a) S; (b) Zn; (c) La; (d) U; (e) Pb.

2. Calculate the resultant temperature when 150 g of water at 75.0°C is mixed with 75.0 g of water at 20.0°C. (Assume no heat loss to container or surroundings.)

3. Calculate the resultant temperature when 50.0 g of silver metal at 150.0°C is mixed with 50.0 g of water at 20.0°C. (Assume no heat loss to container or surroundings.)

4. Suppose 150 ml of water at 50.0°C, 25.0 g of ice at 0°C, and 100 g of Cu at 100.0°C are mixed together. Calculate the resultant temperature, assuming no heat loss, and using the approximate specific heat of Cu.

5. Suppose 150 ml of water at 20.0°C, 50.0 g of ice at 0°C, and 70.0 g of Cu at 100.0°C are mixed together. Calculate the resultant temperature, assuming no heat loss, and using the approximate specific heat of Cu.

6. A Dewar flask (vacuum-jacketed bottle) is used as a calorimeter, and the following data are obtained. Measurements in parts (a) and (b) are made to obtain the heat capacity of the calorimeter, and parts (c) and (d) are performed on an unknown metal.
   (a) Calorimeter with 150.0 g of H_2O has a temperature of 21.3°C.
   (b) When 35.0 ml of H_2O at 99.5°C are added to the calorimeter and water of part (a), the resultant temperature is 35.6°C.
   (c) Calorimeter with 150.0 g of H_2O has a temperature of 22.7°C.
   (d) A 50.3 g sample of metal at 99.5°C added to the calorimeter and water of (c) gives a resultant temperature of 24.3°C.
   Calculate the approximate atomic weight of the metal.

7. A 100.0 g sample of glycerol is put into the calorimeter calibrated in Problem 6, and its temperature is observed to be 20.5°C. Then 45.7 g of iron at 165.0°C are added to the glycerol, giving a resultant temperature of 37.4°C. Calculate the specific heat of the glycerol. (Use the approximate specific heat of iron.)

8. Calculate the heat capacity of a calorimeter (Figure 14-1) that, when containing 300 ml of water, requires a current of 0.840 amp passing for 3 min 41 sec through a 9.05 ohm immersed resistance in order to raise the water temperature from 22.376°C to 23.363°C.

9. Calculate the following quantities, using the calorimeter calibrated in Problem 8.
   (a) The specific heat of a water-insoluble material. First, 38.5 g of a finely divided sample are stirred with 250 ml of water; then a current of 0.695

amp for 8 min 53 sec raises the temperature of the mixture from 24.605°C
to 26.328°C.

(b) The specific heat of a liquid. First, 300 g of it are put in the calorimeter in
place of the water; then a current of 0.742 amp is passed for 3 min 22 sec,
raising its temperature from 21.647°C to 22.406°C.

(c) The enthalpy change per mole of precipitated PbI_2 when 25.0 ml of 6.00 M
NaI are added with stirring to 300 ml of 0.250 M $Pb(NO_3)_2$ contained in the
calorimeter. A current of 0.900 amp passing for 8 min 18 sec through the
heater immersed in the reaction mixture causes a temperature rise only
0.750 of the rise observed for the reaction. Also write the equation for the
reaction involved.

(d) The enthalpy change per mole of $KBrO_3$ when 15.0 g of solid $KBrO_3$ are
dissolved in 200 ml of water in the calorimeter. This is the so-called "heat
of solution." After the salt is dissolved, a current of 0.800 amp passing for
10 min 38 sec is required in order to regain the initial temperature of the
water. Also write the equation for this "reaction."

(e) The enthalpy of fusion per mole of p-iodotoluene, C_7H_7I. When 21.8 g of
p-iodotoluene are added to 200 ml of water in the calorimeter, the crystals
(being immiscible with water and also more dense) sink to the bottom. As
current is passed through the heater, the temperature of the mixture rises
until the p-iodotoluene starts to melt; at this point, all of the electrical
energy is used for fusion and none for raising the temperature. A current
of 0.820 amp passing for 4 min 41 sec is required for the period in which
the temperature stays constant and before the temperature again begins to
rise, as both the water and the *liquid* p-iodotoluene rise above the melting
point of 34.0°C. Write the equation for the "reaction."

10. A metal X, whose specific heat is 0.119 cal/g °C, forms an oxide whose composition is 32.00% O.
    (a) What is the empirical formula of the oxide?
    (b) What is the exact atomic weight of the metal?

11. A metal Y, whose specific heat is 0.0504 cal/g °C, forms two chlorides, whose
compositions are 46.71% Cl and 59.42% Cl.
    (a) Find the formulas of the chlorides.
    (b) What is the exact atomic weight of the metal?

12. (*i*) Use Table 14-1 to determine the standard enthalpy of reaction, in kcal,
for each of the following reactions.
    (*ii*) Give the values of the standard enthalpies of reaction in kjoules.
    (*iii*) Calculate the change in standard internal energy ($\Delta E°$) for each of the
reactions.
    (a) $FeO_{(s)} + H_{2(g)} \rightleftarrows Fe_{(s)} + H_2O_{(l)}$
    (b) $4NH_{3(g)} + 5O_{2(g)} \rightleftarrows 6H_2O_{(g)} + 4NO_{(g)}$
    (c) $Zn_{(s)} + 2HCl_{(aq)} \rightleftarrows H_{2(g)} + ZnCl_{2(aq)}$
    (d) $2FeO_{(s)} + \frac{1}{2}O_{2(g)} \rightleftarrows Fe_2O_{3(s)}$
    (e) $3NO_{2(g)} + H_2O_{(l)} \rightleftarrows 2HNO_{3(aq)} + NO_{(g)}$
    (f) $2N_2O_{(g)} \rightleftarrows 2N_{2(g)} + O_{2(g)}$

13. Calculate the standard enthalpy change (per mole) for the following reactions.
Also write the equation for the reaction in each case.

(a) Dissolving NH_3 in water
(b) The burning (or rusting) of iron to give Fe_2O_3
(c) Combustion of liquid C_6H_6
(d) The reaction of metallic cadmium with dilute HCl
(e) The evolution of H_2S on mixing HCl and Na_2S solutions

14. The enthalpy of combustion of rhombic sulfur is -70.96 kcal/mole. The enthalpy of combustion of monoclinic sulfur is -70.88 kcal/mole. Calculate the standard enthalpy and entropy of transition from rhombic to monoclinic sulfur.

15. Calculate the standard enthalpy and entropy of vaporization for C_2H_5OH at 25°C.

16. Calculate the bond energies in the following gaseous molecules: (a) NO; (b) H_2O; (c) NH_3; (d) PCl_5.

17. The observed heat of combustion at constant volume for sucrose $(C_{12}H_{22}O_{11(s)})$ at 25.0°C is $\Delta E° = -1345$ kcal/mole.
    (a) Write the equation for the combustion reaction and calculate $\Delta H°$ for the reaction.
    (b) Calculate the standard enthalpy of formation of sucrose.
    (c) The overall metabolism of sucrose in your body is the same as the combustion reaction in the bomb calorimeter. Calculate how much heat is produced in your body for every teaspoon of sugar (5.2 g) you eat and metabolize. How much for every pound of sugar you eat?

PROBLEMS B

18. Calculate the approximate specific heat of each of the following elements: (a) Pt; (b) P; (c) Sr; (d) As; (e) Au.

19. Calculate the resultant temperature when 250 g of water at 25.0°C are mixed with 100 g of water at 80.0°C. (Assume no heat loss to container or surroundings.)

20. Calculate the resultant temperature when 100 g of lead metal at 200.0°C are mixed with 200 g of water at 20.0°C. (Assume no heat loss to container or surroundings.)

21. Suppose 200 ml of water at 55.0°C, 35.0 g of ice at 0°C, and 120 g of Zn at 100.0°C are mixed. Calculate the resultant temperature, assuming no heat loss, and using the approximate specific heat of Zn.

22. Suppose 200 ml of water at 15.0°C, 60.0 g of ice at 0°C, and 90.0 g of Cd at 125.0°C are mixed. Calculate the resultant temperature, assuming no heat loss, and using the approximate specific heat of Cd.

23. A metal X, whose specific heat is 0.112 cal/g °C, forms an oxide whose composition is 27.90% O.
    (a) What is the exact atomic weight of the metal?
    (b) What is the formula of the oxide?

24. A metal Y, whose specific heat is 0.0312 cal/g °C, forms two chlorides whose compositions are 35.10% Cl and 15.25% Cl.
    (a) What is the exact atomic weight of the metal?
    (b) What are the formulas of the chlorides?

25. A Dewar flask (vacuum-jacketed bottle) is used as a calorimeter, and the following data are obtained. Measurements in parts (a) and (b) are made to obtain the heat capacity of the calorimeter, and parts (c) and (d) are performed on an unknown metal.
    (a) Calorimeter with 200 g of water has a temperature of 23.7°C.
    (b) When 50.0 g of water at 99.1°C are added to the calorimeter and water in (a), the resultant temperature is 36.0°C.
    (c) Calorimeter with 200 g of water has a temperature of 20.6°C.
    (d) A 91.5 g sample of metal at 99.1°C added to the calorimeter and water in (c) gives a resultant temperature of 21.5°C.
    Calculate the approximate atomic weight of the metal.

26. A 200 g sample of an unknown high-boiling liquid is put into the calorimeter calibrated in Problem 25, and its temperature is observed to be 24.2°C. Then 55.3 g of copper at 180.0°C are added to the liquid in the calorimeter to give a resultant temperature of 28.0°C. Calculate the specific heat of the liquid. Use the approximate specific heat of copper.

27. Calculate the heat capacity of a calorimeter (Figure 14-1) that, when containing 250 ml of water, requires a current of 0.650 amp passing for 5 min 25 sec through the 8.35 ohm immersed resistance in order to raise the water temperature from 25.357°C to 26.213°C.

28. Compute the following quantities, using the calorimeter calibrated in Problem 27.
    (a) The specific heat of an unknown metal. First, 40.0 g of a finely divided sample are stirred with 220 ml of water; then a current of 0.720 amp for 12 min 24 sec raises the temperature from 25.265°C to 27.880°C.
    (b) The specific heat of an unknown liquid. First, 250 g of it are placed in the calorimeter in place of the water; then a current of 0.546 amp is passed for 4 min 45 sec, raising the temperature from 26.405°C to 27.033°C.
    (c) The enthalpy change per mole of precipitated $BaSO_4$ when 125 ml of 0.500 M $BaCl_2$ solution are mixed in the calorimeter with 125 ml of 0.500 M Na_2SO_4 solution. A current of 0.800 amp passing for 5 min 41 sec through the heater immersed in the reaction mixture caused a temperature rise 1.500 times larger than that observed for the reaction. Also write the equation for the reaction involved.
    (d) The enthalpy change per mole of NH_4NO_3 when 16.0 g of solid NH_4NO_3 are dissolved in 250 ml of water in the calorimeter. This is the so-called "heat of solution." After the salt is dissolved, a current of 1.555 amp passing for 4 min 16 sec is required in order to regain the initial temperature of the water. Write the equation for this "reaction."
    (e) The enthalpy of fusion per mole of diiodobenzene, $C_6H_4I_2$. When 20.0 g of diiodobenzene are added to 250 ml of water in the calorimeter, the crystals (being immiscible with water and also more dense) sink to the bottom. As

current is passed through the heater, the temperature of the mixture rises until the diiodobenzene starts to melt; at this point, all of the electrical energy is used for fusion and none for raising the temperature. A current of 0.500 amp passing for 6 min 49 sec is required for the period in which the temperature stays constant and before the temperature again begins to rise, as both the water and the *liquid* diiodobenzene rise above the melting point of 27°C. Write the equation for the "reaction."

29. (*i*) Use Table 14-1 to determine the standard enthalpy of reaction (in kcal) for each of the accompanying reactions.
    (*ii*) Give the values of the standard enthalpies of reaction in kjoules.
    (*iii*) Calculate the change in standard internal energy ($\Delta E°$) for each of the reactions.
    (a) $C_{(s)} + H_2O_{(g)} \rightleftarrows H_{2(g)} + CO_{(g)}$
    (b) $C_6H_{6(l)} + 7\frac{1}{2}O_{2(g)} \rightleftarrows 6CO_{2(g)} + 3H_2O_{(l)}$
    (c) $2K_{(s)} + H_2SO_{4(aq)} \rightleftarrows K_2SO_{4(aq)} + H_{2(g)}$
    (d) $N_2O_{(g)} + H_{2(g)} \rightleftarrows H_2O_{(l)} + N_{2(g)}$
    (e) $NH_{3(g)} \rightleftarrows \frac{1}{2}N_{2(g)} + 1\frac{1}{2}H_{2(g)}$
    (f) $2H_2S_{(g)} + 3O_{2(g)} \rightleftarrows 2H_2O_{(l)} + 2SO_{2(g)}$

30. Calculate the standard enthalpy change (per mole) for the following reactions. Also write the equation for the reaction in each case.
    (a) Dissolving HCl in water
    (b) Combustion of C_4H_{10} vapor
    (c) Combustion of liquid C_2H_5OH
    (d) The production of PCl_5 from PCl_3
    (e) The reaction of metallic zinc with dilute H_2SO_4

31. The enthalpy of combustion of diamond is -94.50 kcal/mole. The enthalpy of combustion of graphite is -94.05 kcal/mole. What is the standard enthalpy and entropy of transition from diamond to graphite?

32. Calculate the standard enthalpy and entropy of vaporization of C_6H_6 at 25.0°C.

33. Calculate the bond energies in the following gaseous molecules: (a) CO; (b) CO_2; (c) SO_2; (d) NaCl. The enthalpy of sublimation of NaCl is $\Delta H = +54.70$ kcal/mole, and the standard enthalpy of formation of $NaCl_{(s)}$ is $\Delta H_f° = -98.23$ kcal/mole.

34. The observed heat of combustion at constant volume of a common fat, glyceryl trioleate ($C_{57}H_{104}O_{6(s)}$) at 25.0°C is $\Delta E° = -7986$ kcal/mole.
    (a) Write the equation for the combustion reaction and calculate $\Delta H°$ for the reaction.
    (b) Calculate the standard enthalpy of formation of glyceryl trioleate.
    (c) The overall metabolism of this fat in your body is the same as the combustion reaction in the bomb calorimeter. Calculate how much heat is produced in your body for every ounce of this fat that you eat and metabolize. How much heat energy would your body produce while getting rid of one pound of this fat?

Chemical Kinetics

THE LAW OF MASS ACTION

The rate of a chemical reaction depends on several important factors but, except for the case where a molecule is unstable and spontaneously decomposes, there is a very basic requirement that two interacting molecules must first collide with each other before any kind of reaction can occur. A collision between two potentially reacting molecules does not necessarily mean that a reaction *will* occur, but at least there's the possibility. The greater the frequency of collision, the faster the reaction is likely to be. The one factor that determines the frequency of collision is concentration: the higher the concentrations of the reactants, the greater the number of molecules per unit volume, and the greater the collision frequency. This basic principle was first put forth quantitatively in 1863 by Guldberg and Waage in their statement of the *law of mass action,* which asserts that, for a general reaction of the type

$$a\,A + b\,B + c\,C + \cdots \rightarrow m\,M + n\,N + \cdots \qquad (15\text{-}1)$$

the rate (V) of the reaction *at a given temperature* will be

$$V = k\,[A]^a[B]^b[C]^c \cdots \qquad (15\text{-}2)$$

where the quantities in brackets are the molar concentrations of the reactants; $a, b, c, m, n, \ldots$ are the coefficients in the balanced chemical equation; and k is a proportionality constant (called the rate constant) that is characteristic of each specific reaction.

One must immediately be aware of the limitations of the law of mass action. Almost every chemical reaction is in actual fact an extremely complicated process, and the familiar balanced chemical equation (which shows the molar relationships between the original reactants and the final products) gives no clue at all to the many intricate sequences of simple intermediate steps that are followed in going from "reactants" to "products." Always bear in mind the following points.

1. The law of mass action really applies *only* to each of the individual intermediate steps in a chemical reaction.

2. The actual overall rate expression for a given chemical reaction can be determined reliably *only* by experiment. It often is a very complicated equation, appearing to bear no relation to the overall chemical equation in the way that Equation 15-2 (the rate equation) is related to Equation 15-1 (the chemical equation).

3. The simple intermediate steps that make up a reaction mechanism invariably involve (a) spontaneous decomposition of one molecule, (b) most commonly a bimolecular collision between two molecules, or (c) an unlikely termolecular collision between three molecules. From a practical standpoint, nothing more complicated is ever observed.

It is important to know the overall rate expression for a reaction, because this permits you to control the reaction and to predict the reaction times needed for different conditions. This expression usually can be determined experimentally without any knowledge of the reaction mechanism itself; in fact, it is a useful aid to working out the mechanism. One objective of the material that follows is an explanation of the manner in which the empirical (experimentally determined) rate expression is obtained for a great many reactions.

We shall not consider those reactions that have very complicated rate expressions. Instead, we shall consider those for which the empirical rate expression *is* of the form given by Equation (15-2),

$$V = k\,[A]^a[B]^b[C]^c \tag{15-2}$$

but where a, b, and c are not necessarily the same as the coefficients in the balanced chemical equation (Equation 15-1).

There are some common terms that are used in studies of reaction rates. One speaks of the *order of the reaction* as being the sum of the exponents ($a + b + c + \cdots$) in the empirical rate expression, and of the *order of a reactant* as being the exponent to which the concentration of the reactant is raised in the empirical rate expression. Thus, if the empirical rate expression is

$$V = k[A]^2[B]$$

we say that it is a third-order reaction ($a + b = 3$) that is second order in A ($a = 2$) and first order in B ($b = 1$). If C is also a reactant but its concentration does not appear in the rate expression, then we say that the reaction is zero order in C ($c = 0$).

HOW CONCENTRATIONS CHANGE WITH TIME

The experimental problem is to find (at a given temperature) the order of the reaction and the reactants and the value of k. Assume for the moment that there is a way to determine [A] as a function of time (t) after mixing the reactants. When the experimental values of [A] are plotted versus t, it is not surprising that the general form of the resulting graph is like that shown in Figure 15-1.

As A is consumed, its concentration drops, and the reaction goes progressively slower and slower as predicted by Equation 15-2. In fact, as shown in Figure 15-2, Equation 15-2 represents the instantaneous rate corresponding to the slope (tangent to the curve, see p 66) at any given point t'—that is, at whatever [A] exists at time t'. The slope (rate) is greatest at the beginning of the reaction and least at the end. This is a problem ideally suited for treatment by the methods of calculus. The slope at t' is $\Delta[A]/\Delta t$ but, in the limit as smaller and smaller increments of [A] and t are chosen, this slope is given by the ratio of the infinitesimals, $d[A]/dt$. The instantaneous rate (V) of Equation 15-2 thus is expressed more profitably in calculus terms by

$$-\frac{d[A]}{dt} = k[A]^a[B]^b[C]^c \cdot \cdot \cdot \tag{15-3}$$

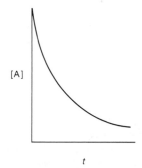

[A]

t

FIGURE 15-1

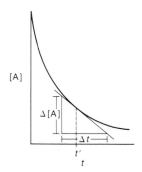

FIGURE 15-2

The negative sign associated with $-d[A]/dt$ indicates that [A] decreases with time.

Experimentally it is just as useful to determine the rate at which [B] or [C] decreases, or the rate at which one of the products [M] or [N] increases. The choice of which compound is used for the rate study will depend on experimental convenience. For example, if M is colored but all other substances colorless, then it would be convenient to study the rate of color increase due to [M]. Perhaps B is a gas and the others are not; in this case the rate of decrease of gas pressure could be used.

Because most reactions involve unequal numbers of molecules of reactants and products, it is necessary to take this into account when comparing the rate of use of a given reactant with the rate of formation of a given product, for example. If the reaction under study is

$$2A + B \rightarrow 3M$$

then it is evident that B is used up just half as fast as A is used up and only one-third as fast as M is produced. The common way to compare these rates is to divide the experimentally-determined rates by the coefficient in the balanced chemical equation to give, in the example above,

$$\frac{1}{3}\left(\frac{d[M]}{dt}\right) = -\frac{d[B]}{dt} = -\frac{1}{2}\left(\frac{d[A]}{dt}\right) \tag{15-4}$$

These interrelated rates and changes in concentration can be illustrated graphically in a single diagram as in Figure 15-3. Unfortunately, in most chemical reactions there is no simple way of following color or pressure changes during the course of the reaction. Instead, one must take samples from the reaction mixture at various times along the way after mixing, do something to stop the reaction in the samples, and then later analyze the samples to find how

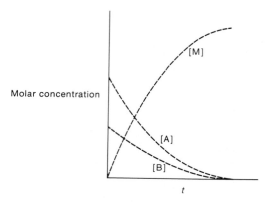

Molar concentration

[M]

[A]

[B]

t

FIGURE 15-3

far the reaction had proceeded in each of the time intervals before sampling. These analyses could be used to provide a few points to construct curves similar to those shown in Figures 15-1 and 15-3, in contrast to the continuous curves that could be obtained by the continuous monitoring of a color change, for example.

DETERMINATION OF RATE CONSTANT AND ORDER

No matter what experimental method is used, constructing tangents to curves is *not* a practical, accurate way to obtain reaction rates. Further, unless precautions are taken to avoid it, it is likely that the concentrations of B and C are also changing at the same time as A, and the observed rate of change of [A] would be a complicated reflection of everything changing at once. In this case it would not be possible to separate out the true dependence of the rate on [A], and it would not be possible to evaluate a or k.

In order to avoid these major complications it is common practice to study reaction rates under conditions such that the concentrations of all of the reactants but one (say A) are used in large excess so that, from a practical standpoint, their concentrations remain constant while [A] alone changes. Thus, if the initial concentrations of B and C are very large compared to A, then for all practical purposes [B] and [C] are constant (and of known value) during the course of the reaction, and Equation 15-3 becomes

$$- \frac{d[A]}{dt} = k[A]^a[B]^b[C]^c = k'[A]^a \qquad (15\text{-}5)$$

an equation involving only two variables, [A] and t. Of course [A] and [B] could have been kept large and constant, or [A] and [C], and the principal result would have been the same because d[A]/dt, d[B]/dt, d[C]/dt, d[M]/dt, and so on are all related to each other through Equation 15-4. There are two general cases of Equation 15-5 to consider: when $a = 1$, and when $a > 1$. We discuss these cases separately, simplifying to consider only two reactants, A and B.

First-Order Reactions ($a = 1$)

A value of $a = 1$ may be obtained in Equation 15-5 under two different situations, as we have noted. First, there is the special circumstance where A is the only reactant (that is, A is unstable and decomposes without any reaction with other substances) and where B and C do not exist (thus, $k' = k$). A common example of this situation is radioactive decay, in which a given radioactive isotope spontaneously decomposes into the isotope of another element at a rate characterized by a rate constant k.

A value of $a = 1$ can also be obtained in some cases where a reactant B is involved, under experimental conditions where $[B] = [B]_e$, a concentration much larger than $[A]$. We can rearrange Equation 15-5 to separate the variables, so that only $[A]$ is on the lefthand side of the equation and only t on the righthand side:

$$-\frac{d[A]}{[A]} = k'dt \tag{15-6}$$

where $k' = k[B]_e^b$. (The same equation applies where A is the only reactant, but in that case $k' = k$.) Using the methods of calculus, we can integrate this equation so as to relate $[A]_0$ (the concentration of A that exists at $t = 0$) to the value of $[A]$ at any later time t:

$$-\int_{[A]_0}^{[A]} \frac{d[A]}{dt} = k'\int_0^t dt \tag{15-7}$$

$$\ln [A] - \ln [A]_0 = -k't \tag{15-8}$$

$$\log [A] = -\frac{k'}{2.30}t + \log [A]_0 \tag{15-9}$$

Equation 15-9 relates the experimentally determined values of $[A]$ to the times t at which the samples were taken. If a reaction is first-order (or pseudo first-order, as would be the case if $[B]$ is much larger than $[A]$), then a plot of log $[A]$ versus t will yield a straight line as in Figure 15-4, with slope equal to $-k'/2.30$, and with y intercept equal to log $[A]_0$.

In order to find the true rate constant k from the slope of this plot, you must know b as well as the large excess concentration $[B]_e$. The value of b can be determined by keeping A at a high known concentration $[A]_e$ and then plotting

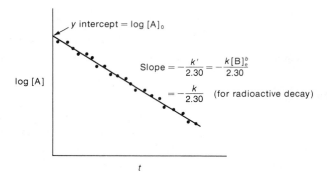

FIGURE 15-4

log [B] versus t as in Figure 15-4 (if $b = 1$), or by making the plots shown in Figure 15-5 (if $b > 1$).

If a reaction is *not* first-order, then the plot shown in Figure 15-4 will *not* be a straight line, and it will be necessary to plot the experimental data in some other way.

Before going on to these alternative procedures, however, we should consider a special way by which true (not pseudo) first-order reactions are often considered. In these cases, $k' = k$. This consideration is especially applicable to radioactive decay processes. It is common practice to describe these true first-order reactions in terms of the time required for one-half of the material to decompose (this time is called the half-life, $t_{\frac{1}{2}}$). In this special circumstance $[A] = \frac{1}{2}[A]_0$ when $t = t_{\frac{1}{2}}$, and Equation 15-9 becomes

$$\log [A] - \log [A]_0 = \log \frac{[A]}{[A]_0} = -\log 2 = -0.301 = -\frac{kt}{2.30} \quad (15\text{-}10)$$

and

$$t_{\frac{1}{2}} = \frac{(2.30)(0.301)}{k} = \frac{0.693}{k} \quad (15\text{-}11)$$

Equation (15-11) shows that, if the half-life of a given first-order reaction is known, it is a simple matter to find the corresponding rate constant, and vice versa. Equation 15-11 also emphasizes the fact that the units of a first-order rate constant are simply reciprocal time: sec^{-1}, min^{-1}, hr^{-1}, and so on. Equation 15-10 illustrates the fact that it doesn't make any difference what concentration units are used; in the log term, the ratio causes the units to cancel. Therefore, it is common practice to use such units as torr, mg, g, or moles/liter. In the case of radioactive decay, the practical unit to use is "counts per minute" (cpm) corrected for background, because this measurement is proportional to the amount (and concentration) of the radioisotope present.

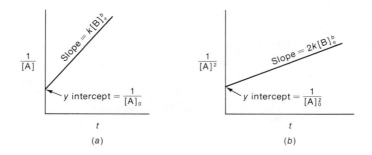

FIGURE 15-5
(a) Second-order plot ($a = 2$). **(b)** Third-order plot ($a = 3$).

Second-Order and Third-Order Reactions ($a > 1$)

If you found that your experimental data did *not* fit a straight-line first-order plot (Figure 15-4), it is natural that you would next consider $a = 2$ or $a = 3$. (It is very unlikely that the order would be greater than 3, because of the implications involving molecular collision processes; we shall not consider such cases.) In this circumstance, using $[B]_e$, a large known excess concentration of B, Equation 15-5 becomes

$$-\frac{d[A]}{[A]^a} = k'dt \tag{15-12}$$

where $k' = k[B]_e^b$.

When Equation 15-12 is integrated so as to relate $[A]_0$ (the concentration of A at $t = 0$) to $[A]$ at time t, we obtain

$$-\int_{[A]_0}^{[A]} \frac{d[A]}{[A]^a} = k' \int_0^t dt \tag{15-13}$$

which corresponds to

$$\frac{1}{[A]^{a-1}} = (a - 1)k't + \frac{1}{[A]_0^{a-1}} \tag{15-14}$$

When experimental values of $[A]$ observed at different times t are plotted as $1/[A]^{a-1}$ versus t as in Figure 15-5, you should get a straight line whose slope is $(a - 1)k' = (a - 1)k[B]_e^b$ and whose y intercept is $1/[A]_0^{a-1}$. Of course, you will have to try different values of a in order to find which value (2 or 3) results in a straight line, as in Figure 15-5a or 5b.

This whole process can be repeated, reversing the roles of A and B, so that a large known initial concentration $[A]_e$, is used. As a result, a set of equations identical to Equations 15-13 and 15-14 would be obtained having B in place of A, and b in place of a. The same is true for Figure 15-5.

Now, having found for both A and B which of the plots in Figures 15-4 or 15-5 gives a straight line, you will know the values of a and b (that is, the order of each reactant). In addition, the slope of whichever plot in Figures 15-4 or 15-5 gives a straight line will give you k (the rate constant) because you know $[B]_e^b$ and $[A]_e^a$. In short,

1. the *type* of plot that gives a straight line tells the *order;*
2. the *slope* of that straight line gives the *rate constant, k.*

IMPORTANT MISCELLANEOUS COMMENTS

1. In the preceding discussion, we emphasized taking a set of data pairs (values of [A] at the corresponding times t) and plotting them as (1) log [A] versus t, (2) $1/[A]$ versus t, and (3) $1/[A]^2$ versus t in order to find which plot gives a straight line. In many ways, a simpler approach is to use your hand calculator or a computer and apply the method of least squares (see p 72) to each of the three possibilities. Whichever combination yields the largest correlation coefficient must be the one that would have given the straight-line plot. At the same time, you would have found the slope and y intercept of that straight line, and your problem would be solved without any graphing.

2. If gaseous reactions are studied, it is customary to express concentration in partial pressures, using atm as the pressure unit. In other words, if A is a gas, then [A] would be expressed as P_A in atm or torr, not moles/liter.

3. The units of k are important. As stated earlier, the units of k for a first-order reaction are reciprocal time: sec^{-1}, min^{-1}, hr^{-1}, and so on. For second-order reactions, k has the units of $M^{-1}sec^{-1}$ (or some other reciprocal time unit); for a gaseous reaction the units would be $atm^{-1}sec^{-1}$. For a third-order reaction, k has the units of $M^{-2}min^{-1}$ (or some other reciprocal time unit) for reactions in solution, or $atm^{-2}min^{-1}$ for gaseous reactions. Other units may be used, but these are the traditional ones; they should *always* be stated.

4. A modification may be needed when dealing with a second-order or third-order plot for a compound involved in a chemical reaction that uses unequal numbers of moles of reactants and products. In this case, any one of the rate terms similar to those given by Equation 15-4 can be set equal to $k[A]^a[B]^b[C]^c$. When you multiply both sides of the equation by the integer (z) needed to cancel the fraction in front of whatever term that you chose, you can see that k' in Equations 15-5, 15-12, 15-13, and 15-14 must *also* include this integer z. As a result, $k' = zk[B]_e^b$, and the slopes of the lines in Figure 15-5 are equal to $zk[B]_e^b$ in part (a) and $2zk[B]_e^b$ in part (b).

5. In the foregoing discussion, we have used the basic assumption that chemical reactions go to completion, or until at least one reactant is completely used up—that they are not reversible. Many reactions do go to completion, or so nearly so as to make no difference. But a huge number of reactions are reversible, and to such an extent that the products form and accumulate and then react with each other to re-form the reactants. The reaction ultimately goes to a position of dynamic equilibrium far from completion where the rate of the forward reaction is the same as the rate of the reverse reaction, and the reaction appears to have ceased. Under these conditions the experimenter observes the *net* rate of reaction, which is simply the difference between the rates of the forward and reverse reactions:

$$V_{net} = -\frac{d[A]}{dt} = k_f[A]^a[B]^b[C]^c - k_r[M]^m[N]^n \qquad (15\text{-}15)$$

[M] and [N] can be expressed in terms of [A] through Equation 15-1, and simpler expressions for use with rate studies can be derived (just as Equations 15-9 and 15-14 were derived from Equation 15-2) for the purpose of finding the values of both the forward and reverse rate constants, k_f and k_r. Such reactions can also be studied by mixing only products together, as well as by mixing only reactants together. Later we examine in some detail the situation that exists at dynamic equilibrium.

ILLUSTRATIVE PROBLEMS

There is no end to the variety of problems that can be found in the area of chemical kinetics. The few given here are fairly typical. To save space, actual plots are not given; we refer to Figure 15-4 or 15-5 for the *type* of curve that would be obtained. In each case the data are converted to the form needed in the plots, so that you can graph them if you wish. Slopes and *y* intercepts of the best-fit lines have been obtained by the method of least squares (see p 72).

PROBLEM:
The activity of a radioactive isotope is studied with the help of a Geiger counter, which counts how many disintegrations occur per minute (counts per min, cpm). The number of cpm is a measure of how much of the isotope is present at any given time. The accompanying data have been corrected for the background cpm always present. Determine the half-life of the isotope.

t (min)	0	2	4	6	8	10	12	14
cpm	3160	2512	1778	1512	1147	834	603	519

SOLUTION:

You know that radioactive decay is first-order, so it is necessary only to find the rate constant for decay, from which the half-life may be calculated by means of Equation 15-11. To make the first-order plot, first convert cpm to log cpm to get

t (min)	0	2	4	6	8	10	12	14
log (cpm)	3.500	3.400	3.250	3.180	3.060	2.921	2.780	2.715

The plot of log (cpm) versus t will look like Figure 15-4, with a correlation coefficient of 0.9974. The slope is -0.0578 min^{-1}, and the y intercept is 3.505. Because the slope is equal to $-k/2.3$,

$$k = -(\text{slope})(2.30) = -(-0.0578 \text{ min}^{-1})(2.30) = 0.133 \text{ min}^{-1}$$

The half-life is given by

$$t_{\frac{1}{2}} = \frac{0.693}{k} = \frac{0.693}{0.133 \text{ min}^{-1}} = 5.21 \text{ min}$$

PROBLEM:

A sample of radioactive sodium-24 ($t_{\frac{1}{2}} = 15.0$ hr) is injected into an animal for studies of Na$^+$ balance. How long will it take for the activity to drop to one-tenth of the original activity?

SOLUTION:

Equation 15-9 shows how the activity [A] at time t is related to the activity $[A]_0$ at the beginning. In this problem $[A] = 0.10[A]_0$ at the time t that you seek. The needed rate constant is obtained from the given half-life:

$$k = \frac{0.693}{t_{\frac{1}{2}}} = \frac{0.693}{15.0 \text{ hr}} = 0.0462 \text{ hr}^{-1}$$

$$\log [A] - \log [A]_0 = \log \frac{[A]}{[A]_0} = \log (0.10) = -1.00 = -\frac{kt}{2.30} = \frac{-0.0462 \text{ hr}^{-1}}{2.30} t$$

$$t = \frac{(2.30)(1.00)}{0.0462 \text{ hr}^{-1}} = 49.8 \text{ hr}$$

PROBLEM:

Cinnamylidene chloride (A) reacts with ethanol (B) to give HCl (C) and a complicated organic molecule (D) by the reaction

$$A + B \rightarrow C + D$$

Because A is the only one of the compounds that absorbs light in the near ultraviolet, the rate is followed by measuring the change in light absorption of a suitable wavelength. The concentrations of A as a function of time after mixing A and B at 23.0°C are the following.

t (min)	0	10	31	67	100	133	178
[A] (moles/liter)	2.11×10^{-5}	1.98×10^{-5}	1.76×10^{-5}	1.33×10^{-5}	1.16×10^{-5}	0.956×10^{-5}	0.743×10^{-5}

The concentration of B remains constant because it also serves as the solvent. Assume that the reaction is zero-order with respect to B, and calculate the rate constant and order with respect to A.

SOLUTION:
Try, in succession, the plots of log [A] versus t, 1/[A] versus t, and 1/[A]^2 versus t until a straight-line plot is obtained. You quickly find that the data converted to

t (min	0	10	31	67	100	133	178
log [A]	-4.676	-4.703	-4.754	-4.876	-4.935	-5.020	-5.129

give an excellent first-order plot (like Figure 15-4) with correlation coefficient of 0.9997. The slope is -2.56×10^{-3} min^{-1} and the y intercept is -4.678. Because the reaction is zero-order with respect to B, the slope is just equal to $-k/2.30$. Therefore

$$k = -(\text{slope})(2.30) = -(-2.56 \times 10^{-3} \text{ min}^{-1})(2.30) = 5.89 \times 10^{-3} \text{ min}^{-1}$$

The overall rate expression is

$$-\frac{d[A]}{dt} = 5.89 \times 10^{-3} \text{ [A]}$$

PROBLEM:
Methyl acetate (A) reacts with water (B) to form acetic acid (C) and methyl alcohol (D):

$$CH_3COOCH_3 + H_2O \rightarrow CH_3COOH + CH_3OH$$
$$A \quad + \quad B \quad \rightarrow \quad C \quad + \quad D$$

The rate of the reaction can be determined by withdrawing samples of known volume from the reaction mixture at different times after mixing, and then quickly titrating the acetic acid that has been formed, using standard NaOH. The initial concentration of A is 0.8500 M. B is used in large excess and remains constant at 51.0 moles/liter. It is known also that reactions like this are first-order with respect to H_2O. From the data given, determine the rate constant for the reaction and the order with respect to A.

t (min)	0	15	30	45	60	75	90
[C] (moles/liter)	0	0.02516	0.04960	0.07324	0.09621	0.1190	0.1404

SOLUTION:
From the concentration of C produced at each time interval, calculate the concentration of A that must remain, knowing that each mole of C resulted from using up

one mole of A. That is, $[A] = 0.850 - [C]$. This calculation gives the following data.

t (min)	0	15	30	45	60	75	90
[A] (moles/liter)	0.8500	0.8248	0.8004	0.7768	0.7538	0.7310	0.7096
log [A]	−0.07058	−0.08365	−0.09669	−0.1097	−0.1227	−0.1361	−0.1490

Try, in succession, the first-order, second-order, and third-order plots until a straight-line plot is obtained. It turns out that the first-order plot gives an excellent fit (correlation coefficient $= 0.9999$), so $a = 1$. The slope is -8.718×10^{-4} min^{-1}, and the y intercept is -0.07054. The slope (see Figure 15-4) is equal to $-\dfrac{k[H_2O]}{2.30}$, and it is given that $[H_2O] = 51.0$ M. Therefore,

$$k = \frac{-(\text{slope})(2.30)}{[H_2O]} = \frac{-(-8.718 \times 10^{-4} \text{ min}^{-1})(2.30)}{\left(51.0 \dfrac{\text{moles}}{\text{liter}}\right)} = 3.93 \times 10^{-5} \text{ M}^{-1}\text{min}^{-1}$$

The overall rate expression for this second order reaction is

$$-\frac{d[A]}{dt} = 3.93 \times 10^{-5}[A][H_2O]$$

PROBLEM:

H_2O_2 is catalytically decomposed by the presence of I$^-$. The accompanying data are obtained at 20.0°C with a 0.250 M H_2O_2 solution in the presence of 0.030 M KI. The reaction flask containing 50.0 ml of reaction mixture is connected to a gas buret, and the rate is followed at a series of time intervals after mixing by measuring the volume (V) of O_2 collected over water at a barometric pressure of 730.0 torr. The reaction is

$$2H_2O_2 \rightarrow 2H_2O + O_2$$

Find the rate constant and order of reaction for the decomposition of H_2O_2.

t (min)	0	5	10	15	25	35	50	65	80
V_{O_2} (ml)	0	7.9	15.5	22.6	35.8	47.8	63.9	77.4	89.4

SOLUTION:

From the volume of O_2 produced at each time interval, calculate the concentration of H_2O_2 that must remain, knowing that 2 moles of H_2O_2 are used up for every mole of O_2 produced.

$$\text{Moles } H_2O_2 \text{ at beginning} = (0.0500 \text{ liters})\left(0.250 \frac{\text{moles}}{\text{liter}}\right) = 0.01250 \text{ moles } H_2O_2$$

The moles of O_2 in V liters at the given conditions are obtained from the ideal gas law:

$$n_{O_2} = \frac{PV}{RT} = \frac{(730 - 18 \text{ torr})(V \text{ liters})}{\left(62.4 \dfrac{\text{torr liter}}{\text{mole K}}\right)(293 \text{ K})} = 0.0389V \text{ moles } O_2$$

The number of moles of H_2O_2 remaining at each time interval will be

$$\text{moles } H_2O_2 \text{ remaining} = (0.01250 \text{ mole } H_2O_2) - \left(2 \frac{\text{moles } H_2O_2}{\text{mole } O_2}\right)(0.0389V \text{ moles } O_2)$$

This number of moles will be in 0.0500 liter, so the concentration will be

$$[H_2O_2] = \frac{(0.01250) - (0.0778V)}{0.0500} = 0.250 - 1.556V$$

Using this equation (with V_{O_2} in liters) to calculate $[H_2O_2]$ gives the following set of data.

t (min)	0	5	10	15	25	35	50	65	80
$[H_2O_2]$ $\left(\dfrac{\text{moles}}{\text{liter}}\right)$	0.2500	0.2377	0.2259	0.2148	0.1943	0.1756	0.1506	0.1296	0.1109
log $[H_2O_2]$	-0.6021	-0.6240	-0.6461	-0.6679	-0.7115	-0.7554	-0.8223	-0.8875	-0.9551

The first-order plot gives an excellent straight line with correlation coefficient of 0.9999, so there is no point in trying second-order and third-order plots. The slope is $-4.41 \times 10^{-3} \text{ min}^{-1}$ and the y intercept is -0.6018. The slope (see Figure 15-4) is equal to $-k/2.30$, so

$$k = -(\text{slope})(2.30) = -(-4.41 \times 10^{-3} \text{ min}^{-1})(2.30)$$
$$= 1.01 \times 10^{-2} \text{ min}^{-1}$$

The overall rate expression is

$$-\frac{d[H_2O_2]}{dt} = 1.01 \times 10^{-2}[H_2O_2]$$

PROBLEM:
The gaseous reaction $2NO + O_2 \rightarrow 2NO_2$ is studied at constant volume by measuring the change in the total pressure with time; this change occurs as 3 gaseous moles are converted to 2 gaseous moles. The *change* in pressure (ΔP) at any given time will be equal to the partial pressure of O_2 *used up*, and to one-half of the partial pressure of NO *used up*. Two series of measurements are made at 27.0°C: (1) with an excess of O_2 where initially $P_{O_2} = 620$ torr and $P_{NO} = 100$ torr, and (2) with an excess of NO where initially $P_{O_2} = 20.0$ torr and $P_{NO} = 315.0$ torr. The variation in total pressure with time is shown in the following data. Determine the rate constant for the reaction and the order with respect to each reactant.

	t (min)	0	1.5	3.2	5.5	8.4	12.7	19.1
(1)	P_{total} (torr)	720.0	715.0	710.0	705.0	700.0	695.0	690.0
	t (min)	0	1.7	3.5	5.6	8.1	11.0	14.5
(2)	P_{total} (torr)	335.0	333.0	331.0	329.0	327.0	325.0	323.0

SOLUTION:

From the total pressure at each time interval, calculate the partial pressures of NO and O_2 that must remain. From the information in the problem, you can see that in series (1) $P_{NO} = 100 - 2(\Delta P)$, and in series (2) $P_{O_2} = 20.0 - \Delta P$. By using the ideal gas law, it would be possible to convert these partial pressures to moles/liter, but this usually is not done with gaseous reactions. Instead we shall simply use the partial pressures (in torr) as the measure of concentration. Try, in succession, the first-order, second-order, and third-order plots until a straight-line graph is obtained for each series.

For series (1):

t (min)	0	1.5	3.2	5.5	8.4	12.7	19.1
P_{NO} (torr)	100.0	90.0	80.0	70.0	60.0	50.0	40.0
$\log P_{NO}$	2.000	1.954	1.903	1.845	1.778	1.699	1.602
$1/P_{NO}$	0.01000	0.01111	0.01250	0.01429	0.01667	0.02000	0.02500

For series (2):

t (min)	0	1.7	3.5	5.6	8.1	11.0	14.5
P_{O_2} (torr)	20.0	18.0	16.0	14.0	12.0	10.0	8.0
$\log P_{O_2}$	1.3010	1.2553	1.2041	1.1461	1.0792	1.0000	0.9031

Series (1) tells us that the reaction is second-order with respect to NO, and series (2) that the reaction is first-order with respect to O_2. The superiority of $1/P_{NO}$ versus t for series (1) is much more apparent from the graphs (the first-order plot has a very pronounced curve, whereas the second-order plot is extremely straight) than from the correlation coefficients (0.9999 for the second versus 0.9908 for the first). We conclude, therefore, that we need the rate constant for the overall rate expression

$$-\frac{dP_{O_2}}{dt} = \frac{1}{2}\left(\frac{-dP_{NO}}{dt}\right) = kP_{NO}^2 P_{O_2}$$

In this case, the slope of the series (1) second-order plot shown in Figure 15-5a is equal to $2k[P_{O_2}] = 7.88 \times 10^{-4}$ torr^{-1} min^{-1}, the factor of 2 being required because of the coefficient of $\frac{1}{2}$ for $\left(\frac{-dP_{NO}}{dt}\right)$ (see miscellaneous note #4, p 236). As a result,

$$k = \frac{7.88 \times 10^{-4}}{2[P_{O_2}]_e} = \frac{7.88 \times 10^{-4} \text{ torr}^{-1} \text{ min}^{-1}}{(2)(620 \text{ torr})} = 6.35 \times 10^{-7} \text{ torr}^{-2} \text{ min}^{-1}$$

The slope of the series (2) first-order plot is -0.0274 min^{-1}, with y intercept $=$ 1.3009. Figure 15-4 shows that the slope $= -k[P_{NO}]_e^2/2.30 = -0.02740$ min^{-1}. Therefore

$$k = \frac{-(2.30)(-0.0274 \text{ min}^{-1})}{[P_{NO}]_e^2} = \frac{(2.30)(0.0274 \text{ min}^{-1})}{(315 \text{ torr})^2} = 6.35 \times 10^{-7} \text{ torr}^{-2} \text{ min}^{-1}$$

The two series agree on the value of k that should be used in the overall rate expression.

THE ENERGY OF ACTIVATION

Molecular collisions are a fundamental requirement for chemical reactions, and concentrations are important because they determine the frequency of these collisions. Nevertheless, most collisions do *not* result in a chemical reaction, because the energy of collision is insufficient to surmount the barrier to reaction. This barrier, the energy required to rearrange atoms in going from reactants to products, is very substantial for most reactions; it is called the *activation energy,* or the enthalpy of activation (ΔH_a).

The *average* kinetic energy of molecules is proportional to the Kelvin temperature ($\sim 3T$ cal/mole), but there is an enormous continuous exchange of energy that takes place between molecules at a given temperature. In fact, there is a *distribution* of kinetic energies similar to that shown in Figure 15-6 for kinetic energies of collisions.

Only an extremely small fraction of molecules are likely to have a series of successive collisions in which they come away with most of the energy, or with almost none of it. The area under the entire distribution curve is unity; it represents all of the molecules. The area lying to the right of any given energy value—say, ΔH_a—represents the fraction of molecules having energies of collision greater than ΔH_a, the minimum amount needed for reaction to occur. This fraction (area) is given by the term

$$e^{-\frac{\Delta H_a}{RT}} = 10^{-\frac{\Delta H_a}{2.3RT}}$$

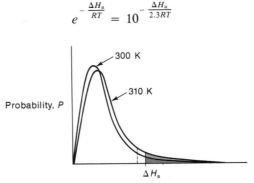

FIGURE 15-6
Distribution of relative kinetic energies of collision in the direction along the line of molecular centers at different temperatures.

For a huge number of reactions, ΔH_a is of the order of 12,800 cal/mole, which at 25.0°C would correspond to a fraction of only 4×10^{-10}, less than a billionth of all the molecules. This fraction increases exponentially with the temperature; an increase of only 10.0°C will *double* the fraction of molecules with enough energy to react. The distribution curve will also be shifted to the right to give a larger area to the right of ΔH_a, as shown in Figure 15-6.

The law of mass action (Equation 15-2) is always stated as applying to a given temperature, and it appears not to have temperature involved in its statement. Yet the rates of chemical reaction invariably increase markedly with increase in temperature. Because concentrations will be negligibly affected by temperature, the temperature-sensitive factor in the law of mass action must be the rate constant, k. As a good approximation, we say that k is proportional to the fraction of molecules (or collisions) that have the required enthalpy of activation:

$$k = Ae^{-\frac{\Delta H_a}{RT}} = A \cdot 10^{-\frac{\Delta H_a}{2.3RT}} \qquad (15\text{-}16)$$

where A is just a proportionality constant. This equation provides a method of finding the enthalpy of activation for a given chemical reaction. By taking the logarithm of each side and rearranging, we get

$$\log k = \left(\frac{-\Delta H_a}{2.30R}\right) \cdot \frac{1}{T} + \log A \qquad (15\text{-}17)$$

If you determine the rate constant for a reaction (as in the first part of this chapter) at several different temperatures, you can plot your data as $\log k$ versus $1/T$ (Figure 15-7) and obtain a straight-line graph, preferably by the method of least squares.

The y intercept corresponds to $\log A$, and the slope corresponds to $-\Delta H_a/2.3R$, from which we can obtain the activation enthalpy in cal/mole as

$$\Delta H_a = -(2.3)\left(1.987 \frac{\text{cal}}{\text{mole K}}\right)(\text{slope, K})$$

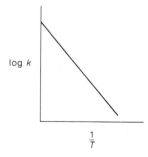

FIGURE 15-7

PROBLEM:
The rate constants at several different temperatures are obtained for the rearrangement of cyclopropane (A) to propylene (B):

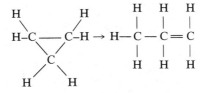

The data are the following.

t (°C)	470	485	500	510	519	530
k (sec^{-1})	1.10×10^{-4}	2.61×10^{-4}	5.70×10^{-4}	10.21×10^{-4}	16.4×10^{-4}	28.6×10^{-4}

(a) Determine the enthalpy of activation for this reaction, and (b) calculate the fraction of molecules having the requisite enthalpy of activation for entering into reaction at 500°C.

SOLUTION:
(a) Transform the data to the form needed for the application of Equation 15-17 and a straight-line plot, as follows (with T in K).

$1/T$	1.346×10^{-3}	1.319×10^{-3}	1.294×10^{-3}	1.277×10^{-3}	1.263×10^{-3}	1.245×10^{-3}
$\log k$	-3.959	-3.583	-3.244	-2.991	-2.785	-2.544

When these data are plotted as in Figure 15-7, you get a straight line whose slope is -1.407×10^4 K, and whose y intercept is 14.98, with a correlation coefficient of 0.9998. The slope is equal to $-\Delta H_a/2.3R$. Therefore,

$$\Delta H_a = -(2.30)(R)(\text{slope})$$
$$= -(2.30)\left(1.987 \frac{\text{cal}}{\text{mole K}}\right)(-1.407 \times 10^4 \text{ K}) = 6.43 \times 10^4 \frac{\text{cal}}{\text{mole}}$$
$$= 64.3 \text{ kcal/mole}$$

(b) The fraction of molecules with any given energy is obtained from $10^{-\frac{\Delta H_a}{2.3RT}}$. Therefore, at 500°C (773 K), the fraction is

$$10^{-\frac{64,300 \text{ cal}}{(2.30)\left(1.987 \frac{\text{cal}}{\text{mole K}}\right)(773 \text{ K})}} = 6.29 \times 10^{-19}$$

This is a very small fraction indeed!

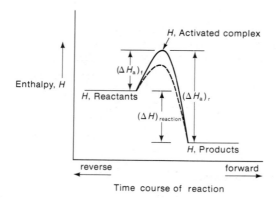

FIGURE 15-8

Reversible Reactions

For reactions that are reversible, the concept of an activation energy barrier applies to the reverse reaction just as it does to the forward reaction. Determining the rate constants for the reverse reaction at several different temperatures by simply mixing the products together would enable you to find ΔH_a for the reverse reaction. Figure 15-8 shows the relationship between the enthalpies of activation for the forward and reverse reactions.

The energy content (enthalpy, H) of the substances is shown on the vertical scale. On collision, the reactants contribute their kinetic energy to the formation of the activated complex and, if this is equal to or greater than $(\Delta H_a)_f$, the complex can be formed and subsequently can decompose into the products, which take up energy $(\Delta H_a)_r$ for use in kinetic energy. In Figure 15-8, the products contain less energy than the reactants, and the difference is given off as the energy (enthalpy) of reaction, $(\Delta H)_{reaction}$. Just the reverse would be true if you started with products. From the diagram you can see that these three quantities are related by

$$(\Delta H)_{reaction} = (\Delta H_a)_f - (\Delta H_a)_r \qquad (15\text{-}18)$$

This relationship makes it possible for you to determine $(\Delta H_a)_f$ and combine it with the enthalpy of reaction determined in a separate experiment to find $(\Delta H_a)_r$ without having to study the rates of the reverse reaction.

Catalysts

A catalyst provides an alternative pathway by which reactants can proceed to form the products, a pathway that involves a lower energy barrier and a different intermediate activated complex. This alternative pathway is shown by the dotted line in Figure 15-8. With a smaller activation energy required, there will be a larger fraction of molecules that can react (the area to the right of the dotted line in Figure 15-6), and the reaction will go faster. Note that a catalyst lowers the activation for the forward and reverse reactions by exactly the same amount, and that $(\Delta H)_{reaction}$ remains unchanged.

PROBLEMS A

1. The following counts per minute (cpm) are obtained for a sample of an un-known isotope at 10 min intervals, beginning with $t = 0$:   10,000, 8166, 7583, 6464, 5381, 5023, 4466, 3622, 2981, 2690, 2239, 2141, 1775, 1603, 1348, 1114, 1048. Determine the half-life of the isotope.

2. As part of a radio-iodine treatment for a thyroid problem, a sample of radioac-tive NaI is injected into a person at 9:00 A.M. on a certain day. The iodide goes almost immediately to the thyroid gland. If the sample has 10,000 cpm when injected, how many cpm will there be at 3:00 P.M. on the same day? The half-life of the iodine-128 isotope is 25.08 min.

3. How many milligrams of a 100 mg sample of radon-222 will disintegrate in 30 days, given that the half-life is 3.82 days?

4. The Dead Sea scrolls were found by radioactive carbon-dating techniques to have 11.9 cpm per gram of carbon. Living material similar to that from which they were made has 15.3 cpm per gram of carbon. How old must the scrolls be, knowing that the half-life of carbon-14 is 5730 yr?

5. The p-nitrophenyl ester of N-carbobenzoxylysine reacts with water according to the following equation (B and P are just symbols used to simplify descrip-tion of this compound; H and O represent hydrogen and oxygen).

$$BOOP + H_2O \rightarrow BOOH + POH$$

There is a very great difference in the ultraviolet absorption of the products and the reactants, so that the rate of the reaction can be followed easily by means of a spectrophotometer at a wavelength of 347.6 nm. The following data are obtained for a very dilute (3.388×10^{-5}M) solution of BOOP. C refers to the molality of BOOP at time t; in each case the value should be multiplied by 10^{-5}. Consider the concentration of water to remain constant at 55.5 moles/liter, and the reaction to be first order with respect to water. Determine the order of the reaction with respect to BOOP, and the rate constant for the reaction.

t (min)	0	5	10	15	20	25	30	35	40	45
C (moles/liter)	3.388	3.081	2.783	2.553	2.290	2.065	1.892	1.710	1.538	1.406

6. Cyclopentadiene (C) reacts with itself to give a dimer (D):

$$2C_5H_6 \rightarrow C_{10}H_{12}$$

When this reaction is carried out at constant volume in the gas phase, the total pressure drops because two gaseous moles are converted to one. The *change* in total pressure at any given time interval will be equal to one-half of the partial pressure of C *used up* in that time. The following experiment is run at 127.0°C, using an initial pressure of 200 torr of C. Use partial pressures in torr as a measure of concentration (instead of moles/liter), and determine the rate constant and order of the reaction.

t (min)	0	10	20	30	40	50	60
P_{total} (torr)	200.0	177.5	163.4	153.6	146.4	140.9	136.6

7. NO_2 decomposes at high temperature according to the gaseous reaction

$$NO_2 \rightarrow NO + \tfrac{1}{2}O_2$$

When this reaction is carried out at constant volume, the total pressure increases because one gaseous mole is converted to one and one-half mole. The *change* in total pressure at any given time will be equal to one-half of the partial pressure of NO_2 *used up* in that time. The following experiment is run at 277.0°C, using an initial pressure of 100 torr of NO_2. Use partial pressures in torr as a measure of concentration (instead of moles/liter), and determine the rate constant and order of the reaction.

t (min)	0	10	20	30	40	50	60	80
P_{total} (torr)	100.0	104.2	107.8	110.8	113.4	115.7	117.7	121.1

8. Ditertiarybutylperoxide (D) decomposes to acetone (A) and ethane (E) by the gaseous reaction

$$D \rightarrow 2A + E$$

The accompanying data are obtained at 155.0°C by observing the pressure change that occurs at constant volume as one gaseous mole is converted to three gaseous moles. The *change* in total pressure at any given time is just equal to twice the partial pressure of D *used up* in that time. Use partial pressures in torr as a measure of concentration (instead of moles/liter), and determine the rate constant and order of the reaction.

t (min)	0	2.0	3.0	5.0	6.0	8.0	9.0	11.0	12.0	14.0	15.0	17.0	18.0	20.0	21.0
P_{total} (torr)	173.5	187.3	193.4	205.3	211.3	222.9	228.6	239.8	244.4	254.5	259.2	268.7	273.9	282.0	286.8

9. (a) Calculate the average molar kinetic energies for He, H_2, O_2, and CO_2 at 27.0°C.

   (b) How many molecules of each gas are present in 1.00 ml at 760 torr and 27.0°C?

10. Calculate the fraction of CH_4 molecules having kinetic energies greater than 13.0 kcal/mole at each of the following temperatures.
    - (a) 200 K
    - (b) 300 K
    - (c) 400 K
    - (d) 500 K
    - (e) 1000 K

11. How much faster will a reaction be at 50.0°C than at 20.0°C, when its activation energy is 15.0 kcal?

12. Two reactants for a given reaction are M and N.
    - (a) If, for a given initial concentration of M, a five-fold increase in the concentration of N causes a 25-fold increase in the initial rate of reaction, what can be said about the order with respect to N?
    - (b) If M is not only a reactant but also the solvent, what is the probable order with respect to M in the experimentally determined rate expression?

13. The accompanying values of the rate constant k have been obtained at different temperatures for the reaction

$$CO + NO_2 \rightarrow CO_2 + NO$$

t (°C)	267	319	365	402	454
k ($M^{-1}sec^{-1}$)	0.0016	0.021	0.12	0.63	2.70

   - (a) Determine the energy of activation for this reaction.
   - (b) What fraction of the molecules possess the necessary enthalpy of activation at 350.0°C?
   - (c) Given that the enthalpy change for this reaction is −54.2 kcal, calculate the enthalpy of activation for the reverse reaction.
   - (d) Draw an enthalpy diagram for this reaction that shows the relationship between the enthalpy of the reaction and the enthalpies of activation for the forward and reverse reactions.

14. The accompanying values of the rate constant k have been determined as a function of temperature for the reaction

$$NO_2 \rightarrow NO + \tfrac{1}{2}O_2$$

t (°C)	319	330	354	379	383
k ($M^{-1}sec^{-1}$)	0.522	0.755	1.700	4.02	5.03

   - (a) Determine the energy of activation for this reaction.
   - (b) What fraction of the molecules possess the necessary energy of activation at 250.0°C?
   - (c) Given that the enthalpy change for this reaction is +13.5 kcal, calculate the enthalpy of activation for the reverse reaction.
   - (d) Draw an enthalpy diagram for this reaction that shows the relationship between the enthalpy of the reaction and the enthalpies of activation for the forward and reverse reactions.

PROBLEMS B

15. The radioactivity of an isotope is studied, and the following counts per minute (cpm) are obtained at five-minute intervals, starting with $t = 0$:   4500, 3703, 2895, 2304, 1507, 1198, 970, 752, 603, 496, 400, 309, 250, 199. Determine the half-life of this isotope.

16. You buy a 250 mg sample of radioactive Na_3PO_4 in order to carry out some experiments with the radioactive ^{32}P, but unforeseen problems delay the initiation of the experiments for 60 days. What percentage of the original sample remains active for your use if the half-life of ^{32}P is 14.3 days?

17. Free, isolated neutrons have a half-life of 12.8 min. If you produce 10^4 neutrons in a pulsed nuclear target experiment, how long will you have to wait until only 10 neutrons remain (statistically)?

18. A piece of wood from the doorway of a Mayan temple is subjected to radioactive carbon-dating techniques and found to have 12.7 cpm per gram of carbon. The living trees from which the wood came had 15.3 cpm per gram of carbon. How old must this temple be, knowing that the half-life of ^{14}C is 5730 yr?

19. Benzene diazonium chloride (B) decomposes in water according to the equation

$$C_6H_5N_2Cl \rightarrow C_6H_5Cl + N_2$$

It is possible to follow the rate of the reaction by measuring the volume V of N_2 produced at a given temperature and pressure. The following data are obtained at 50.0°C using 35.0 ml of a 0.0672 M solution of B. The N_2 gas is collected over water in a gas buret where the temperature is 14.0°C. The barometric pressure is 745 torr. From these data, determine the order of the reaction and rate constant.

t (min)	0	6.0	7.0	8.0	9.0	10.0	12.0	13.0	17.0	19.0	21.0	23.0
V (ml)	0.0	20.2	22.6	25.4	27.8	29.9	33.6	35.4	41.0	43.1	45.0	46.5

20. Acetaldehyde (A) decomposes in the gas phase according to the equation

$$CH_3CHO \rightarrow CH_4 + CO$$

The rate of decomposition can be measured by observing the rate of change of gas pressure at a given volume and temperature, because one gaseous mole is converted to two gaseous moles. The *change* in total pressure at any given time interval will be equal to the partial pressure of A *used up* in that time. In the accompanying set of data obtained at 518.0°C, the total pressure of the gas mixture is given as a function of time. Determine the order of the reaction and the rate constant. Use pressure in torr as a measure of concentration, not moles/liter.

t (sec)	0	42	73	105	190	242	310	384	480	665	840	1070	1440
P_{total} (torr)	363	397	417	437	477	497	517	537	557	587	607	626	645

21. Acetyl fluoride (A) reacts with water (B) to form acetic acid (C) and hydro-fluoric acid (D). The reaction is carried out in acetone (an inert solvent) so that the concentrations can be kept low and the reaction under control. The equation is

$$CH_3COF + H_2O \rightarrow CH_3COOH + HF$$
$$A \quad + \quad B \quad \rightarrow \quad C \quad + \quad D$$

Samples of the reaction mixture are taken at 10 min intervals and quickly titrated with standard NaOH to obtain the combined concentration of C and D. Two series of measurements are made: (1) with an excess of H_2O where $[H_2O] = 1.000$ M and $[A] = 0.0100$ M at the outset, and (2) with an excess of A where $[A] = 0.800$ M and $[H_2O] = 0.0200$ M at the outset.

t (min)	0	10.0	20.0	30.0	40.0	50.0	60.0
(1) [C + D] (moles/liter)	0	0.00286	0.00530	0.00740	0.00920	0.01076	0.01208
(2) [C + D] (moles/liter)	0	0.004805	0.008800	0.01236	0.01561	0.01842	0.02098

Determine the order of this reaction with respect to each reactant, and the rate constant for the reaction.

22. The gas-phase reaction

$$2NO + H_2 \rightarrow N_2O + H_2$$

is studied at 826.0°C under two sets of conditions: (1) with a large excess of NO, and (2) with a large excess of H_2. Because 3 gaseous moles are converted to 2 gaseous moles, the rate can be followed by measuring the total gas pressure as a function of time. The *change* in pressure at any given time will be equal to the partial pressure of H_2 *used up* and to one-half of the partial pressure of NO *used up* in that time. The following data are obtained for series (1), where the initial mixture contains 400 torr of NO and 100 torr of H_2.

(1)	t (sec)	0	20.0	40.0	60.0	80.0	100.0	120.0
	P_{total} (torr)	500.0	466.1	445.7	429.5	420.9	413.2	409.1

For series (2), the initial mixture contains 600 torr of H_2 and 100 torr of NO, and the following data are obtained.

(2)	t (sec)	0	15.0	30.0	45.0	60.0	75.0	90.0
	P_{total} (torr)	700.0	693.1	688.8	684.7	681.0	678.4	676.5

Determine the order of this reaction with respect to each reactant, and the rate constant for the reaction. Express concentrations as partial pressures in torr, not moles per liter.

23. (a) Calculate average molar kinetic energies for Ne, Ar, CO, and C_2H_6 at 127.0°C.
    (b) How many molecules of each gas are present in 1.00 ml at 730 torr and 127.0°C?

24. Calculate the fraction of N_2 molecules having kinetic energies greater than 15.0 kcal/mole at each of the following temperatures.
    (a) 27.0°C   (d) 327.0°C
    (b) 127.0°C   (e) 427.0°C
    (c) 227.0°C

25. How much faster will a reaction be at 40.0°C than at 25.0°C, when its activation energy is 18.0 kcal?

26. The initial rate of reaction is observed by mixing given concentrations of A and B. First doubling the concentration of A (keeping B constant) and then doubling B (keeping A constant) causes a doubling of the initial rate in the first instance but a quadrupling in the second. What can be said about the order of the reaction with respect to A and B?

27. The accompanying values of the rate constant k are obtained for the gaseous reaction

$$2NO + Cl_2 \rightarrow 2NOCl$$

Each of the values of k should be multiplied by 10^{12}.

t (°C)	0	40	82	128	178	225	293
k ($M^{-2}sec^{-1}$)	1.38	7.5	27.2	72.2	182	453	1130

(a) Determine the energy of activation for this reaction.
(b) What fraction of the molecules possess the necessary activation energy at 100.0°C (to enter into reaction)?
(c) Given the enthalpy change for this reaction as -18.1 kcal, calculate the enthalpy of activation for the reverse reaction.
(d) Draw an enthalpy diagram for this reaction that shows the relationship between the enthalpy of the reaction and the enthalpies of activation for the forward and reverse reactions.

28. The following values of the rate constant k are determined for the gaseous reaction

$$N_2O_5 \rightarrow 2NO_2 + \tfrac{1}{2}O_2$$

t (°C)	0	25	35	45	55	65
k (sec^{-1})	7.87×10^3	3.46×10^5	1.35×10^6	4.98×10^6	1.50×10^7	4.87×10^7

(a) Calculate the enthalpy of activation for this reaction.
(b) What fraction of the molecules possess the necessary energy of activation to decompose at 50.0°C?
(c) Given the enthalpy change for this reaction as $+12.6$ kcal, calculate the enthalpy of activation for the reverse reaction.
(d) Draw an enthalpy diagram for this reaction that shows the relationship between the enthalpy of the reaction and the enthalpies of activation for the forward and reverse reactions.

Chemical Equilibrium in Gases

GENERAL PRINCIPLES

When the reversible chemical reaction

$$a\,A + b\,B + c\,C + \cdots \rightleftarrows m\,M + n\,N + \cdots$$

reaches equilibrium, the rates of the forward and reverse reactions are exactly equal, and the net rate of reaction (V_{net}) is zero. Under these conditions, Equation 15-15 can be rearranged and cast into the form

$$K_e = \frac{k_f}{k_r} = \frac{[M]^m[N]^n}{[A]^a[B]^b[C]^c} \tag{16-1}$$

The constant K_e, which was not present in Equation 15-15, is a new constant (the ratio of the forward and reverse rate constants) called the *equilibrium constant*. Each of the quantities in brackets is the *equilibrium* concentration of the substance shown. At any given temperature, the value of K_e remains constant no matter whether you start with A, B, and C or with M and N, and regardless of the proportions in which they are mixed. K_e varies with temperature because k_f and k_r vary with temperature, but not by exactly the same amount. This dependence on temperature is discussed at the end of this chapter. Equation 16-1 applies to *any* chemical reaction at equilibrium, no matter how many or how complicated the intermediate steps in going from reactants to products.

A good understanding of the principles of chemical equilibrium is important for the prediction of the yield of a reaction at a given temperature and for knowing how to change the yield to your advantage. You will also be able to predict the effect of temperature changes on the yield. Every living biological system contains thousands of reversible reactions whose shifting equilibria must be carefully controlled for the health of the system. Every industrial chemical process is optimized by using the principles of chemical equilibrium.

Special cases of chemical equilibrium in solution are considered in several later chapters, so here we deal only with gaseous reactions. When concentrations are expressed in moles/liter (as they usually are in solution) in Equation 16-1, the equilibrium constant is designated as K_c, whereas for concentrations expressed as partial pressures in atm (as they usually are for gases) the equilibrium constant is designated as K_p. For gaseous equilibria, then, Equation 16-1 becomes

$$K_p = \frac{P_M^m \cdot P_N^n}{P_A^a \cdot P_B^b \cdot P_C^c} \tag{16-2}$$

The relationship between K_c and K_p is easily established by means of the ideal gas law because

$$M \frac{\text{moles}}{\text{liter}} = \frac{n}{V} = \frac{P}{RT} \tag{16-3}$$

Substitution of P/RT for each molar concentration in Equation 16-1 gives

$$K_c = \frac{P_M^m \cdot P_N^n}{P_A^a \cdot P_B^b \cdot P_C^c} \cdot \frac{1}{(RT)^{(m+n)-(a+b+c)}} = \frac{K_p}{(RT)^{\Delta n}} \tag{16-4}$$

where Δn is the difference in the number of moles of gaseous products and reactants. If $\Delta n = 0$, then $K_c = K_p$; otherwise you can convert one to the other by means of Equation 16-4, using 0.08206 liter atm/mole K for R if partial pressures are expressed in atm.

DETERMINATION OF K_p

Values of the equilibrium constant may be obtained by allowing the reactants to come to equilibrium at a given temperature, analyzing the equilibrium mixture, and then substituting the equilibrium concentrations into Equation 16-2.

PROBLEM:

Starting with a $3:1$ mixture of H_2 and N_2 at 450.0°C, the equilibrium mixture is found to be 9.6% NH_3, 22.6% N_2, and 67.8% H_2 by volume. The total pressure is 50.0 atm. Calculate K_p and K_c. The reaction is $N_2 + 3H_2 \rightleftarrows 2NH_3$.

SOLUTION:

According to Dalton's law of partial pressures (see p 163), the partial pressure of a gas in a mixture is given by the product of its volume fraction and the total pressure. Therefore the equilibrium pressure of each gas is

$$P_{NH_3} = (0.096)(50.0 \text{ atm}) = 4.8 \text{ atm}$$
$$P_{N_2} = (0.226)(50.0 \text{ atm}) = 11.3 \text{ atm}$$
$$P_{H_2} = (0.678)(50.0 \text{ atm}) = \underline{33.9 \text{ atm}}$$
$$\text{Total pressure} = 50.0 \text{ atm}$$

By substitution in Equation 16-2,

$$K_p = \frac{P_{NH_3}^2}{P_{N_2} \cdot P_{H_2}^3} = \frac{(4.80 \text{ atm})^2}{(11.3 \text{ atm})(33.9 \text{ atm})^3} = 5.23 \times 10^{-5} \text{ atm}^{-2}$$

Use Equation 16-4 to calculate K_c, noting that $\Delta n = (2 - 4) = -2$.

$$K_c = \frac{K_p}{(RT)^{\Delta n}} = \frac{5.23 \times 10^{-5} \text{ atm}^{-2}}{\left[\left(0.08206 \dfrac{\text{liter atm}}{\text{mole K}}\right)(723 \text{ K})\right]^{-2}}$$

$$= 0.184 \left(\frac{\text{mole}}{\text{liter}}\right)^{-2} = 0.184 \text{ M}^{-2}$$

Note that the *starting* composition does not enter into the calculations, only the *equilibrium* composition.

Table 16-1 gives equilibrium constants for a number of gaseous reactions.

THE EFFECT OF CHANGE IN TOTAL PRESSURE ON K_p AND EQUILIBRIUM POSITION

When the preceding experiment with H_2 and N_2 is conducted at 450.0°C and 100 atm total pressure, an equilibrium mixture is obtained that, on analysis, proves to be 16.41% NH_3, 20.89% N_2, and 62.70% H_2. Following the same line of reasoning as in the preceding problem, we can determine that the equilibrium pressures are $P_{NH_3} = 16.41$ atm, $P_{N_2} = 20.89$ atm, and $P_{H_2} = 62.70$ atm, and the calculated value of $K_p = 5.23 \times 10^{-5}$ atm^{-2}. Note that doubling the total pressure does *not* double the pressure of each gas, as it would have done in a mixture of nonreacting gases. Instead, the chemical equilibrium shifts to

TABLE 16-1
Equilibrium Constants for Selected Gaseous Reactions

Equilibrium	Temp (°C)	K_p
$H_2 \rightleftarrows 2H$	1000	7.0×10^{-18} atm
	2000	3.1×10^{-6} atm
$Cl_2 \rightleftarrows 2Cl$	1000	2.45×10^{-7} atm
	2000	0.570 atm
$N_2O_4 \rightleftarrows 2NO_2$	25	0.143 atm
	45	0.671 atm
$2H_2O \rightleftarrows 2H_2 + O_2$	1000	6.9×10^{-15} atm
	1700	6.4×10^{-8} atm
$2H_2S \rightleftarrows 2H_2 + S_2$	1130	0.0260 atm
	1200	0.0507 atm
$H_2 + Cl_2 \rightleftarrows 2HCl$	1200	2.51×10^4
	1800	1.12×10^3
$SO_2 + \frac{1}{2}O_2 \rightleftarrows SO_3$	900	6.55 atm$^{-\frac{1}{2}}$
	1000	1.86 atm$^{-\frac{1}{2}}$
$CO_2 + H_2 \rightleftarrows CO + H_2O$	700	0.534
	1000	0.719

make more NH_3 at the expense of N_2 and H_2 in such a way that the new equilibrium pressures give exactly the same value of K_e.

We can generalize from this example and say that, for any chemical equilibrium involving a different number of moles of gas on each side of the balanced equation, the equilibrium position will always shift with an increase in total pressure toward the side with the smaller number of gaseous moles; the value of K_p will remain unchanged.

THE EFFECT OF CHANGE IN PARTIAL PRESSURE OF ONE GAS ON K_p AND EQUILIBRIUM POSITION

It often is important to know the yield of a chemical reaction—that is, the percentage of reactants converted to products. The following example shows how this yield may be calculated, and how conditions may be altered to increase the yield.

PROBLEM:
$K_p = 54.4$ at 355.0°C for the reaction $H_2 + I_2 \rightleftarrows 2HI$. What percentage of I_2 will be converted to HI if 0.20 mole each of H_2 and I_2 are mixed and allowed to come to equilibrium at 355.0°C and a total pressure of 0.50 atm?

SOLUTION:
Assume that X moles each of H_2 and I_2 are used up in reaching equilibrium to give $2X$ moles of HI, in accordance with the chemical equation, leaving

$0.20 - X$ moles each of H_2 and I_2. The partial pressure of each gas is given by the product of its mole fraction and the total pressure (see p 163).

$$\text{Moles of } H_2 \text{ at equilibrium} = 0.20 - X$$

$$\text{Moles of } I_2 \text{ at equilibrium} = 0.20 - X$$

$$\text{Moles of HI at equilibrium} = 2X$$

$$\overline{\text{Total moles at equilibrium} = 0.40 - 2X + 2X = 0.4}$$

$$P_{HI} = \left(\frac{2X}{0.40}\right)(0.50 \text{ atm})$$

$$P_{H_2} = P_{I_2} = \left(\frac{0.20 - X}{0.40}\right)(0.50 \text{ atm})$$

$$K_p = \frac{P_{HI}^2}{P_{H_2} \cdot P_{I_2}} = \frac{\left[\left(\frac{2X}{0.40}\right)(0.50 \text{ atm})\right]^2}{\left[\left(\frac{0.20 - X}{0.40}\right)(0.50 \text{ atm})\right]^2}$$

$$54.4 = \left(\frac{2X}{0.20 - X}\right)^2$$

Taking the square root of each side, we obtain

$$7.4 = \frac{2X}{0.20 - X}$$

$$X = \frac{1.48}{9.4} = 0.157 = \text{moles of } H_2 \text{ and } I_2 \text{ used up}$$

$$\text{Percentage conversion (yield)} = \frac{0.157}{0.200} \times 100 = 78.5\%$$

PROBLEM:

What percentage of I_2 will be converted to HI at equilibrium at 355.0°C, if 0.200 mole of I_2 is mixed with 2.00 moles of H_2 at total pressure of 0.50 atm?

SOLUTION:

In this problem, it is advantageous first to assume that the large excess of H_2 will use almost the entire amount of I_2, leaving only X moles of it unused. In general, it is always advantageous to let X represent the smallest unknown entity because it often simplifies the mathematical solution. If X moles of I_2 are *not* used, then $0.20 - X$ moles *are* used. For every mole of I_2 used up, one of H_2 is used up, and two of HI are formed. Proceeding as in the last problem, the number of moles of each component at equilibrium is

$$\text{moles of } H_2 = 2.00 - (0.20 - X) = 1.80 + X$$

$$\text{moles of } I_2 = \qquad\qquad X = \qquad X$$

$$\underline{\text{moles of HI} = 2(0.20 - X) \qquad = 0.40 - 2X}$$

$$\text{Total moles} = 2.20$$

The partial pressure of each component will be the mole fraction of each times the total pressure, as follows.

$$P_{H_2} = \left(\frac{1.80 + X}{2.20}\right) (0.50 \text{ atm})$$

$$P_{I_2} = \left(\frac{X}{2.20}\right) (0.50 \text{ atm})$$

$$P_{HI} = \left(\frac{0.40 - 2X}{2.20}\right) (0.50 \text{ atm})$$

When we substitute these partial pressures into the expression for K_p, we get an expression that will be tedious to solve unless we make a reasonable approximation: we assume that X is negligible in comparison with 0.20 and 1.80.

$$K_p = 54.4 = \frac{\left[\left(\dfrac{0.40 - 2X}{2.20}\right)(0.50)\right]^2}{\left[\left(\dfrac{1.80 + X}{2.20}\right)(0.50)\right]\left[\left(\dfrac{X}{2.20}\right)(0.50)\right]} \cong \frac{(0.40)^2}{(1.80)(X)}$$

$$X = \frac{(0.40)^2}{(1.80)(54.4)} = 0.0016 \text{ moles of } I_2 \text{ not used}$$

$$0.200 - 0.0016 = 0.1984 \text{ moles of } I_2 \text{ used}$$

$$\text{Percentage of } I_2 \text{ used} = \frac{0.1984}{0.200} \times 100 = 99.2\%$$

Note that the wise decision to let X = the amount of I_2 *not* used instead of the amount of I_2 that *was* used really did simplify the solution by making it possible to neglect X when added to or subtracted from larger numbers. If we had solved the quadratic equation instead, we would have found that 99.197% of the I_2 had been used up. This is a *common* method of simplifying a math problem, and at the end you can always check to see whether your answer really is negligible compared to what you said it was. Many chemists say that if X is less than 10.0% of what it is added to or subtracted from, it is okay to neglect it.

The preceding problem illustrates the fact that, although the value of K_p does not change with changes in concentration, the *equilibrium position* will change to use up part of the excess of any one reagent that has been added. In this problem, the large excess of H_2 shifts the equilibrium position to the right, causing more of the I_2 to be used up (99.2% compared to 78.5%) than when H_2 and I_2 are mixed in equal proportions. Advantage may be taken of this principle by using a large excess of a cheap chemical to convert the maximum amount of an expensive chemical to a desired product. In this case I_2, the more expensive chemical, is made to yield more HI by using more of the cheaper H_2.

THE PERCENTAGE DECOMPOSITION OF GASES

Many gases decompose into simpler ones at elevated temperatures, and it often is important to know the extent to which decomposition takes place.

PROBLEM:

$K_p = 1.78$ atm at 250.0°C for the decomposition reaction $PCl_5 \rightleftarrows PCl_3 + Cl_2$. Calculate the percentage of PCl_5 that dissociates if 0.0500 mole of PCl_5 is placed in a closed vessel at 250.0°C and 2.00 atm pressure.

SOLUTION:

Although you are told that you are starting with 0.0500 mole PCl_5, this piece of information is not needed to find the percentage dissociation at the given pressure and temperature. If you were asked for the volume of the reaction vessel, then you would need to know the actual number of moles; otherwise not. To answer the question that is asked, it is simpler to just start with one mole (don't worry about the volume) and assume that X moles of PCl_5 dissociate to give X moles each of PCl_3 and Cl_2 and $1 - X$ moles of PCl_5 at equilibrium.

$$\text{Moles of } PCl_5 = 1.00 - X$$
$$\text{Moles of } PCl_3 = \quad X$$
$$\text{Moles of } Cl_2 = \quad \underline{\quad X \quad}$$
$$\text{Total moles} = 1.00 + X$$

The partial pressures are given by the mole fractions times the total pressure, and are substituted into the K_p expression, to give

$$K_p = 1.78 \text{ atm} = \frac{\left[\left(\frac{X}{1+X}\right)(2.00 \text{ atm})\right]\left[\left(\frac{X}{1+X}\right)(2.00 \text{ atm})\right]}{\left[\left(\frac{1-X}{1+X}\right)(2.00 \text{ atm})\right]}$$

$$1.78 = \frac{2X^2}{(1-X)(1+X)} = \frac{2X^2}{1-X^2}$$

$$1.78 - 1.78X^2 = 2X^2$$

$$X^2 = \frac{1.78}{3.78} = 0.471$$

$$X = 0.686 \text{ moles } PCl_5 \text{ dissociate}$$

$$\text{Percentage of } PCl_5 \text{ dissociated} = \frac{0.686}{1.00} \times 100 = 68.6\%$$

This was not a difficult quadratic equation to solve but, even if it had been, it would not be possible to neglect X compared to 1.00; it is too large. If we had neglected X, we would have obtained the extremely erroneous answer of 94.3%

dissociated. If K_p is very large (or very small), it means that the equilibrium position lies far to the right (or to the left). In either of these cases it is possible to choose X so that it will be very small and amenable to a simplified math solution. The value of K_p for the PCl_5 equilibrium is neither very large nor very small, and hence it never will be possible to neglect X.

THE EFFECT OF TEMPERATURE ON K_p
AND EQUILIBRIUM POSITION

Equation 16-1 shows that

$$K_e = \frac{k_f}{k_r}$$

and Equation 14-16 shows that a rate constant is equal to $A \cdot 10^{-\frac{\Delta H_a}{2.3RT}}$. Substitution shows the temperature dependence of K_e, as follows:

$$K_e = \frac{A_f 10^{-\frac{(\Delta H_a)_f}{2.3RT}}}{A_r 10^{-\frac{(\Delta H_a)_r}{2.3RT}}} = \frac{A_f}{A_r} 10^{-\frac{(\Delta H_a)_f - (\Delta H_a)_r}{2.3RT}} \tag{16-5}$$

Equation 15-18 and Figure 15-8 show that $(\Delta H_a)_f - (\Delta H_a)_r = \Delta H$, the enthalpy (energy) change for the reaction. Therefore, we can write Equation 16-5 as

$$K_e = Z \cdot 10^{-\frac{\Delta H}{2.3RT}} \tag{16-6}$$

where Z is a constant, the ratio of the two constants A_f and A_r. By taking the logarithm of both sides of this equation, we get

$$\log K_e = \left(-\frac{\Delta H}{2.3R}\right) \cdot \frac{1}{T} + \log Z \tag{16-7}$$

Equation 16-7 not only shows the simple way that K_e depends on temperature, it also shows a simple way to determine the enthalpy change for a reaction. By determining the value of K_e at several different temperatures, and then plotting $\log K_e$ versus $1/T$, we should get a straight line whose slope is $-\Delta H/2.3R$. If the reaction is exothermic (ΔH is negative), the slope will be positive; if the reaction is endothermic (ΔH is positive), the slope will be negative (Figure 16-1). Equation 16-7 applies to all chemical equilibria and is *independent* of the concentration units used; either K_p or K_c can be used equally

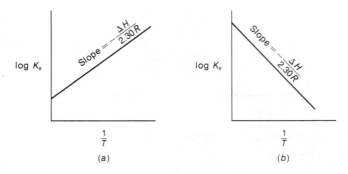

FIGURE 16-1
(a) An exothermic reaction. **(b)** An endothermic reaction.

well. The value of R normally will be 1.987 cal/mole K to go along consistently with ΔH expressed in *calories* per mole.

We can also conclude that for exothermic reactions K_e always decreases with increasing temperature, while for endothermic reactions K_e always increases with increasing temperature. This, in turn, means that an increase in temperature always shifts the position of equilibrium to the left for an exothermic reaction. The reverse is true for endothermic reactions.

PROBLEM:
The following values of K_p are found for the gas-phase reaction

$$H_2 + CO_2 \rightleftarrows CO + H_2O$$

t (°C)	600	700	800	900	1000
K_p	0.39	0.64	0.95	1.30	1.76

Determine the enthalpy of this reaction from these data.

SOLUTION:
We need to transform the data into a form ($1/T$ and log K_p) that will give a straight-line plot with Equation 16-7.

$1/T$	0.001145	0.001028	0.000932	0.000853	0.000786
log K_p	−0.4089	−0.1938	−0.0223	0.1139	0.2455

Plotting these data as in Figure 16-1, and using the method of least squares (see p 72), we obtain a best-fit straight line whose slope is −1809.5 K and whose y

intercept is 1.6637, with a correlation coefficient of 0.9999. Because the slope is equal to $-\Delta H/2.30R$, we can find ΔH.

$$\Delta H = -(\text{slope})(2.30)\left(1.987 \frac{\text{cal}}{\text{mole K}}\right)$$

$$= -(-1809.5 \text{ K})(2.30)\left(1.987 \frac{\text{cal}}{\text{mole K}}\right) = 8270 \frac{\text{cal}}{\text{mole}}$$

We see that the reaction is endothermic, in keeping with the fact that the equilibrium constant increases with increasing temperature.

There are times when you know the enthalpy of the reaction and the equilibrium constant at some temperature, but would like to know the value of K_p at a different temperature. Equation 16-6 is easily adapted to this situation. Let's say that the equilibrium constants K_1 and K_2 correspond to the Kelvin temperatures T_1 and T_2. Substituting these values into Equation 16-7, we obtain $\log K_2 = -\Delta H/2.30RT_2 + \log Z$, and $\log K_1 = -\Delta H/2.30RT_1 + \log Z$. If we subtract the second equation from the first we get

$$\log K_2 - \log K_1 = \log \frac{K_2}{K_1} = -\frac{\Delta H}{2.30R}\left[\frac{1}{T_2} - \frac{1}{T_1}\right] \qquad (16\text{-}8)$$

PROBLEM:
The reaction $PCl_3 + Cl_2 \rightleftarrows PCl_5$ is exothermic with $\Delta H = -22.2$ kcal/mole. The value of K_p is 0.562 atm^{-1} at 250.0°C. Calculate the value of K_p at 200.0°C.

SOLUTION:
We know the value of ΔH, and K_p at 250.0°C. What we want is K_p at 200.0°C. Equation 16-8 is ideal for this, as follows.

$$\log \frac{K_2}{0.562} = \frac{-\left(-22,200 \frac{\text{cal}}{\text{mole}}\right)}{(2.30)\left(1.987 \frac{\text{cal}}{\text{mole K}}\right)}\left[\frac{1}{473 \text{ K}} - \frac{1}{523 \text{ K}}\right]$$

$$= 0.98183$$

Taking the antilog of each side, we obtain

$$\frac{K_2}{0.562} = 9.59$$

$$K_2 = 5.39 \text{ at } 200.0°C$$

The equilibrium constant is larger at the lower temperature, as expected for an exothermic reaction. Note also that this reaction, which is written in the reverse order from that in an earlier problem (p 260), has an equilibrium constant (0.562

atm^{-1}) that is just the reciprocal of the other (1.78 atm). This reciprocity will always be the case.

CATALYSTS

A catalyst affects only the *rate* of a chemical reaction; it has no effect on K_e or on the position of equilibrium at a given temperature. You cannot, therefore, increase the yield of a chemical reaction at a given temperature by adding a catalyst to the reaction mixture. Catalysts are, however, of great practical value because they may make an impractically slow reaction reach equilibrium at a practical rate, or may permit such a reaction to go at a practical rate at a lower temperature, where a more favorable equilibrium position exists.

SUMMARY

The *principle of Le Châtelier* summarizes the conclusions that may be drawn from the illustrative examples in this chapter: "Whenever a stress is placed on a system at equilibrium, the equilibrium position shifts in such a way as to relieve that stress." If the stress is an increase in the partial pressure (concentration) of one component, the equilibrium shifts toward the opposite side in order to use up part of the increase. If the stress is an increase in the total pressure, the stress may be partially relieved by a shift toward the side with the smaller number of gaseous moles; if there are the same number of gaseous moles on each side, no shift will occur, and no stress will be relieved. If the stress is an increase in temperature, the stress is partly relieved because, for an endothermic reaction, the equilibrium constant increases and the equilibrium shifts to the right; for an exothermic reaction, the equilibrium constant decreases and the equilibrium shifts to the left. A catalyst places no stress on the system and causes no shift in the equilibrium position.

PROBLEMS A

1. In the following table, columns A, B, and C refer to these gaseous equilibria:

A: $N_2O_4 \rightleftarrows 2NO_2$ ΔH is positive
B: $CO_2 + H_2 \rightleftarrows CO + H_2O$ ΔH is positive
C: $SO_2 + \tfrac{1}{2}O_2 \rightleftarrows SO_3$ ΔH is negative

For each reaction, tell the effect on K_e and on the equilibrium position of each change listed in the table. Assume that each change affects the reaction mixture only after it has already once reached equilibrium. Use the following symbols in completing the table: $+$, $-$, 0, R, and L to mean increase, decrease, no effect, shift right, and shift left, respectively.

Change	A. Effect on		B. Effect on		C. Effect on	
	position	K_p	position	K_p	position	K_p
Decrease in total pressure						
Increase in temperature						
Decrease in partial pressure of last-named gas						
A catalyst added						
Argon gas added, keeping total pressure constant						

2. A 0.50 mole sample of $SbCl_5$ is put into a closed container and heated to 248.0°C at 1 atm. At equilibrium, analysis shows 42.8% by volume of Cl_2 in the mixture. Calculate K_p at this temperature for the dissociation reaction $SbCl_5 \rightleftarrows SbCl_3 + Cl_2$.

3. Analysis of the equilibrium mixture that results from heating 0.25 mole of CO_2 to a temperature of 1100°C at a pressure of 10 atm shows the presence of 1.40 $\times$ 10^{-3}% O_2 by volume. Calculate the value of K_p for the dissociation reaction $2CO_2 \rightleftarrows 2CO + O_2$ at this temperature.

4. Calculate the value of K_c for (a) the dissociation of H_2S at 1130°C and 1200°C, and (b) the reaction of $H_2 + Cl_2$ at 1200°C and 1800°C, using the K_p values given in Table 16-1.   (c) Calculate the values of K_p for the dissociation of HCl at 1200°C and 1800°C.

5. Calculate the percentage of H_2 that dissociates to atoms at a total pressure of 0.1 atm H_2 and (a) 1000°C and (b) 2000°C.

6. Calculate the percentage of H_2S that dissociates at a pressure of 100 atm and 1130°C.

7. Calculate the composition of the equilibrium gas mixture that results when (a) 0.500 mole each of CO_2 and H_2 are mixed at 1000°C and a total pressure of 2.00 atm, and (b) 1.00 mole of CO and 5.00 moles of H_2O are mixed at 1000°C and a total pressure of 2.00 atm.

8. What percentage of Br_2 will be converted to $COBr_2$ if (a) an equimolar mixture of CO and Br_2 is heated to 70.0°C at a total pressure of 10.0 atm, and (b) a 9 : 1 mixture of CO to Br_2 is heated to 70.0°C at a total pressure of 10.0 atm? The value of K_p is 0.0250 atm^{-1} for the reaction $CO_{(g)} + Br_{2(g)} \rightleftarrows COBr_{2(g)}$.   (c) What is the value of K_c for this reaction?

9. The equilibrium composition of mixtures of H_2 and I_2 is studied at various temperatures in order to get the following values of K_p. Use these data, preferably with the method of least squares, to determine the enthalpy of the reaction $H_2 + I_2 \rightleftharpoons 2HI$.

t (°C)	340	360	380	400	420	440	460	480
K_p	70.8	66.0	61.9	57.7	53.7	50.5	46.8	43.8

10. Values of K_p are obtained at a variety of temperatures for the reaction $2SO_2 + O_2 \rightleftharpoons 2SO_3$. Use these data, preferably with the method of least squares, to determine the enthalpy of this reaction.

t (°C)	627	680	727	789	832	897
K_p (atm^{-1})	42.9	10.5	3.46	0.922	0.397	0.130

11. Using the values of K_p given in Table 16-1, calculate the enthalpy of reaction for (a) the dissociation of H_2S, and (b) the reaction of H_2 with Cl_2.

12. For the reaction $N_2 + 3H_2 \rightleftharpoons 2NH_3$, it is known that the enthalpy of the reaction is -22.1 kcal and that $K_p = 5.23 \times 10^{-5}$ atm^{-1} at 450.0°C. Calculate the value of K_p at 550.0°C.

PROBLEMS B

13. In the following table, columns A, B, and C refer to the following gaseous equilibria:
    A: $H_2 + Cl_2 \rightleftharpoons 2HCl$ ΔH is negative
    B: $2H_2O \rightleftharpoons 2H_2 + O_2$ ΔH is positive
    C: $H_2 + C_2H_4 \rightleftharpoons C_2H_6$ ΔH is negative
    Read the instructions given in Problem 1.

	A. Effect on		B. Effect on		C. Effect on	
Change	position	K_p	position	K_p	position	K_p
Decrease in temperature						
Increase in total pressure						
A catalyst added						
Decrease in partial pressure of first-named gas						
Helium gas added, keeping the volume constant						

14. Four moles of $COCl_2$ are put into a sealed vessel and heated to 395.0°C at a pressure of 0.200 atm. At equilibrium, analysis shows 30.0% by volume of CO in the mixture. Calculate K_p for the dissociation reaction $COCl_2 \rightleftarrows CO + Cl_2$ at this temperature.

15. An equilibrium mixture results from heating 2.00 moles of NOCl to a temperature of 225.0°C at a pressure of 0.200 atm. Analysis shows the presence of 34.0% NO by volume. Calculate the value of K_p for the dissociation reaction $2NOCl \rightleftarrows 2NO + Cl_2$ at this temperature.

16. Calculate the values of K_c for (a) the reaction of SO_2 with O_2 at 900.0°C and 1000.0°C, and (b) the reaction of CO_2 with H_2 at 700.0°C and 1000.0°C, using the K_p values given in Table 16-1. (c) Calculate the values of K_p for the reaction of CO with H_2O at 700.0°C and 1000.0°C.

17. Calculate the percentage of Cl_2 that dissociates to atoms at a total pressure of 1.00 atm and (a) 1000.0°C and (b) 2000.0°C.

18. Calculate the percentage of H_2O that dissociates at a pressure of 0.500 atm and (a) 1000°C and (b) 1700°C.

19. What percentage of H_2 will be converted to HCl at 1800°C using (a) an equimolar mixture of H_2 and Cl_2 at a total pressure of 0.900 atm, and (b) an $8:1$ mixture of Cl_2 to H_2 at a total pressure of 0.900 atm.

20. Calculate the composition of the equilibrium gas mixture that results when (a) 2.00 moles of NO and 1.00 mole of O_2 are mixed at 210.0°C and a total pressure of 0.800 atm, and (b) 10.0 moles of NO and 1.00 mole of O_2 are mixed at 210.0°C and a total pressure of 5.00 atm. The value of K_p is 3.36×10^3 atm^{-1} for the reaction $2NO_{(g)} + O_{2(g)} \rightleftarrows 2NO_{2(g)}$. (c) Calculate the value of K_c for this reaction.

21. The equilibrium that results from mixing SO_2 and NO_2 is studied at a variety of temperatures, and the accompanying values of K_p are obtained. Use these data, preferably with the method of least squares, to determine the enthalpy of the reaction $SO_2 + NO_2 \rightleftarrows SO_3 + NO$.

t (°C)	477	527	577	627	677
K_p	282	198	145	110	86.0

22. The dissociation of NO_2 is studied at several different temperatures and the accompanying values of K_p corresponding to the reaction $2NO_2 \rightleftarrows 2NO + O_2$ are obtained. Use these data, preferably with the method of least squares, to determine the enthalpy of this reaction.

t (°C)	136	150	184	210	226	239
K_p (atm)	1.82×10^{-6}	5.47×10^{-6}	6.82×10^{-5}	2.98×10^{-4}	7.35×10^{-4}	1.47×10^{-3}

23. Using the values of K_p given in Table 16-1, calculate the enthalpy of reaction for (a) the reaction of SO_2 with O_2, and (b) the reaction of CO_2 with H_2.

24. For the reaction $N_2O_4 \rightleftarrows 2NO_2$, it is known that the enthalpy of reaction is $+14.53$ kcal, and that K_p at 25.0°C is 0.143 atm. Calculate the value of K_p at 0.0°C.

25. Derive a general mathematical equation that relates K_p, equilibrium pressure (P), and fraction (α) of A dissociated in the gaseous equilibrium $A \rightleftarrows B + C$. Assume that B and C come only from the dissociation of A, and that the temperature remains constant.

Electrochemistry I:
Batteries and Free Energy

Many of the simplest chemical reactions involve only an interchange of atoms or ions between reactants, or perhaps only the dissociation of one reactant into two parts. In such reactions, there is no change in the electrical charge of any of the atoms involved. This chapter deals with another type of reaction, in which one or more electrons are transferred between atoms, with the result that some of the atoms involved do have their electrical charges changed. These reactions are known as electron-transfer reactions. You can appreciate their importance when you realize that every battery used in electronic devices and machines, every impulse involved in nerve transmission, every metabolic reaction that produces energy in biological systems, photosynthesis, and combustion processes (to mention but a few examples) requires electron-transfer reactions.

ELECTRON-TRANSFER POTENTIAL

Oxidation and Reduction

In the transfer of electrons between atoms, an electron donor is called a *reducing agent;* an electron acceptor is called an *oxidizing agent*. Whenever a reducing agent donates electrons, we say that it has been oxidized; we say that an oxidizing agent, on accepting electrons, has been reduced. Oxidation and reduction always occur simultaneously; if one atom or ion donates an electron, another atom or ion must accept it.

The process of donating and accepting electrons is reversible. For example, under one set of conditions metallic cadmium may donate electrons and become Cd^{2+} ions, as it does when immersed in HCl,

$$Cd + 2H^+ \rightarrow Cd^{2+} + H_2\uparrow$$

or, under another set of conditions, the Cd^{2+} ion may accept electrons and be reduced to Cd metal, as it is when it comes in contact with metallic Zn:

$$Cd^{2+} + Zn \rightarrow Zn^{2+} + Cd$$

Reducing agents differ in their ability to donate electrons. For example, metallic Zn can donate electrons to Cd^{2+} to produce metallic Cd and Zn^{2+}, but metallic Cd is unable to donate electrons to Zn^{2+} to produce metallic Zn and Cd^{2+}. This illustrates the fact that metallic Zn is a stronger reducing agent than metallic Cd.

A reversible electron-transfer reaction written in the form

$$ne^- + \text{oxidizing agent} \rightleftarrows \text{reducing agent}$$

is called a *half-reaction,* because it cannot occur unless it is coupled with another half-reaction going in the opposite direction. If we use half-reactions in the manner described in the next few paragraphs, we can assign a number to each reducing agent to describe its strength, or ability to donate electrons.

Galvanic Cells

An electric current is a flow of electrons through a conductor. Many electron-transfer reactions can be arranged so that the electrons donated by the reducing agent are forced to flow through a conducting wire to reach the oxidizing agent. Such an arrangement is called a *battery,* or *electrochemical (galvanic) cell:* a simple form is shown in Figure 17-1. The electron-transfer reaction that produces the current is

$$Zn + Cu^{2+} \rightarrow Zn^{2+} + Cu$$

Note that *all* the components of one half-reaction are placed in one beaker, and *all* the components of the second half-reaction are placed in the other beaker. One must *not* put the materials of the lefthand side of the equation in one beaker, and those of the righthand side in the other! If one did, electron transfer could occur on contact of the two reactants, and there would be no flow of current from one electrode to the other. The *negative electrode* always involves the half-reaction with the greatest reducing strength; in this case it is a strip of zinc dipping into a solution of Zn^{2+} (any soluble zinc salt). The *positive*

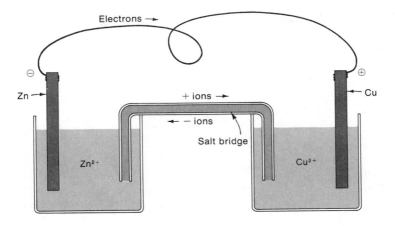

FIGURE 17-1
Simple galvanic cell.

electrode, the one to which the electrons flow in the connecting wire, is a strip of copper dipping into a solution of Cu^{2+} (any soluble copper salt). The two solutions are connected by an inverted U-tube filled with a salt solution (held in place by a gel such as agar-agar), to permit ions to pass from one beaker to the other. Each beaker with its contents is called a half-cell. Traditionally, the negative electrode is shown at the left.

Now, let us trace the reaction that occurs when one atom of Zn donates its electrons according to the half-reaction

$$Zn \rightarrow Zn^{2+} + 2e^-$$

The Zn^{2+} ion goes into solution, while the electrons pass through the wire to the Cu electrode. Here, the electrons combine with a Cu^{2+} ion from solution, according to the half-reaction

$$Cu^{2+} + 2e^- \rightarrow Cu$$

and the atom of Cu is deposited at the metal surface where the electron transfer occurs. The net result of this pair of half-reactions is to put a zinc ion into solution in the lefthand beaker and to remove a copper ion from the righthand beaker. This intolerable situation would quickly lead to the accumulation of an excess of positive ions in one beaker and an excess of negative ions in the other. The reaction would stop immediately if it were not for the "salt bridge," which permits negative ions to migrate into the lefthand beaker and positive ions to migrate into the righthand beaker, in order to maintain electroneutrality in each beaker at all times. The completed circuit thus involves the *uni*direc-

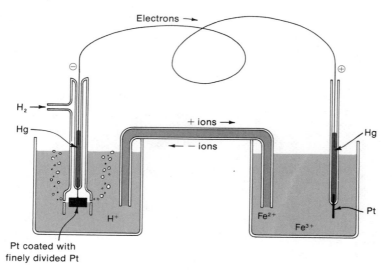

FIGURE 17-2
Galvanic cell that uses the hydrogen electrode
and the Fe^{2+}, Fe^{3+} electrode.

tional flow of electrons through the copper wire, but the *bi*directional flow of ions in solution.

If, in the construction of a galvanic cell, a combination of half-reactions is used that does not involve metals that can be used for electrodes, then a piece of platinum or other inert conducting substance is used to transfer the electrons to and from the solution. Figure 17-2 shows a galvanic cell in which one half-reaction involves a gas (H_2) that must make electrical contact with both the H^+ ions in solution and the platinum electrode. The other half-reaction involves only *ions* (Fe^{2+} and Fe^{3+}) in solution that make contact with a platinum-wire electrode. The liquid mercury provides a simple electrical connection to the wires.

A standard shorthand notation is used to describe the construction of galvanic cells and avoid the necessity of drawing pictures. It is based on the convention that the negative electrode is shown at the left. The notation for the galvanic cell shown in Figure 17-1 is

$$Zn/Zn^{2+}(M_1)//Cu^{2+}(M_2)/Cu$$

The galvanic cell of Figure 17-2 is described as

$$Pt/H_2(P),\ H^+(M_1)//Fe^{2+}(M_2),\ Fe^{3+}(M_3)/Pt$$

where M_1, M_2, and M_3 are the numerical values of the concentrations of the ions, and P is the pressure of the H_2 gas. A single / indicates the interface between the solid electrode and the solution, and a double // indicates the salt bridge or some other junction between the two half-cells.

Standard Electrode Potentials

If, in Figures 17-1 and 17-2, we cut the wire connecting the electrodes and connect the two ends to a voltmeter that draws essentially no current (such as a vacuum-tube voltmeter) or to a potentiometer, the observed voltage (E_{cell}) reading will be an accurate measure of the *difference* in the reducing strengths of the two half-reactions involved. If the voltage were zero, we would know that they were of equal reducing strength. By definition,

$$
\begin{aligned}
E_{cell} &= E_{half-cell\ accepting\ electrons} - E_{half-cell\ donating\ electrons} \\
&= E_{oxidizing\ half-reaction} - E_{reducing\ half-reaction} \\
&= E_{ox} - E_{red}
\end{aligned}
\tag{17-1}
$$

Note that this definition of cell potential involves the basic assumption that the reducing strength of a half-reaction can be represented by a potential (a voltage), the *half-cell* potential. In order to use these cell voltages to measure the strength of each individual reducing agent, we must (a) compare voltages under conditions that eliminate the effect of temperature and concentration, and (b) know the value of E for at least one half-reaction.

To make a fair comparison we choose a set of reference conditions called the *standard state*. For a *pure* substance, this is taken as the physical form stable at 1 atm and 25.0°C; under these conditions, it is said to be at unit activity. For practical purposes we shall also assume that the water in a dilute solution is at unit activity. The *solute* in solution is said to be at unit activity when it behaves as though it were a fictitious *ideal* one-molar solution in which there are no electrical interactions between ions or molecules. The *actual* solution concentration required to produce unit activity varies considerably from solute to solute: for HCl it is 1.20 M; for LiCl it is 1.26 M. We shall not dwell at this time on the problems connected with finding the actual concentrations of solutions associated with unit activity. If we restrict our solution concentrations to 0.1 M or less, our computational errors generally will be less than 5.0% if we use molar concentrations instead of activities; the more dilute the solution, the less the error.

For a variety of reasons, it is impossible to find the absolute value of E for the strength of any reducing agent, even though the difference between any two of them can be measured very accurately by Equation 17-1. Instead, we *arbitrarily* select a voltage of zero for the half-reaction

$$
2e^- + 2H^+ \rightleftarrows H_{2(g)}
$$

when all of the components are at unit activity. This special electrode is called the "standard hydrogen electrode." Letting a superscript ° indicate the standard state of unit activity, our fundamental convention is that, for the standard hydrogen electrode,

$$E^\circ_{H_2-H^+} = 0.00$$

If we now make a whole series of galvanic cells with solutions at unit activity and use the standard hydrogen electrode as one-half of every cell, then the measured cell voltage (E°_{cell}) for each cell will be given, according to Equation 17-1, by either

$$E^\circ_{cell} = E^\circ_{ox} - E^\circ_{H_2-H^+} = E^\circ_{ox}$$

or

$$E^\circ_{cell} = E^\circ_{H_2-H^+} - E^\circ_{red} = -E^\circ_{red}$$

depending on whether the standard hydrogen electrode is a stronger or weaker reducing agent then the other half-reaction. If we measure the voltage of the cell shown in Figure 17-2, with all components at unit activity, we find $E^\circ_{cell} = 0.77$ volts. From our fundamental equation and the assumption for the standard hydrogen electrode,

$$E^\circ_{cell} = E^\circ_{Fe^{2+}-Fe^{3+}} - E^\circ_{H_2-H^+} = E^\circ_{Fe^{2+}-Fe^{3+}}$$
$$E^\circ_{Fe^{2+}-Fe^{3+}} = +0.77 \text{ volts}$$

This direct determination of the E° of every reducing agent by comparison with another to which an arbitrary value of zero has been assigned is the way, in principle, in which all the E° values (or *standard electrode potentials*) in Table 17-1 were found. Reducing agents that are stronger than H_2 have a negative value of E°, and those that are weaker have a positive value. The more negative the E° value, the stronger the reducing agent. Table 17-1 can always be used to calculate the voltage of any galvanic cell whose concentrations are at unit activity.

PROBLEM:
Calculate the voltage of the cell shown in Figure 17-1 if the solutions are at unit activity.

SOLUTION:
We get the appropriate values for $E^\circ_{Zn-Zn^{2+}}$ and $E^\circ_{Cu-Cu^{2+}}$ from Table 17-1 and substitute them into Equation 17-1:

$$E^\circ_{cell} = E^\circ_{ox} - E^\circ_{red} = E^\circ_{Cu-Cu^{2+}} - E^\circ_{Zn-Zn^{2+}}$$
$$= +0.34 - (-0.76) = +1.10 \text{ volts}$$

TABLE 17-1
Standard Electrode Potentials ($E°$) at 25°C.

Half-reaction	$E°$ (volts)	Half-reaction	$E°$ (volts)
$e^- + Li^+ \rightleftarrows Li$	−3.04	$2e^- + Hg_2Cl_2 \rightleftarrows 2Cl^- + 2Hg$	0.27
$e^- + K^+ \rightleftarrows K$	−2.92	$2e^- + Cu^{2+} \rightleftarrows Cu$	0.34
$2e^- + Ca^{2+} \rightleftarrows Ca$	−2.87	$e^- + Cu^+ \rightleftarrows Cu$	0.52
$e^- + Na^+ \rightleftarrows Na$	−2.71	$2e^- + I_2 \rightleftarrows 2I^-$	0.54
$3e^- + La^{3+} \rightleftarrows La$	−2.52	$2e^- + 2H^+ + O_2 \rightleftarrows H_2O_2$	0.68
$3e^- + Ce^{3+} \rightleftarrows Ce$	−2.48	$e^- + Fe^{3+} \rightleftarrows Fe^{2+}$	0.77
$2e^- + Mg^{2+} \rightleftarrows Mg$	−2.36	$2e^- + Hg_2^{2+} \rightleftarrows 2Hg$	0.79
$3e^- + Lu^{3+} \rightleftarrows Lu$	−2.26	$e^- + Ag^+ \rightleftarrows Ag$	0.80
$3e^- + Al^{3+} \rightleftarrows Al$	−1.66	$2e^- + 3H^+ + NO_3^- \rightleftarrows H_2O + HNO_2$	0.94
$2e^- + Zn^{2+} \rightleftarrows Zn$	−0.76	$3e^- + 4H^+ + NO_3^- \rightleftarrows 2H_2O + NO$	0.96
$2e^- + 2H^+ + 2CO_2 \rightleftarrows H_2C_2O_4$	−0.49	$e^- + H^+ + HNO_2 \rightleftarrows H_2O + NO$	1.00
$2e^- + Fe^{2+} \rightleftarrows Fe$	−0.44	$2e^- + Br_{2(l)} \rightleftarrows 2Br^-$	1.07
$2e^- + Cd^{2+} \rightleftarrows Cd$	−0.40	$4e^- + 4H^+ + O_2 \rightleftarrows 2H_2O$	1.23
$2e^- + Co^{2+} \rightleftarrows Co$	−0.28	$6e^- + 14H^+ + Cr_2O_7^{2-} \rightleftarrows 7H_2O + 2Cr^{3+}$	1.33
$2e^- + Ni^{2+} \rightleftarrows Ni$	−0.25	$2e^- + Cl_2 \rightleftarrows 2Cl^-$	1.36
$e^- + AgI \rightleftarrows I^- + Ag$	−0.15	$3e^- + Au^{3+} \rightleftarrows Au$	1.50
$2e^- + Sn^{2+} \rightleftarrows Sn$	−0.14	$5e^- + 8H^+ + MnO_4^- \rightleftarrows 4H_2O + Mn^{2+}$	1.51
$2e^- + Pb^{2+} \rightleftarrows Pb$	−0.13	$e^- + Ce^{4+} \rightleftarrows Ce^{3+}$	1.61
$2e^- + 2H^+ \rightleftarrows H_2$	0.00	$2e^- + 2H^+ + 2HClO \rightleftarrows 2H_2O + Cl_2$	1.63
$e^- + AgBr \rightleftarrows Br^- + Ag$	0.07	$e^- + Au^+ \rightleftarrows Au$	1.69
$2e^- + S_4O_6^{2-} \rightleftarrows 2S_2O_3^{2-}$	0.08	$2e^- + 2H^+ + H_2O_2 \rightleftarrows 2H_2O$	1.78
$2e^- + 2H^+ + S \rightleftarrows H_2S$	0.14	$6e^- + 6H^+ + XeO_3 \rightleftarrows 3H_2O + Xe$	1.80
$2e^- + Sn^{4+} \rightleftarrows Sn^{2+}$	0.15	$2e^- + S_2O_8^{2-} \rightleftarrows 2SO_4^{2-}$	2.01
$e^- + Cu^{2+} \rightleftarrows Cu^+$	0.15	$2e^- + 2H^+ + O_3 \rightleftarrows H_2O + O_2$	2.07
$e^- + AgCl \rightleftarrows Cl^- + Ag$	0.22	$2e^- + F_2 \rightleftarrows 2F^-$	2.87

Variation of Cell Voltage with Concentration

When concentrations of reactants are not at unit activity, the half-cell potential may be calculated by the relation

$$E = E° - \frac{RT}{nF} \ln Q_{\frac{1}{2}} \qquad (17\text{-}2)$$

which is known as the *Nernst equation*. This equation is derived and discussed in more advanced courses. Here n is the number of moles of electrons as shown in the half-reaction equation, and $Q_{\frac{1}{2}}$ is the "ion product" that has the same form as the usual equilibrium-constant expression but that uses the actual ionic concentrations, *not* the equilibrium values. It also differs because half-reactions by themselves are fictitious, and isolated electrons do not exist in aqueous solution—thus, there is no factor in the $Q_{\ddagger}$ expression for $[e^-]^n$, the "electron concentration." If we express the gas constant R in fundamental units (using the value of 1.013×10^6 dynes/cm² for 1 atm that we determined on p 158, along with the volume of 22,400 cm³/mole at 1 atm and 273.2 K), we get $R = 8.314 \times 10^7$ ergs/mole K. Reexpressed in different units, $R = 8.314$ volt coulombs/mole K. The total electrical charge F on one mole of electrons is 96,487 coulombs/mole of electrons. If we use these values of R and F, limit ourselves to 25.0°C for convenient use of Table 17-1, and convert the logarithm to base 10, then the general expression for the *voltage of a half-cell* is

$$E = E° - \frac{\left(8.314 \dfrac{\text{volt coulombs}}{\text{mole K}}\right)(298.2 \text{ K})(2.303)}{\left(n \dfrac{\text{mole of electrons}}{\text{mole}}\right)\left(96,487 \dfrac{\text{coulombs}}{\text{mole of electrons}}\right)} \log Q_{\ddagger}$$

$$= E° - \frac{0.0591}{n} \log Q_{\ddagger} \qquad\qquad (17\text{-}3)$$

PROBLEM:

What is the voltage of a cell constructed as in Figure 17-1, but with a 0.1 M $ZnSO_4$ solution in the lefthand beaker and a $10^{-4} M$ $CuSO_4$ solution in the righthand beaker?

SOLUTION:

The voltage of the cell is

$$E_{\text{cell}} = E_{\text{ox}} - E_{\text{red}} = E_{\text{Cu}-\text{Cu}^{2+}} - E_{\text{Zn}-\text{Zn}^{2+}}$$

Each of the half-cell potentials is given by Equation 17-3, with $E°$ values taken from Table 17-1:

$$E_{\text{Cu}-\text{Cu}^{2+}} = E°_{\text{Cu}-\text{Cu}^{2+}} - \frac{0.0591}{2} \log \frac{[\text{Cu}]}{[\text{Cu}^{2+}]}$$

$$= +0.34 - \frac{0.0591}{2} \log \frac{1}{10^{-4}} = +0.22 \text{ volt}$$

$$E_{\text{Zn}-\text{Zn}^{2+}} = E°_{\text{Zn}-\text{Zn}^{2+}} - \frac{0.0591}{2} \log \frac{[\text{Zn}]}{[\text{Zn}^{2+}]}$$

$$= -0.76 - \frac{0.0591}{2} \log \frac{1}{10^{-1}} = -0.79 \text{ volt}$$

$$E_{\text{cell}} = +0.22 - (-0.79) = +1.01 \text{ volt}$$

In case you wonder why [Cu] and [Zn] appear to have magically disappeared, you must recall that solids such as metallic Cu and Zn have constant invariant concentrations and are said to be at "unit activity"—that is, their values are set equal to 1. The same will always be done for $[H_2O]$ and concentrations of other liquids that appear in half-reaction equations.

We could have combined the equations for E_{ox} and E_{red} into one equation before solving for E_{cell}, as follows.

$$E_{cell} = \left[E_{ox}^{\circ} - \frac{0.0591}{n} \log (Q_{\frac{1}{2}})_{ox} \right] - \left[E_{red}^{\circ} - \frac{0.0591}{n} \log (Q_{\frac{1}{2}})_{red} \right]$$

$$= [E_{ox}^{\circ} - E_{red}^{\circ}] - \frac{0.0591}{n} \log \frac{(Q_{\frac{1}{2}})_{ox}}{(Q_{\frac{1}{2}})_{red}}$$

$$= E_{cell}^{\circ} - \frac{0.0591}{n} \log \frac{(Q_{\frac{1}{2}})_{ox}}{(Q_{\frac{1}{2}})_{red}}$$

Of particular importance here is the fact that the *ratio* $(Q_{\frac{1}{2}})_{ox}/(Q_{\frac{1}{2}})_{red}$ is identical to the ion product for the overall cell reaction, which we shall simply call Q. In other words, we can write our expression for *the voltage of the whole cell* as

$$E_{cell} = E_{cell}^{\circ} - \frac{0.0591}{n} \log Q \qquad (17\text{-}4)$$

or the general form of the Nernst equation for the whole cell as

$$E_{cell} = E_{cell}^{\circ} - \frac{2.30RT}{n} \log Q \qquad (17\text{-}5)$$

For the cell reaction

$$Zn + Cu^{2+} \rightleftarrows Zn^{2+} + Cu$$

the corresponding calculation is

$$E_{cell} = [(+0.34) - (-0.76)] - \frac{0.0591}{2} \log \frac{[Zn^{2+}]}{[Cu^{2+}]}$$

$$= +1.10 - \frac{0.0591}{2} \log \frac{10^{-1}}{10^{-4}}$$

$$= +1.10 - \frac{(0.0591)(3)}{2} = +1.01 \text{ volts}$$

You should note that Equations 17-4 and 17-5 apply only to chemical equations that are *complete* and *balanced* and that, for a given electron-transfer reaction, n

is the number of electrons lost by the reducing agent (or gained by the oxidizing agent) in the balanced chemical equation. Equation 17-3, on the other hand, applies only to a *balanced half*-reaction. When Equation 17-3 is applied to two different half-reactions that are to be combined to make one complete reaction, it is essential that the half-reactions be balanced to give the *same* value of n, the value that will be used in Equation 17-4 and that will refer to the complete reaction.

CAUTION: We must call your attention to a complication that may result from the unthinking application of Equation 17-4 to a reaction such as

$$Hg + Hg^{2+} \rightleftarrows Hg_2^{2+}$$

where one ion (in this case Hg_2^{2+}) is common to *both* half-reactions. Application of Equation 17-4 to this overall reaction would give the *erroneous* result

$$E_{cell} = E_{cell}^\circ - \frac{0.0591}{1} \log \frac{[Hg_2^{2+}]}{[Hg^{2+}]}$$

because it loses track of the fact that there are *two different* Hg_2^{2+} concentrations, one in each half-cell, that can be varied independently. However, the method of the last problem can be applied to the separate half-reactions:

(oxidizing) $2e^- + 2Hg^{2+} \rightleftarrows Hg_2^{2+}$
(reducing) $2e^- + Hg_2^{2+} \leftrightarrows 2Hg$

Using this method, we obtain a *correct* expression for the cell voltage:

$$E_{cell} = E_{cell}^\circ - \frac{0.0591}{2} \log \frac{[Hg_2^{2+}]_{ox} [Hg_2^{2+}]_{red}}{[Hg^{2+}]^2}$$

ELECTRON-TRANSFER EQUILIBRIUM

If you permit a battery to completely "run down" or become "dead," its E_{cell} becomes zero, and the cell reaction reaches an equilibrium condition. Making a battery and letting it go dead is not the only way to let an electron-transfer reaction reach equilibrium; the components could just as well be mixed together in a single beaker. The important thing to realize is that, no matter how you reach equilibrium, the ion product Q at equilibrium is now equal to the equilibrium constant. Thus, *for any electron-transfer reaction at equilibrium* at 25.0°C, we could write

$$E_{cell} = 0 = E_{cell}^\circ - \frac{0.0591}{n} \log K_e$$

or $$E_{cell}^\circ = \frac{0.0591}{n} \log K_e \qquad (17\text{-}6)$$

or, for the general case from Equation 17-5, we could write

$$E^{\circ}_{cell} = \frac{2.30RT}{n} \log K_e \qquad (17\text{-}7)$$

The most important thing about Equations 17-6 and 17-7 is that the equilibrium constant for electron-transfer reactions can be calculated from standard electrode potentials without ever having to make experimental measurements.

PROBLEM:

Calculate the equilibrium constant at 25.0°C for the reaction

$$Fe^{2+} + Ag^+ \rightleftarrows Fe^{3+} + Ag$$

SOLUTION:

From Table 17-1, we get

$$E^{\circ}_{cell} = E^{\circ}_{Ag-Ag^+} - E^{\circ}_{Fe^{2+}-Fe^{3+}}$$
$$= (+0.80) - (+0.77) = +0.030 \text{ volt}$$

Only 1 mole of electrons is transferred in the equation as written, so

$$\log K_e = \frac{nE^{\circ}_{cell}}{0.0591} = \frac{(1)(0.030)}{0.0591} = 0.51$$

$$K_e = 3.2$$

Electrical Energy from Chemical Energy

Up to this point we have emphasized the *voltage* of a galvanic cell. We are also in a position to consider the conversion of chemical energy to electrical energy in a galvanic cell. Electrical energy is calculated as the product of the voltage of a cell and the total electrical charge (in coulombs) that passes from the battery:

$$\text{electrical energy} = \text{volts} \times \text{coulombs}$$

$$= \text{joules (the SI-approved unit of energy)}$$

The total charge that passes from a battery will be determined by the number of moles of electrons (n) that pass through the circuit. By definition,

$$1 \text{ faraday } (F) = 96{,}487 \frac{\text{coulombs}}{\text{mole of electrons}}$$

and

$$1 \text{ cal} = 4.184 \text{ joules} = 4.184 \text{ volt coulombs}$$

Therefore,

electrical energy

$$= (\text{volts})(\text{moles of electrons})\left(\frac{\text{coulombs}}{\text{mole of electrons}}\right)\left(\frac{\text{calories}}{\text{volt coulomb}}\right)$$

$$= (E)(n)(F)\left(\frac{1}{4.184}\right) = nE \times \frac{96,487}{4.184}$$

$$= 23,060 \times nE \text{ cal} \tag{17-8}$$

The cell voltages we have been talking about in this chapter are all maximal voltages, voltages measured with a potentiometer that just matches the voltage of the cell without actually draining any current from the cell, or with a vacuum-tube voltmeter whose resistance is so high that the result is essentially the same. If a battery is used to do work, the voltage will be less; not all the energy that is produced can be employed usefully because of partial dissipation as heat. In fact, *all* the electrical energy could be wasted if you so wished. Nevertheless, our potentiometrically-measured (*maximal*) voltages can be used for the calculation of the *maximal* amount of available electrical energy that can be obtained from a chemical reaction, *regardless of whether it actually is used for work*. For a given chemical reaction, we can equate maximal available electrical energy with maximal available work, and write

$$\text{maximal available work} = \text{maximal available electrical energy}$$
$$= 23,060 \times nE \text{ cal} \tag{17-9}$$

THE CONCEPT OF FREE ENERGY

The preceding statements imply that every substance has an amount of energy, called its *free energy* (G), that could be used for useful work. If, as a result of a chemical reaction that occurs at constant temperature and pressure, the sum of the products ("state 2") possesses an amount of free energy G_2 whereas the sum of the reactants ("state 1") possesses an amount of free energy G_1, then the change in free energy (ΔG) for the reaction will be

$$\Delta G = G_2 - G_1 = G_{\text{products}} - G_{\text{reactants}} \tag{17-10}$$

If the *sign* of ΔG is negative, it means that the products have less free energy than the reactants. The *magnitude* of the change is the maximal amount of work that might have been obtained from the reaction. The maximal obtainable work is associated with a *decrease* in free energy $(-\Delta G)$—that is, the work is obtained at the expense of the chemical system (Figure 17-3). In equation form, we would write

$$\text{maximal available work} = -\Delta G \tag{17-11}$$

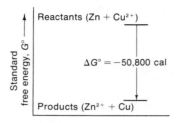

FIGURE 17-3
The decrease in standard free energy (ΔG°)
that occurs when Zn reacts with Cu^{2+}.

and, for maximal available electrical work,

$$\Delta G = -nFE$$
$$= -23,060 \times nE \text{ cal} \tag{17-12}$$

Free Energies of Reaction

If the reactants and products were in their standard states of unit activity, then of course the voltage (E°) is the "standard cell potential," and the change in free energy is the change in "standard free energy" (ΔG°):

$$\Delta G^\circ = -nFE^\circ \tag{17-13}$$

For the reaction

$$Zn + Cu^{2+} \rightleftarrows Zn^{2+} + Cu$$

the change in standard free energy that accompanies the reaction is

$$\Delta G^\circ = -23,060 \times nE^\circ \text{ cal}$$
$$= -(23,060)(2)(1.10) = -50,800 \text{ cal/mole of Zn or Cu}$$

The change in standard free energy is $-50,800$ cal/mole whether metallic Zn is wastefully put into a beaker of $CuSO_4$ or whether the reaction is usefully employed as a battery as in Figure 17-1. Either way, the products end up being less capable of doing work than they were before starting the reaction.

The Relationship Between ΔG° and K_e

There is a very important extension of the concepts associated with ΔG, which up to this point has been closely linked with electron-transfer reactions and the production of electrical energy. Equations 17-7 and 17-13 can be combined to give

$$\Delta G^\circ = -2.30RT \log K_e \qquad\qquad (17\text{-}14)$$

$$= -(2.30)\left(1.987 \ \frac{\text{cal}}{\text{mole K}}\right)(T \text{ K})\log K_e$$

$$= -4.57 \ T \log K_e \ \text{cal/mole} \qquad\qquad (17\text{-}15)$$

where T is the temperature at which the equilibrium constant is known. With this last expression, you can see that the concept of electron-transfer has been eliminated ("n" is no longer involved), and you see that *the change in standard free energy for a reaction is associated with its equilibrium constant*. In other words, for *any* chemical reaction, regardless of whether it involves electron-transfer, it is possible to calculate the change in standard free energy at a given temperature if its equilibrium constant is known for that temperature. In eliminating "n" to obtain Equation 17-14, we must not overlook the fact that the value of ΔG° still depends on the amount of material used and the way the chemical equation is written. For the reaction written as 20 H_2 + 10 $O_2 \rightleftarrows$ 20 H_2O, the value of the equilibrium constant is $(K_e)^{10}$, compared with K_e for the same reaction written as $2H_2 + O_2 \rightleftarrows 2H_2O$; the value of ΔG° for the first is 10 times that for the second, as expected from Equation 17-14.

The values of K_e used in Equations 17-6 or 17-7 and 17-14 or 17-15 must correspond to K_c (see p 255), where concentrations are expressed in moles/liter. Values of K_p with concentrations expressed in atm or torr can be converted to K_c by means of Equation 16-4.

Because no reversible reaction, in principle, ever goes to completion, we need an expression that indicates our expectation of whether the reaction "goes" (with the final equilibrium more to the right), or "doesn't go" (with the final equilibrium more to the left). We summarize all this in one word, "spontaneous." We say that a reaction is *spontaneous* if ΔG° is negative, corresponding to a value of K_e greater than one and an equilibrium position that is more to the right than to the left. All of this is consistent with Equation 17-14. If ΔG° is positive, we say that the reaction is *not* spontaneous. The reaction between Zn and Cu^{2+} is very spontaneous.

It is not always easy to harness non-electron-transfer reactions to do useful work, but the *potential* to do the work is still present. To see how this wider concept ties in with other equilibrium reactions studied in this book, consider the following examples.

1. At 25.0°C for the reaction (p 351)

$$HNO_2 \rightleftarrows H^+ + NO_2^-$$

$$\Delta G^\circ = -(4.57)(298)\log K_e = -(4.57)(298)\log(4.5 \times 10^{-4})$$

$$= +4550 \ \text{cal/mole}$$

2. At 25.0°C for the reaction (p 374)

$$AgCl \rightleftarrows Ag^+ + Cl^-$$

$$\Delta G^\circ = -(4.57)(298)\log K_e = -(4.57)(298)\log(1.6 \times 10^{-10})$$
$$= +13,350 \text{ cal/mole}$$

3. At 355.0°C for the reaction (p 257)

$$H_{2(g)} + I_{2(g)} \rightleftarrows 2HI_{(g)}$$

$$\Delta G^\circ = -(4.57)(628)\log K_e = -(4.57)(628)\log(54.4)$$
$$= -4994 \text{ cal per 2 moles of HI}$$
$$= -2497 \text{ cal/mole}$$

Standard Free Energies of Formation

Equation 17-13 may be used to calculate standard free energies of reaction from E°_{cell} values derived from Table 17-1, and Equation 17-14 may be used if values of K_e are known. There is also a third way by which ΔG° may be calculated. Just as for enthalpy (H, see pp 215–219), it is impossible to know the actual free energy content (G) of any substance; only changes can be measured. Also, as for enthalpy, it is possible to construct a table of standard free energies of formation for each substance, based on the arbitrary assumption that the elements in their standard states (physical forms stable at 25.0°C and 1 atm) have zero free energy of formation. You recall that these "formation" reactions correspond to the formation of compounds directly from the elements. Table 17-2 shows a few selected values of ΔG_f°. Once such a table has been constructed, it is possible to *calculate* the ΔG° or K_e for any reaction for which data are available—even though the reaction might not actually occur because the activation energy is too high or because other reactions occur instead. Just as with enthalpy, we can say

$$\Delta G^\circ = \Sigma(\Delta G_f^\circ)_{products} - \Sigma(\Delta G_f^\circ)_{reactants} \qquad (17\text{-}16)$$

and apply it in the following typical manner.

PROBLEM:
What is the change in standard free energy, and what is the value of K_e, at 25.0°C for the gaseous dissociation of PCl_5 into PCl_3 and Cl_2?

SOLUTION:
Take the needed standard free energies of formation from Table 17-2 and use them with the balanced chemical equation.

$$PCl_{5(g)} \quad\rightleftharpoons\quad PCl_{3(g)} \quad + \quad Cl_{2(g)}$$

$$(1 \text{ mole}) \left(-77.59 \ \frac{kcal}{mole}\right) \quad (1 \text{ mole}) \left(-68.42 \ \frac{kcal}{mole}\right) \quad (1 \text{ mole}) \left(0.00 \ \frac{kcal}{mole}\right)$$

$$\Sigma(\Delta G_f^\circ)_{products} = (1)(-68.42) + (1)(0.00) \qquad = -68.42 \text{ kcal}$$

$$\Sigma(\Delta G_f^\circ)_{reactants} = (1)(-77.59) \qquad\qquad\qquad = -77.59 \text{ kcal}$$

$$(\Delta G^\circ)_{reaction} = (-68.42 \text{ kcal}) - (-77.59 \text{ kcal}) = +9.17 \text{ kcal}$$

$$\log K_e = \frac{-\Delta G^\circ}{2.30RT} = \frac{-9170 \text{ cal}}{(2.30)\left(1.987 \ \dfrac{cal}{mole \ K}\right)(298 \text{ K})} = -6.73$$

$$K_e = 1.85 \times 10^{-7} \text{ moles/liter}$$

Because K_e is relatively small, you would conclude that PCl_5 is only slightly dissociated at room temperature. Also note that, if you wanted the value of K_p (with concentrations expressed in atm), you would have to use Equation 16-4 with $\Delta n = +1$, to give

$$K_p = (RT)^{\Delta n} \cdot K_c = \left(0.08206 \ \frac{liter \ atm}{mole \ K}\right)(298 \text{ K})\left(1.85 \times 10^{-7} \ \frac{moles}{liter}\right)$$

$$K_p = 4.52 \times 10^{-6} \text{ atm}$$

ENTROPIES OF REACTION

We are now in a position to see the relationship between H, G, and S. We say that the energy content or enthalpy (H) of a substance is composed of two parts; one that can do work and is called free energy (G), and another that cannot do work and is the product of temperature (T) and entropy (S). The product of T and S is necessary in order to obtain the units of energy, because S has the units of cal/K. In equation form we write

$$H = G + TS \tag{17-17}$$

In chemical and physical changes, these properties of H, G, and S change for each substance, so that for a *change* (Δ) we would write

$$\Delta H = \Delta G + \Delta(TS) \tag{17-18}$$

and, if the change takes place at constant temperature and pressure,

$$\Delta H = \Delta G + T(\Delta S) \tag{17-19}$$

TABLE 17-2
Standard Free Energies of Formation at 25°C (ΔG_f° in kcal/mole)

Substance	State	ΔG_f°	Substance	State	ΔG_f°	Substance	State	ΔG_f°
Any element	standard	0.00	CO	g	−32.81	N (atom)	g	81.47
Ag^+	aq	18.43	CO_2	g	−94.26	Na (atom)	g	18.67
AgCl	s	−26.22	Cu^{2+}	aq	15.53	Na^+	aq	−62.59
Br (atom)	g	19.69	F (atom)	g	14.20	NH_3	g	−3.98
Br^-	aq	−24.57	F^-	aq	−66.08	NH_3	aq	−6.37
BrCl	g	−0.21	FeO	s	−58.40	NH_4^+	aq	−19.00
C (atom)	g	160.84	Fe_2O_3	s	−177.10	NO	g	20.72
C (diamond)	s	0.68	H (atom)	g	48.58	NO_2	g	12.39
Ca^{2+}	aq	−132.18	H^+	aq	0.00	NO_3^-	aq	−26.41
Cd^{2+}	aq	−18.58	HBr	g	−12.72	N_2O	g	24.76
CH_3OH	l	−39.73	HCl	g	−22.77	O (atom)	g	54.99
CH_3OH	g	−38.69	HI	g	0.31	OH^-	aq	−37.60
C_2H_5OH	l	−41.77	H_2O	l	−56.69	P (atom)	g	66.71
C_2H_5OH	g	−40.30	H_2O	g	−54.64	PCl_3	g	−68.42
C_4H_{10}	g	−3.75	H_2S	g	−7.89	PCl_5	g	−77.59
C_6H_6	l	29.76	I (atom)	g	16.77	S (atom)	g	43.57
C_6H_6	g	30.99	I^-	aq	−12.35	S^{2-}	aq	20.00
Cl (atom)	g	25.19	ICl	g	−1.32	SO_2	g	−71.79
Cl^-	aq	−31.35	K^+	aq	−67.47	SO_4^{2-}	aq	−177.34
ClO_4^-	aq	−2.57	Li^+	aq	−70.22	Zn^{2+}	aq	−35.18

When changes occur with all materials in their standard states, the superscript °
would be used. For example, for $\Delta S°$ we would have

$$\Delta S° = \frac{\Delta H° - \Delta G°}{T} \qquad (17\text{-}20)$$

PROBLEM:
Calculate the change in standard entropy for the dissociation of PCl_5 into PCl_3 and
Cl_2 at 25.0°C.

SOLUTION:
Take the value of $\Delta G° = +9170$ cal as obtained in the preceding problem, and
the value of $\Delta H° = +22,130$ cal as obtained on p. 218 in Chapter 14. Substitute
these values into Equation 17-20 and get

$$\Delta S° = \frac{(22,130 \text{ cal}) - (9170 \text{ cal})}{298.2 \text{ K}} = +43.48 \frac{\text{cal}}{\text{K}} = +43.48 \text{ e.u.}$$

The positive sign of the 43.48 e.u. indicates that the reaction goes with an increase in entropy—that the products represent a state of greater molecular disorder than do the reactants.

Entropies of Transition

When a substance changes from one physical state to another at a temperature where the two states can coexist (such as at the melting point or the normal boiling point), the two phases are at equilibrium and $\Delta G_T^\circ = 0$ for the transition. Under these conditions

$$\Delta S_T^\circ = \frac{\Delta H_T^\circ - \Delta G_T^\circ}{T} = \frac{\Delta H_T^\circ}{T} \tag{17-21}$$

It is for this reason (that $\Delta G_T^\circ = 0$) that we were able to calculate standard entropies of transition by simply dividing the enthalpies of transition by the Kelvin temperature of transition as we did on p 214. Entropies of reaction *cannot* be calculated this way; the more general expression must be used.

Absolute Entropies

In addition to calculating ΔS from values of ΔH and ΔG, there is another independent source of information that demonstrates that the arbitrary values for ΔH_f° and ΔG_f° are self-consistent. This source is based on the reasonable assumption (known as the *third law of thermodynamics*) that, at the absolute zero of temperature, the entropy of perfect crystals of all pure elements and compounds is zero. This assumption is reasonable because it embodies the concepts that these elements and compounds have minimal energy, no "molecular chaos," and perfect order. By measuring the amount of energy required to raise the temperature of a mole of a given substance from absolute zero (that is, by measuring its molar heat capacity at a series of different temperatures) to some temperature such as 25.0°C, it is possible to calculate the *absolute* (or actual) *entropy* at 25.0°C. This calculation includes any entropies of transition $(\Delta H_T/T)$ that occur in going from 0.0 K to 25.0°C. Some selected values of absolute entropies (S°) are shown in Table 17-3. Because it is impossible to know the absolute values of H and G for any substance, it was necessary to make an arbitrary assignment of zero for the enthalpies and free energies of formation of *elements* in their standard states in Tables 14-1 and 17-2. Note that this is *not* true for the entropies of the elements in their standard states in Table 17-3.

These values of S° are measures of the energy that a substance requires at 25.0°C in order to maintain its characteristic variety of internal atomic and molecular motions (vibrations and rotations), and its random movement in

TABLE 17-3
Absolute Entropies at 25°C ($S°$ in cal/mole °C)

Substance	State	$S°$	Substance	State	$S°$	Substance	State	$S°$
Ag^+	aq	17.67	Cu	s	7.96	N_2	g	45.77
AgCl	s	22.97	Cu^{2+}	aq	−23.60	Na	s	12.20
Br (atom)	g	41.81	F (atom)	g	37.92	Na (atom)	g	36.72
Br^-	aq	19.29	F^-	aq	−2.30	Na^+	aq	14.40
Br_2	l	36.40	F_2	g	48.60	NH_3	g	46.01
BrCl	g	57.30	Fe	s	6.49	NH_3	aq	26.30
C (atom)	g	37.76	FeO	s	12.90	NH_4^+	aq	26.97
C (diamond)	s	0.58	Fe_2O_3	s	21.50	NO	g	50.34
C (graphite)	s	1.36	H (atom)	g	27.39	NO_2	g	57.47
Ca	s	9.95	H^+	aq	0.00	NO_3^-	aq	35.00
Ca^{2+}	aq	−13.20	H_2	g	31.21	N_2O	g	52.58
Cd	s	12.30	HBr	g	47.44	O (atom)	g	38.47
Cd^{2+}	aq	−14.60	HCl	g	44.62	O_2	g	49.00
CH_3OH	l	30.30	HI	g	49.31	OH^-	aq	−2.52
CH_3OH	g	56.80	H_2O	l	16.72	P (atom)	g	38.98
C_2H_5OH	l	38.40	H_2O	g	45.11	P_4 (per P)	s	10.60
C_2H_5OH	g	67.40	H_2S	g	49.15	PCl_3	g	74.49
C_4H_{10}	g	74.10	I (atom)	g	43.18	PCl_5	g	84.30
C_6H_6	l	48.50	I^-	aq	26.14	S (atom)	g	40.08
C_6H_6	g	64.34	I_2	s	27.90	S^{2-}	aq	5.30
Cl (atom)	g	39.46	ICl	g	59.12	S_8 (per S)	s	7.62
Cl^-	aq	13.17	K	s	15.20	SO_2	g	59.40
Cl_2	g	53.29	K^+	aq	24.50	SO_4^{2-}	aq	4.10
ClO_4^-	aq	43.50	Li	s	6.70	Zn	s	9.95
CO	g	47.30	Li^+	aq	3.40	Zn^{2+}	aq	−25.45
CO_2	g	51.06	N (atom)	g	36.61			

space if it is a gas. The terms "perfect" and "pure" are required in the basic definition of zero entropy at absolute zero, because any imperfections would represent a degree of disorder, as would the presence of any impurities. A solid solution of Ar and Kr at absolute zero would have a positive value of entropy at absolute zero, because the *separate* crystals of Ar and Kr, representing a greater state of order at absolute zero, have zero entropy.

In using Table 17-3 for calculating changes in entropy accompanying reactions, we can follow our usual custom of saying that the change in entropy equals the difference between "state 2" (products) and "state 1" (reactants):

$$\Delta S°_{\text{reaction}} = \Sigma S°_{\text{products}} - \Sigma S°_{\text{reactants}} \qquad (17\text{-}22)$$

PROBLEM:

Calculate the change in standard entropy for the reaction

$$PCl_{5(g)} \rightleftarrows PCl_{3(g)} + Cl_{2(g)}$$

SOLUTION:

Take the needed information from Table 17-3, and use it with the balanced chemical equation

$$PCl_{5(g)} \qquad\qquad \rightleftarrows \qquad\qquad PCl_{3(g)} \qquad + \qquad Cl_{2(g)}$$

$$(1 \text{ mole}) \left(84.30 \ \frac{cal}{mole \ °C} \right) \quad (1 \text{ mole}) \left(74.49 \ \frac{cal}{mole \ °C} \right) \quad (1 \text{ mole}) \left(53.29 \ \frac{cal}{mole \ °C} \right)$$

$$\Sigma S°_{products} = (1)(74.49) + (1)(53.29) \qquad = 127.78 \ \frac{cal}{°C}$$

$$\Sigma S°_{reactants} = (1)(84.30) \qquad\qquad\qquad = \quad 84.30 \ \frac{cal}{°C}$$

$$\Delta S_{reaction} = \left(127.78 \ \frac{cal}{°C} \right) - \left(84.30 \ \frac{cal}{°C} \right) = \quad 43.48 \ \frac{cal}{°C}$$

PROBLEMS A

1. Make a simple sketch to show how you would arrange the materials used in constructing batteries that utilize the chemical reactions given below. For each battery show also the polarity, the directions of flow of electrons and ions, and the half-reaction that occurs in each cell.
   (a) $Mg + 2Ag^+ \rightleftarrows Mg^{2+} + 2Ag$
   (b) $Cu + Hg_2^{2+} \rightleftarrows Cu^{2+} + 2Hg$
   (c) $Sn^{2+} + S_2O_8^{2-} \rightleftarrows 2SO_4^{2-} + Sn^{4+}$
   (d) $Sn^{2+} + Br_2 \rightleftarrows Sn^{4+} + 2Br^-$
   (e) $Fe + 2Fe^{3+} \rightleftarrows 3Fe^{2+}$

2. Describe each of the cells in Problem 1 in simplified shorthand notation.

3. Calculate the cell voltage for each of the cells constructed in Problem 1, assuming (i) all components at unit activity, or (ii) the negative half-cell *ionic* concentrations to be 10^{-4} molar and the positive half-cell *ionic* concentrations to be 10^{-2} molar.

4. Calculate the equilibrium constant for each of the reactions listed in Problem 1.

5. Calculate the change in standard free energy that accompanies each of the reactions listed in Problem 1.

6. The Edison cell is a rugged storage battery that may receive hard treatment and yet give good service for years. It may even be left uncharged indefinitely and still be recharged. It gives 1.3 volts. Its electrolyte is a 21.0% KOH

solution to which a small amount of LiOH is added. The chemical reaction that occurs on discharge is

$$Fe + 2Ni(OH)_3 \rightleftarrows Fe(OH)_2 + 2Ni(OH)_2$$

Write the half-cell reactions for the electrodes. Which pole must be the negative pole?

7. Equilibrium constants for the following reactions may be found in the tables cited. Calculate $\Delta G°$ for each of the reactions at 25.0°C.
   (a) $HC_2H_3O_2 \rightleftarrows H^+ + C_2H_3O_2^-$ (Table 23-1)
   (b) $NH_3 + H_2O \rightleftarrows NH_4^+ + OH^-$ (Table 23-1)
   (c) $Ag_2CrO_4 \rightleftarrows 2Ag^+ + CrO_4^{2-}$ (Table 24-1)
   (d) $Ag(NH_3)_2^+ \rightleftarrows Ag^+ + 2NH_3$ (Table 25-1)
   (e) $N_2O_{4(g)} \rightleftarrows 2NO_{2(g)}$ (Table 16-1)

8. Calculate $\Delta G°$ and K_e at 25.0°C for each of the reactions listed. Tell which ones are spontaneous.
   (a) $H_2S_{(g)} + 3O_{2(g)} \rightarrow 2H_2O_{(l)} + 2SO_{2(g)}$
   (b) $2CH_3OH_{(l)} + 3O_{2(g)} \rightarrow 2CO_{2(g)} + 4H_2O_{(l)}$
   (c) $2NO_{2(g)} \rightarrow 2NO_{(g)} + O_{2(g)}$
   (d) $I_{2(s)} + 2HBr_{(g)} \rightarrow 2HI_{(g)} + Br_{2(l)}$

9. Calculate $\Delta S°$ for each of the reactions in Problem 8.

10. Calculate $\Delta H°$ for each of the reactions in Problem 8, but do not use data given in Table 14-1.

11. Calculate values of K_e at 25.0°C for the following equilibria, using data from Table 17-1 exclusively.
    (a) $AgBr_{(s)} \rightleftarrows Ag_{(aq)}^+ + Br_{(aq)}^-$
    (b) $2H_{2(g)} + O_{2(g)} \rightleftarrows 2H_2O_{(l)}$
    (c) $2H_2O_{2(l)} \rightleftarrows 2H_2O_{(l)} + O_{2(g)}$

12. Use the value of $\Delta H°$ obtained in Problem 28(c), p. 226, for the reaction $Ba_{(aq)}^{2+} + SO_{4\,(aq)}^{2-} \rightleftarrows BaSO_{4(s)}$, along with information given in Table 24-1; to calculate the value of $\Delta S°$ for this reaction.

PROBLEMS B

13. Make a simple sketch to show how you would arrange the materials used in constructing batteries that utilize the following chemical reactions. For each battery, also show the polarity, the directions of flow of electrons and ions, and the half-reaction that occurs in each cell.
    (a) $2La + 3Cu^{2+} \rightleftarrows 2La^{3+} + 3Cu$
    (b) $Zn + Hg_2^{2+} \rightleftarrows Zn^{2+} + 2Hg$
    (c) $Fe + Cl_2 \rightleftarrows Fe^{2+} + 2Cl^-$
    (d) $Sn + Sn^{4+} \rightleftarrows 2Sn^{2+}$
    (e) $2Hg + Cl_2 \rightleftarrows Hg_2Cl_2$

14. Describe each of the cells in Problem 13 in simplified shorthand notation.

15. Calculate the cell voltage for each of the cells constructed in Problem 13, assuming (i) all components at unit activity, or (ii) the negative half-cell *ionic* concentrations to be 10^{-1} molar and the positive half-cell *ionic* concentrations to be 10^{-5} molar.

16. Calculate the equilibrium constant for each of the reactions listed in Problem 13.

17. Calculate the change in standard free energy that accompanies each of the reactions listed in Problem 13.

18. One-hundredth mole each of Ag_2SO_4 and $FeSO_4$ are mixed in one liter of water at 25.0°C and permitted to reach equilibrium according to the reaction

$$Ag^+ + Fe^{2+} \rightleftarrows Ag + Fe^{3+}$$

The silver that is formed is filtered off, washed, dried, and weighed, and found to weigh 0.0645 g. From these data, calculate $\Delta G°$ for the reaction.

19. Equilibrium constants for the following reactions may be found in the tables cited. Calculate $\Delta G°$ for each of the reactions at 25.0°C.
    (a) $HCN \rightleftarrows H^+ + CN^-$ (Table 23-1)
    (b) $(CH_3)_3N + H_2O \rightleftarrows (CH_3)_3NH^+ + OH^-$ (Table 23-1)
    (c) $Tl_2S \rightleftarrows 2Tl^+ + S^{2-}$ (Table 24-1)
    (d) $Hg(CN)_4^{2-} \rightleftarrows Hg^{2+} + 4CN^-$ (Table 25-1)
    (e) $Br_{2(l)} + Cl_{2(g)} \rightleftarrows 2BrCl_{(g)}$ ($K_p = 42.5$)

20. Calculate $\Delta G°$ and K_e at 25.0°C for each of the following reactions. Tell which ones are spontaneous.
    (a) $Cl_{2(g)} + 2HBr_{(g)} \rightarrow 2HCl_{(g)} + Br_{2(l)}$
    (b) $2C_6H_{6(l)} + 7O_{2(g)} \rightarrow 12CO_{2(g)} + 6H_2O_{(l)}$
    (c) $NH_{4(aq)}^+ + NO_{3(aq)}^- \rightarrow N_2O_{(g)} + 2H_2O_{(l)}$
    (d) $I_{2(s)} + Cl_{2(g)} \rightarrow 2ICl_{(g)}$

21. Calculate $\Delta S°$ for each of the reactions in Problem 20.

22. Calculate $\Delta H°$ for each of the reactions in Problem 20, but do not use data given in Table 14-1.

23. Calculate values of K_e at 25.0°C for the following equilibria, using data from Table 17-1 exclusively.
    (a) $Hg_2Cl_2 \rightleftarrows Hg_2^{2+} + 2Cl^-$
    (b) $3HNO_2 \rightleftarrows H^+ + NO_3^- + 2NO + H_2O$
    (c) $Cl_2 + H_2O \rightleftarrows H^+ + Cl^- + HClO$

24. Use the value of $\Delta H°$ obtained in Problem 9(c), p 224, for the reaction $Pb_{(aq)}^{2+} + 2I_{(aq)}^- \rightleftarrows PbI_{2(s)}$ along with information given in Table 24-1 to calculate the value of $\Delta S°$ for this reaction.

18

Electrochemistry II: Balancing Equations

The basic principles discussed at the beginning of Chapter 17 (in connection with the construction of simple electrochemical cells) are exactly the ones used to write and balance chemical equations for electron-transfer reactions. These principles also enable you to predict whether or not a given electron-transfer reaction will actually take place.

Let us review these principles.

1. Every electron-transfer reaction may be considered to be composed of two half-reactions, with each half-reaction written in the form

$$n e^- + \text{oxidizing agent} \rightleftarrows \text{reducing agent}$$

2. Electron-transfer half-reactions may be listed in order of decreasing tendency for the reducing agents to give up electrons, as in Table 17-1.

3. In any electron-transfer reaction, there must be the same number of electrons gained by the oxidizing agent as are lost by the reducing agent.

4. An electron-transfer reaction will occur spontaneously only if the reducing half-reaction lies above the oxidizing half-reaction in a table such as Table 17-1. This results in a positive E°_{cell}, a negative ΔG°, and $K_e > 1$.

This chapter concerns the application of these four principles to the balancing of equations for electron-transfer reactions.

BALANCING EQUATIONS, WITH HALF-REACTIONS GIVEN

The simplest situation that exists for balancing electron-transfer equations is the one in which a table of standard electrode potentials is at hand, and the two needed half-reactions are included in it. The following problem illustrates this situation.

PROBLEM:
Write a balanced ionic equation for the reaction between MnO_4^- and H_2S in acid solution.

SOLUTION:
The reaction is spontaneous because, in Table 17-1, the reducing half-reaction

$$2e^- + 2H^+ + S \rightleftarrows H_2S$$

lies above the oxidizing half-reaction

$$5e^- + 8H^+ + MnO_4^- \rightleftarrows 4H_2O + Mn^{2+}$$

In order to combine these two half-reactions to give the complete reaction, we must multiply each one by a factor that will yield the same number of electrons lost as gained. The factor 5 is needed for the H_2S half-reaction, and the factor 2 is needed for the MnO_4^- half-reaction, in order to provide a loss of 10 electrons by H_2S and a gain of 10 by MnO_4^-:

$$10e^- + 16\ H^+ + 2MnO_4^- \rightleftarrows 8H_2O + 2Mn^{2+}$$
$$10e^- + 10\ H^+ + 5S \rightleftarrows 5H_2S$$

Now if we subtract the second (reducing) half-reaction from the first (oxidizing) half-reaction, the 10 electrons cancel to give

$$5H_2S + 2MnO_4^- + 16\ H^+ \rightarrow 5S + 2Mn^{2+} + 10H^+ + 8H_2O$$

This equation can be simplified by subtracting 10 H^+ from each side to give

$$5H_2S + 2MnO_4^- + 6H^+ \rightarrow 5S + 2Mn^{2+} + 8H_2O$$

We can make the following general statement. *To balance any electron-transfer equation, you must subtract the reducing half-reaction equation from the oxidizing half-reaction equation after the two equations have first been written to show the same number of electrons.* Simplify the final equation, if needed.

DETERMINATION OF OXIDATION STATE

The general method of balancing electron-transfer equations requires that half-reaction equations be available. Short lists of common half-reactions, similar to Table 17-1, are given in most textbooks, and chemistry handbooks have extensive lists. However, no list can provide all possible half-reactions, and it is not practical to carry lists in your pocket for instant reference. The practical alternative is to learn to make your own half-reaction equations. There is only one prerequisite for this approach: you must know the oxidation states of the oxidized and reduced forms of the substances involved in the electron-transfer reaction. In Chapter 8 you learned the charges on the ions of the most common elements; now we review the method of determining the charge (the oxidation state) of an element when it is combined in a radical.

PROBLEM:
What is the charge of Cr in the $Cr_2O_7^{2-}$ ion?

SOLUTION:
It will be convenient to remember that, whenever oxygen is combined with other elements, it *always* has a charge of -2 unless it is in a peroxide (in which case it is -1). Similarly, it will be useful to know that, whenever hydrogen is combined with other elements, it *always* has a charge of $+1$ unless it is a hydride (in which case it is -1). Peroxides and hydrides are not common.

The total charge (C) on an ion is the sum of the charges of the atoms that compose it. If we let z be the charge of a given element in the ion, and n be the number of atoms of that element in the ion, then

$$C = n_1z_1 + n_2z_2 + \cdots = \Sigma n_iz_i$$

If we apply this equation to the $Cr_2O_7^{2-}$ ion whose charge is -2, we have

$$-2 = n_{Cr}z_{Cr} + n_Oz_O$$
$$-2 = (2)(z_{Cr}) + (7)(-2)$$
$$2z_{Cr} = 14 - 2 = 12$$
$$z_{Cr} = +6 = \text{charge on Cr}$$

WRITING YOUR OWN HALF-REACTIONS

The second simplest situation that exists for balancing electron-transfer equations is the one in which the principal oxidation and reduction products are given, and you know (or are told) that the reaction actually takes place. All you need do is to write your own half-reactions, and then proceed as illustrated in the first problem.

Often the oxidized form of an atom is combined with oxygen, whereas in the reduced state it is combined with less oxygen, or none. Chromium is a typical example; in the Cr^{6+} state it is combined as $Cr_2O_7^{2-}$, but in the Cr^{3+} state it is uncombined (except for hydration). In working out a suitable half-reaction equation, you must decide what to do with the oxygen atoms. The answer is simple: you may use H^+, OH^-, and H_2O on either side of the equations for balancing, so long as you comply with the actual state of acidity of the solutions. If a solution is acidic, you must not use OH^- on either side of an equation for balancing; you must use H^+ and/or H_2O.

PROBLEM:
Write a balanced half-reaction for the oxidation of metallic gold to its highest oxidation state.

SOLUTION:
Like all elements in the uncombined state, metallic gold (Au) has a charge of 0. There is no simple way in which you can *reason* out the fact that gold's highest oxidation state is $+3$; presumably you *learned* this in Chapter 8. Now, knowing the two oxidation states involved, you can write your half-reaction as

$$3e^- + Au^{3+} \rightleftarrows Au$$

As a check we note that the sum of the electrical charges on each side is zero.

PROBLEM:
Write a balanced half-reaction equation for the oxidation of Mo^{3+} to MoO_2^+ in acid solution.

SOLUTION:
You are not expected to know about the chemistry of Mo but, once you are given the reactant and product, there is no difficulty. For the MoO_2^+ ion,

$$C = +1 = (1)(z_{Mo}) + (2)(z_0)$$
$$= z_{Mo} + (2)(-2)$$
$$z_{Mo} = +5$$

Because the oxidized form contains oxygen and the reduced form does not, and because the solution is acidic, the most direct approach here is to add sufficient H^+ to combine with the oxygen atoms to form water; in this case $4H^+$ would combine with $2\ O^{2-}$ to make $2\ H_2O$. Note that the H^+ and O^{2-} do not involve any changes in charge. We write

$$2e^- + 4H^+ + MoO_2^+ \rightleftarrows 2H_2O + Mo^{3+}$$

As a check, we note that the sum of the electrical charges on each side is $+3$.

BALANCING EQUATIONS WITHOUT USING HALF-REACTIONS

A simple alternative to writing half-reactions and taking the difference between them is the following stepwise procedure.

1. Write down only the oxidizing and reducing agents on the left side of the equation, and their reduced and oxidized forms on the right. If the oxidized and reduced forms of a given agent differ in the number of atoms of the element responsible for electron exchange, make these numbers equal by using a preliminary integer as a coefficient for the form with the smaller number of atoms.

2. Determine, for the elements responsible for electron exchange, the number of e^- gained by the oxidizing agent and the number of e^- lost by the reducing agent, as they go to their respective products.

3. Multiply each of these numbers by integers that will give the same (and smallest possible) number of e^- lost as gained. Use these integers as coefficients for the oxidizing agent (and its corresponding product) and the reducing agent (and its corresponding product) as written down in step 1. These integers must be multiplied by any preliminary coefficients used in step 1.

4. Determine the sum of the ionic charges on each side of the equation. If they are not equal, calculate the number of H^+ ions or OH^- ions that must be added to one side or the other in order to make the sums equal. If the reaction conditions are acidic, you must use H^+; if basic, use OH^-. You cannot have H^+ on one side of the equation and OH^- on the other; both sides are in the same container with the same acidity.

5. If necessary, balance the H's *or* the O's by adding the proper number of H_2O molecules to whichever side of the equation needs them. The equation will now be balanced.

PROBLEM:
Balance the equation for the reaction between $Cr_2O_7^{2-}$ (from $Na_2Cr_2O_7$) and Br^- (from NaBr) in acid solution.

SOLUTION:
We follow the steps just outlined.

(1) Write down the oxidizing and reducing agents, and their corresponding reduced and oxidized forms as products:

$$Cr_2O_7^{2-} + 2Br^- \rightarrow 2Cr^{3+} + Br_2$$

Preliminary coefficients of 2 are used in front of Br^- and Cr^{3+} because, no matter what the final coefficients, there will always be $2Br^-$ for every Br_2 and $2Cr^{3+}$ for every $Cr_2O_7^{2-}$. Cr and Br are the elements involved in electron transfer.

(2) Determine the loss and gain of e^- as Cr goes from $+6$ to $+3$, and Br goes from -1 to 0.

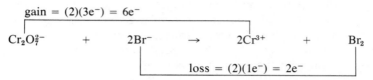

$$gain = (2)(3e^-) = 6e^-$$

$$Cr_2O_7^{2-} \quad + \quad 2Br^- \quad \longrightarrow \quad 2Cr^{3+} \quad + \quad Br_2$$

$$loss = (2)(1e^-) = 2e^-$$

(3) The least common multiple for 6 and 2 is 6. Provide an equal loss and gain of e^- by multiplying $Cr_2O_7^{2-}$ and $2Cr^{3+}$ by 1, and $2Br^-$ and Br_2 by 3, to give

$$gain = (2)(3e^-) = 6e^-$$

$$Cr_2O_7^{2-} \quad + \quad 6Br^- \quad \longrightarrow \quad 2Cr^{3+} \quad + \quad 3Br_2$$

$$loss = (3)(2)(1e^-) = 6e^-$$

(4) The sum of the ionic charges on the left side is -8; the sum on the right is $+6$. Addition of 14 H^+ to the left or 14 OH^- to the right would balance the ionic charges, but we must use 14 H^+ because the solution is acidic. This gives

$$Cr_2O_7^{2-} + 6Br^- + 14\ H^+ \rightarrow 2Cr^{3+} + 3Br_2$$

(5) There are 7 O atoms on the left, and none on the right, so we must add $7H_2O$ to the right to give the balanced equation

$$Cr_2O_7^{2-} + 6Br^- + 14\ H^+ \rightarrow 2Cr^{3+} + 3Br_2 + 7H_2O$$

PROBLEM:
Balance the equation for the reaction between MnO_4^- (from $KMnO_4$) and AsO_3^{3-} (from Na_3AsO_3) in basic solution.

SOLUTION:
We follow the stepwise procedure as before.

(1) Write down the oxidizing and reducing agents, and their corresponding reduced and oxidized forms:

$$MnO_4^- + AsO_3^{3-} \rightarrow MnO_2 + AsO_4^{3-}$$

No preliminary coefficients are needed.

(2) Determine the loss and gain of e^- as Mn goes from $+7$ to $+4$, and As goes from $+3$ to $+5$.

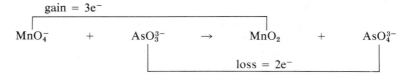

$$2MnO_4^- + 3AsO_3^{3-} \rightarrow 2MnO_2 + 3AsO_4^{3-}$$

(above with annotations: gain = (2)(3e⁻) = 6e⁻ and loss = (3)(2e⁻) = 6e⁻)

(3) The least common multiple for 2 and 3 is 6. Provide an equal loss and gain of e⁻ by multiplying MnO_4^- and MnO_2 by 2, and AsO_3^{3-} and AsO_4^{3-} by 3, to give

(4) The sum of the ionic charges on the left is −11; the sum on the right is −9. Balance the ionic charges by adding 2 OH⁻ to the right (the solution is basic) to give

$$2MnO_4^- + 3AsO_3^{3-} \rightarrow 2MnO_2 + 3AsO_4^{3-} + 2\,OH^-$$

(5) There are 2 Hs on the right and none on the left, so we must add 1 H_2O to the left to give the balanced equation

$$2MnO_4^- + 3AsO_3^{3-} + H_2O \rightarrow 2MnO_2 + 3AsO_4^{3-} + 2\,OH^-$$

BALANCING EQUATIONS CONTAINING ORGANIC COMPOUNDS

Balancing equations involving the oxidation and reduction of organic compounds appears to be much more difficult than balancing those for inorganic compounds, because you seem to have no idea at all about the valence charge on each of the atoms in the organic molecule. As a matter of fact it is not correct to think of the atoms as having charges, because they usually are involved in covalent bonds, not ionic bonds. However, *for purposes of balancing equations,* it makes no difference what charges are assigned to the atoms, so long as the net charge on the whole molecule remains unchanged. Here are some simplifying steps to take; the result is to assign all electron-transfer properties exclusively to the carbon atom.

1. Assign charges of +1 to H and −2 to O in all organic compounds.

2. Assign to all other atoms (except C) in the organic molecule the same charges that they have in the products of the reaction.

3. Assign to C whatever charge is needed (along with the assignments in 1 and 2) to give the overall charge on the organic molecule (usually zero). This charge frequently will be a fraction.

4. Proceed to balance the equation either (a) by making half-reactions and taking the difference, or (b) by using the stepwise procedure of the preceding problems.

PROBLEM:

Write a balanced ionic equation for the reaction of $Cr_2O_7^{2-}$ with C_2H_3OCl in acid solution, given that CO_2 and Cl_2 are the two main products from the organic compound.

SOLUTION:

We use the stepwise procedure.

(1) Write down the oxidizing and reducing agents, and their corresponding reduced and oxidized forms as products:

$$Cr_2O_7^{2-} + 2C_2H_3OCl \rightarrow 2Cr^{3+} + 4CO_2 + Cl_2$$

Preliminary coefficients are used in front of Cr^{3+} and C_2H_3OCl because, no matter what the final coefficients, there will always be $2Cr^{3+}$ for each $Cr_2O_7^{2-}$, and $2C_2H_3OCl$ will be needed for every Cl_2. Also, there will always be $4CO_2$ produced for every $2C_2H_3OCl$.

(2) Determine the loss and gain of e^- for C and Cr. As before, Cr goes from $+6$ to $+3$. For C, we must first find the charge that we will assign it. In C_2H_3OCl, H is assigned $+1$ and O is assigned -2; Cl is assigned 0, the same valence as Cl in Cl_2, the product. All together, H_3OCl has a charge of $+1$, which means that each C has a charge of $-\frac{1}{2}$ to give a total charge of zero. Each C goes from $-\frac{1}{2}$ in C_2H_3OCl to $+4$ in CO_2, a loss of $4\frac{1}{2}e^-$.

$$gain = (2)(3e^-) = 6e^-$$

$$Cr_2O_7^{2-} \quad + \quad 2C_2H_3OCl \quad \rightarrow \quad 2Cr^{3+} \quad + \quad 4CO_2 \quad + \quad Cl_2$$

$$loss = (4)(4\tfrac{1}{2}e^-) = 18e^-$$

(3) The least common multiple for 6 and 18 is 18. Provide an equal loss and gain of e^- by multiplying $Cr_2O_7^{2-}$ and $2Cr^{3+}$ by 3, and $2C_2H_3OCl$, $4CO_2$, and Cl_2 by 1, to give

$$3Cr_2O_7^{2-} + 2C_2H_3OCl \rightarrow 6Cr^{3+} + 4CO_2 + Cl_2$$

(4) The sum of the ionic charges on the left is -6; the sum on the right is $+18$. In an acid solution, we can balance the ionic charge by adding 24 H^+ to the left to give

$$3Cr_2O_7^{2-} + 2C_2H_3OCl + 24H^+ \rightarrow 6Cr^{3+} + 4CO_2 + Cl_2$$

(5) There are 30 H atoms on the left, and none on the right, so we must add 15 H_2O to the right to give the balanced equation

$$3Cr_2O_7^{2-} + 2C_2H_3OCl + 24H^+ \rightarrow 6Cr^{3+} + 4CO_2 + Cl_2 + 15H_2O$$

BALANCING MOLECULAR EQUATIONS

After we have developed an ionic equation for an electron-transfer reaction, we frequently need to show the molecules involved in the solutions—that is, the substances that are initially put into the solution, and those that are obtained from it after the reaction has occurred. We must have such molecular equations if stoichiometric calculations are to be made.

We can use the ionic equation to write the molecular equation of the same reaction, keeping in mind that every ion of the original substances was obtained from some acid, base, or salt, and that every ion in the products must be shown as the salt, base, or acid that would be obtained if the solution were evaporated to dryness.

PROBLEM:
Metallic copper is oxidized by dilute nitric acid. Write ionic and molecular equations for the reaction.

SOLUTION:
The two half-reactions (from Table 17-1) are

$$3e^- + 4H^+ + NO_3^- \rightleftarrows NO + 2H_2O$$
$$2e^- + Cu^{2+} \rightleftarrows Cu$$

Balancing electrons and subtracting the second half-reaction from the first, we obtain

$$6e^- + 8H^+ + 2NO_3^- \rightleftarrows 2NO + 4H_2O$$
$$6e^- + 3Cu^{2+} \rightleftarrows 3Cu$$
$$3Cu + 2NO_3^- + 8H^+ \rightleftarrows 3Cu^{2+} + 2NO + 4H_2O$$

To write the molecular equation, we use 8 HNO_3 to furnish the required 8 H^+. In the products Cu^{2+} appears as $Cu(NO_3)_2$. We have

$$3Cu + 8HNO_3 \rightleftarrows 3Cu(NO_3)_2 + 2NO + 4H_2O$$

SOME GENERAL GUIDELINES

It should be evident that with a little practice you can very quickly, efficiently, and infallibly balance the most complicated electron-transfer equations. It is a straightforward mechanical process. This statement is true *IF* you know what the products of oxidation and reduction are. The most difficult situation that exists for balancing equations is the one characterized by the following request: "Write a balanced ionic equation for the reaction, if any, that occurs when you mix A and B." You know the potential reactants because they are given, but that is all.

If you are faced with such a request, there is one question that must be answered before any other: "Will it be an electron-transfer reaction?" The answer depends on two basic requirements for an electron-transfer reaction.

1. The reactants must include *both* an oxidizing and a reducing agent.

2. In terms of Table 17-1, the reducing agent (on the righthand side of the half-reaction) must lie *above* the oxidizing agent (on the lefthand side of the half-reaction).

If both of these requirements are met, the reaction will be electron-transfer, and the equation will be balanced by the principles outlined in this chapter. If only one (or neither) of these requirements is met, the reaction (if any) will be limited to such reactions as double decomposition, association, or dissociation as described in Chapter 27.

Lacking a table of standard electrode potentials, or one that is adequate, what guidelines can be used to identify oxidizing and reducing agents, and to estimate their relative strengths? Here are a few.

1. If uncombined elements are among the reactants, an electron-transfer reaction is the only possibility. Metals can react *only* as reducing agents; nonmetallic elements *only* as oxidizing agents.

2. If a reactant is an -ous acid, an -ite ion, or an -ous metal ion, there is implied the existence of a higher valence form and thus the possibility that the reactant is a reducing agent.

3. If a reactant is an -ic metal ion, there is implied the existence of a lower valence form and thus the possibility that the reactant is an oxidizing agent.

4. If the reactants include substances that are well-known strong oxidizing and reducing agents, then the reaction will be an electron-transfer reaction.

TABLE 18-1
Electron Transfer Table by Chemical Groups

Group number	Chemical group	$E°$ (volts)
1	Alkali metals ($e^- + M^+ \rightleftarrows M$) Li, Cs, Rb, K, Na	-3.04 to -2.71
2	Alkaline earth metals ($2e^- + M^{2+} \rightleftarrows M$) Ba, Sr, Ca, Mg	-2.90 to -2.36
3	Active metals ($ne^- + M^{n+} \rightleftarrows M$) Al, Zn, Cr(3+), Fe(2+), Cd	-0.76 to -0.40 (Al is -1.66)
4	Medium active metals ($ne^- + M^{n+} \rightleftarrows M$) Co(2+), Ni(2+), Sn(2+), Pb(2+)	-0.40 to -0.13
5	Midrange: $2e^- + 2H^+ \rightleftarrows H_2$ $2e^- + 2H^+ + S \rightleftarrows H_2S$	0.00 0.14
6	Second stage oxidation ($ne^- + -ic \rightleftarrows -ous$) Sn^{2+}, Fe^{2+}, Hg_2^{2+} H_2SO_3, H_3AsO_3, HNO_2	0.15, 0.77, 0.91 0.17, 0.56, 0.94
7	Jewelry metals ($ne^- + M^{n+} \rightleftarrows M$) Cu(2+), Ag(1+), Hg(2+), Pt(2+), Au(3+)	0.34, 0.80, 0.85, 1.20, 1.50
8	Halogens ($2e^- + X_2 \rightleftarrows 2X^-$) I^-, Br^-, Cl^-, F^-	0.54, 1.07, 1.36, 2.87
9	Oxidizing negative ions and others	

$$ne^- + H^+ + \begin{bmatrix} NO_3^- \rightleftarrows NO \\ IO_3^- \rightleftarrows I^- \\ O_2 \rightleftarrows H_2O \\ MnO_2 \rightleftarrows Mn^{2+} \\ Cr_2O_7^{2-} \rightleftarrows Cr^{3+} \\ BrO_3^- \rightleftarrows Br^- \\ ClO_3^- \rightleftarrows Cl^- \\ PbO_2 \rightleftarrows Pb^{2+} \\ MnO_4^- \rightleftarrows Mn^{2+} \\ H_2O_2 \rightleftarrows H_2O \end{bmatrix} + H_2O$$

	$E°$
$NO_3^- \rightleftarrows NO$	0.96
$IO_3^- \rightleftarrows I^-$	1.20
$O_2 \rightleftarrows H_2O$	1.23
$MnO_2 \rightleftarrows Mn^{2+}$	1.28
$Cr_2O_7^{2-} \rightleftarrows Cr^{3+}$	1.33
$BrO_3^- \rightleftarrows Br^-$	1.44
$ClO_3^- \rightleftarrows Cl^-$	1.45
$PbO_2 \rightleftarrows Pb^{2+}$	1.45
$MnO_4^- \rightleftarrows Mn^{2+}$	1.51
$H_2O_2 \rightleftarrows H_2O$	1.78

NOTE: In each group, the elements are listed from left to right in decreasing strength as reducing agents. Half-cell potentials correspond to unit activity and 25°C. The number in parentheses after the symbol of some elements denotes the electrical charge of the oxidized form.

Most students have some *dis*organized knowledge that can be usefully organized for assistance in these guidelines. Table 18-1 helps in this organization and helps to determine what the expected products of reaction will be.

If you study this table carefully, you will note that the first five groups correspond to reactivities and general information with which you are familiar; the association of approximate $E°$ values with each group is very helpful. Group 7 contains all the common metals that do not displace H^+ (as H_2 gas) from acids. Again, the range and degree of reactivity is related to common experience; from a chemist's standpoint it is very helpful to associate an approximate value of the electrode potential with each.

Likewise, group 8 contains elements often associated with each other because of position in the periodic table and similarity in chemical properties. The
emphasis in this group is on the oxidized form of the element (the halogen X_2).
The same reduced forms of the halogen occur again in group 9, but associated
in a different way. There is no problem with F_2 because it is the strongest
oxidizing agent of all chemicals; it lies at the bottom of every list. You need
concern yourself only with the approximate $E°$ values of the first three; the
order is the same as that in the periodic table.

Groups 6 and 9 lie farthest from the previous experience of students and
require more study than the others. Most chemists think of the oxidizing agents
of group 9 as being *very strong* oxidizing agents and would classify a large
number of them as having $E°$ values of about 1.35 to 1.50. It pays to know that
the $E°$ of HNO_3 (giving NO) is approximately 1.0, because it is so common.
Note that almost none of the half-reactions in the first eight groups overlap with
group 9. Another help with group 9 is to note that all the halogen-containing
oxidizing agents are reduced to the halide forms. You should probably think of
the first seven groups in terms of the relative strengths of the *reducing* agents
(those chemicals on the righthand side of the half-reaction), and think of groups
8 and 9 in terms of the relative strengths of the *oxidizing* agents (those chemicals
on the lefthand side of the half-reaction).

Two major oversimplifications are involved in Table 18-1: (a) only metals are
shown lying above H_2, and (b) no reactions in basic solution are shown. You
will probably spend far less time memorizing this table, or something similar to
it, than you would spend trying to get comparable information in some other
intuitive, haphazard manner, and your work in memorizing it will be rewarded
by superior predictions. This table is not practical for making predictions about
reactions composed of half-reactions whose $E°$ values lie very close together.

EFFECT OF ACID CONCENTRATION

The $E°$ values of all the half-reactions involving H^+ depend on having H^+
concentration of unit activity; for all other half-reactions, the $E°$ values are
independent of the H^+ concentration. For those that do involve H^+, the actual
half-reaction potentials (and therefore the strengths as oxidizing or reducing
agents) can be greatly affected by the H^+ concentration. By means of the Nernst
equation (p 275) we can calculate that, in pure water, where $[H^+] = 10^{-7}$ M,

$$E_{H_2-H^+} = -0.41 \text{ volts}$$

Immediately we see that none of the medium active metals (group 4) are
able to react with water to displace H_2, and most of those in the active metal
group (group 3) would react with little energy change.

PROBLEM:
Calculate the electrode potential for the Cr^{3+}–$Cr_2O_7^{2-}$ half-reaction if the $Cr_2O_7^{2-}$ and Cr^{3+} ions are kept at unit activity, and the H^+ concentration is (a) raised to an activity of 5 M or (b) lowered to 10^{-7} M as it is in distilled water.

SOLUTION:
Our fundamental equation for the electrode potential (p 275) gives

$$E_{Cr^{3+}-Cr_2O_7^{2-}} = E^\circ_{Cr^{3+}-Cr_2O_7^{2-}} - \frac{0.0591}{6} \log \frac{[Cr^{3+}]^2}{[Cr_2O_7^{2-}][H^+]^{14}}$$

With unit activity for Cr^{3+} and $Cr_2O_7^{2-}$, and an E° value of 1.36 volts from Table 17-1, we get

$$E_{Cr^{3+}-Cr_2O_7^{2-}} = 1.36 + \frac{(0.0591)(14)}{6} \log[H^+]$$

(a) If $[H^+] = 5$ M, then $E_{Cr^{3+}-Cr_2O_7^{2-}} = +1.46$ volts. Only a modest increase in oxidizing power results from a fivefold increase in the activity of the H^+.

(b) If $[H^+] = 10^{-7}$ M, then $E_{Cr^{3+}-Cr_2O_7^{2-}} = +0.39$ volts. This result shows that one of the potentially strongest oxidizing agents, $Cr_2O_7^{2-}$, is *greatly reduced* in potency when it is merely dissolved in water (with the H^+ activity decreased 10 million-fold); it is about as strong an oxidizing agent as Cu^{2+} at unit activity.

PROBLEM:
Calculate the electrode potential for the NO–NO_3^- half-reaction for an aqueous solution of KNO_3 in which the NO_3^- is at unit activity, the NO pressure is 1 atm, and $[H^+] = 10^{-7}$ M (distilled water).

SOLUTION:

$$E_{NO-NO_3^-} = E^\circ_{NO-NO_3^-} - \frac{0.0591}{3} \log \frac{[NO]}{[H^+]^4[NO_3^-]}$$

$$= +0.96 + \frac{(0.0591)(4)}{3} \log(10^{-7})$$

$$= +0.41 \text{ volts}$$

This result shows that, *in a neutral aqueous solution*, the NO_3^- ion is a relatively weak oxidizing agent. It is a common error to attempt to make an electron-transfer reaction using the NO_3^- ion in a neutral aqueous solution; you should not confuse the weak oxidizing power of the NO_3^- ion under these conditions with its strong oxidizing power in strong acid solution. Nitrate salts are commonly used in the laboratory because they are so soluble and readily available, and because they are not appreciably oxidizing in neutral aqueous solution.

We have already seen that strongly acid solutions will make the oxygen-containing negative ions even stronger oxidizing agents; the nitrate ion in an acid solution is no exception. When concentrated HNO_3 is used, the half-cell potential is comparable to that of MnO_4^-, but the gaseous product is no longer NO; it is NO_2. For concentrated HNO_3, the half-reaction that must be used is

$$e^- + 2H^+ + NO_3^- \rightleftarrows H_2O + NO_2$$

It is a common error to think that a metal, if it reacts with an acid, will produce H_2 gas. This is not true if the acid is HNO_3; the gaseous products will be NO or NO_2, depending on the concentration. Under certain circumstances (usually involving dilute solution and a strong reducing agent such as Zn), the reduction products of HNO_3 may actually be N_2 or NH_3.

Likewise, it is a common error to think that a sulfide, if it reacts with an acid, will always produce H_2S. This is not true if the acid is HNO_3; the gaseous products will be NO or NO_2 (depending on the concentration), and sulfur will be formed. The NO_3^- half-reaction is below the H_2S (or S^{2-}) half-reaction. Concentrated HNO_3 is so powerful an oxidizing agent that almost all the really difficult soluble sulfides can be dissolved through oxidation, even though the same sulfides remain untouched by those strong acids that would lead to the formation of H_2S.

It is worth noting that a large number of reducing agents (most of those in the first eight groups of the abbreviated electron-transfer table!) are unstable in solution if not protected from the air. Even in neutral water solution, the effect of air oxidation may be very marked, because O_2 is a strong oxidizing agent. In some cases the effects may be complicated because of low O_2 pressure or unusual hydration effects.

By now it may have become a matter of some concern to you that aqueous solutions of such strong oxidizing agents as $KMnO_4$ and $Na_2Cr_2O_7$ are stable for indefinite periods of time. In fact, it would appear that no oxidizing agent that lies below the half-reaction

$$4e^- + 4H^+ + O_2 \rightleftarrows 2H_2O \qquad\qquad E^\circ = 1.23 \text{ volts}$$

could exist in water without decomposing the water to liberate O_2 gas. Because water is *not* decomposed by these substances, it is evident that, although the equilibrium position of such reactions does lie far in the direction favoring the evolution of O_2, the activation energy must be extremely high and the *rate* must be vanishingly small. The detailed reasons why this should be so are still unknown, despite enormous research effort.

PROBLEMS A

1. Find the electrical charge on each atom in the following molecules and ions.
   (a) CO_2 (c) BaO_2 (e) $Ca_2P_2O_7$ (g) $C_2O_4^{2-}$ (i) $B_4O_7^{2-}$
   (b) $AgNO_3$ (d) $Fe_2(SO_4)_3$ (f) LiH (h) $PtCl_6^{2-}$ (j) UO_2^{2+}

2. Write balanced ionic half-reaction equations for the oxidation of each of the following reducing agents in acid solution.
   (a) HNO_2 (c) Al (e) Hg_2^{2+} (g) I^-
   (b) H_3AsO_3 (d) Ni (f) H_2O_2

3. Write balanced ionic half-reaction equations for the reduction of each of the following oxidizing agents in acid solution.
   (a) PbO_2 (c) Co^{3+} (e) BrO^- (g) F_2
   (b) NO_3^- (d) ClO_4^- (f) Ag^+ (h) Sn^{2+}

4. Write balanced ionic equations for the following reactions.
   (a) $Zn + Cu^{2+} \rightarrow$ (i) $PbO_2 + Sn^{2+} + H^+ \rightarrow$
   (b) $Zn + H^+ \rightarrow$ (j) $MnO_4^- + H_2C_2O_4 + H^+ \rightarrow$
   (c) $Cr_2O_7^{2-} + I^- + H^+ \rightarrow$ (k) $H_2O_2 + HNO_2 \rightarrow$
   (d) $MnO_4^- + Cl^- + H^+ \rightarrow$ (l) $Fe + Cu^{2+} \rightarrow$
   (e) $ClO_3^- + Br^- + H^+ \rightarrow$ (m) $Fe^{3+} + I^- \rightarrow$
   (f) $MnO_4^- + H_2O_2 + H^+ \rightarrow$ (n) $ClO_4^- + H_3AsO_3 \rightarrow$
   (g) $MnO_2 + H^+ + Cl^- \rightarrow$ (o) $H_2S + ClO^- \rightarrow$
   (h) $Ag + H^+ + NO_3^-$ (dilute) $\rightarrow$

5. Write balanced molecular equations for the reactions in Problem 4. Use potassium salts of the negative ions and sulfate salts of the positive ions for the reactants given.

6. Complete and balance the following ionic equations for reactions that occur in aqueous solution. The nature of the solution (acidic or basic) is indicated for each reaction.
   (a) $CH_2O + Ag_2O \rightarrow Ag + HCO_2^-$ (basic)
   (b) $C_2H_2 + MnO_4^- \rightarrow CO_2$ (acidic)
   (c) $C_2H_3OCl + Cr_2O_7^{2-} \rightarrow CO_2 + Cl_2$ (acidic)
   (d) $Ag^+ + AsH_3 \rightarrow Ag + H_3AsO_3$ (acidic)
   (e) $CN^- + Fe(CN)_6^{3-} \rightarrow CNO^- + Fe(CN)_6^{4-}$ (basic)
   (f) $C_2H_4O + NO_3^- \rightarrow NO + C_2H_4O_2$ (acidic)

7. Write balanced equations for the reactions that occur in each of the following situations.
   (a) Metallic zinc and dilute nitric acid produce ammonium nitrate as one of the products.
   (b) Sodium thiosulfate, $Na_2S_2O_3$, is used to titrate iodine in quantitative analysis; one of the products is sodium tetrathionate, $Na_2S_4O_6$.
   (c) A sample of potassium iodide contains some potassium iodate as impurity. When sulfuric acid is added to a solution of this sample, iodine is

produced, as shown by a blue color that appears when a little starch solution is added. Give the equation for the formation of the iodine.

(d) When copper is heated in concentrated sulfuric acid, an odor of sulfur dioxide is noted.

(e) A classical operation in quantitative analysis is the use of a Jones reductor, a column of granulated zinc. A solution of ferric salts is passed through this column prior to titration with potassium permanganate. Give the equation for the reaction in the column.

(f) Pure hydriodic acid cannot be prepared by adding concentrated sulfuric acid to sodium iodide and distilling off the hydriodic acid, because of side reactions. One side reaction yields hydrogen sulfide, as noted by the odor. Give the equation for this side reaction.

(g) A solution of sodium hypochlorite is heated. One of the products is sodium chlorate.

8. Balance the equations given in Problem 3, p 423.

PROBLEMS B

9. Find the oxidation number of each atom in the following molecules and ions.

(a) SO_3 (c) $Zn(IO_3)_2$ (e) $(NH_4)_3PO_4$ (g) $V_2O_7^{4-}$ (i) $S_2O_3^{2-}$
(b) H_2SO_3 (d) Na_2O_2 (f) NaH (h) SiO_3^{2-} (j) BiO^+

10. Write balanced ionic half-reactions for the oxidation of each of the following reducing agents in acid solution.

(a) Sn (c) $H_2C_2O_4$ (e) Br^-
(b) Sn^{2+} (d) H_2S (f) Ba

11. Write balanced ionic half-reaction equations for the reduction of each of the following oxidizing agents in aqueous acid solution.

(a) MnO_2 (solid)   (c) ClO_2^- (e) H_2O_2 (g) Fe^{2+}
(b) $Cr_2O_7^{2-}$ (d) IO_3^- (f) Br_2 (h) Cd^{2+}

12. Write balanced ionic equations for the following reactions.

(a) $Fe + Ag^+ \rightarrow$ (i) $H_2O_2 + ClO_3^- \rightarrow$
(b) $Cd + H^+ \rightarrow$ (j) $H_3AsO_3 + Ce^{4+} \rightarrow$
(c) $Cd + H^+ + NO_3^-$ (dilute) $\rightarrow$ (k) $Sn^{2+} + ClO_2^- + H^+ \rightarrow$
(d) $Cl_2 + HNO_2 \rightarrow$ (l) $Sn^{2+} + Mg \rightarrow$
(e) $MnO_4^- + H_2S + H^+ \rightarrow$ (m) $PbO_2 + HNO_2 + H^+ \rightarrow$
(f) $ClO_3^- + Sn^{2+} + H^+ \rightarrow$ (n) $Cr_2O_7^{2-} + H_2C_2O_4 + H^+ \rightarrow$
(g) $Br_2 + Fe^{2+} \rightarrow$ (o) $Cl_2 + I^- \rightarrow$
(h) $H_2O_2 + H_2SO_3 \rightarrow$

13. Write balanced molecular equations for the reactions in Problem 12. Use potassium salts of the negative ions and sulfate salts of the positive ions for the reactants given.

14. Complete and balance the following ionic equations for reactions that occur in aqueous solution. The nature of the solution (acidic or basic) is indicated for each reaction.

(a) $C_3H_8O + MnO_4^- \rightarrow CO_2$ (acidic)
(b) $CN^- + MnO_4^- \rightarrow CNO^- + MnO_2$ (basic)
(c) $CH_2O + Ag(NH_3)_2^+ \rightarrow Ag + HCO_2^-$ (basic)
(d) $CHCl_3 + MnO_4^- \rightarrow Cl_2 + CO_2$ (acidic)
(e) $MnO_4^- + C_2H_3OCl \rightarrow CO_2 + Cl_2$ (acidic)
(f) $C_2H_6O + Cr_2O_7^{2-} \rightarrow HC_2H_3O_2$ (acidic)

15. Write balanced equations for the reactions that occur in each of the following situations.
    (a) Chlorine gas is bubbled through a solution of ferrous bromide.
    (b) In the final step of producing bromine from sea water, a mixture of sodium bromide and sodium bromate is treated with sulfuric acid.
    (c) A microchemical procedure uses a cadmium amalgam (cadmium dissolved in metallic mercury) to reduce iron salts to their lowest valence state prior to titration with standard ceric sulfate. Give the equation for the reaction involving the cadmium.
    (d) When zinc is heated with concentrated sulfuric acid, hydrogen sulfide is evolved.
    (e) Pure hydrobromic acid cannot be prepared by treating sodium bromide with concentrated sulfuric acid and distilling off the hydrobromic acid, because some sulfur dioxide is produced at the same time, as noted by the odor. Give the equation for the production of sulfur dioxide.
    (f) Sodium perchlorate is prepared by carefully heating solid sodium chlorate.
    (g) The Marsh test for the detection of arsenic depends on the reaction of an arsenic compound, such as H_3AsO_4, with metallic zinc in acid solution to give arsine, AsH_3.

16. Balance the equations given in Problem 9, page 425.

Electrochemistry III: Electrolysis

Chapter 17 emphasizes the principles associated with *obtaining* electrical energy from electron-transfer reactions in solution. This chapter emphasizes what happens when electrical energy is *applied* to solutions in the operation of *electrolytic cells*. The oxidation and reduction processes that take place in an electrolytic cell are called *electrolysis*. We focus on determining what products are obtained and how much energy is required.

Unlike a galvanic cell, an electrolytic cell needs only a single beaker; both electrodes are immersed in it. The electrode *to* which the electrons are furnished by the power supply is the negative electrode; the other electrode is positive. In solution, the positive ions are attracted to the negative electrode, and the negative ions to the positive electrode. Thus, conduction of current in solution is a *bi*directional flow of ions, just as in the salt bridge of a galvanic cell. The electron-transfer process that occurs at each electrode (removal of electrons *from* the negative electrode and supply of electrons *to* the positive electrode) depends on the material between the electrodes, as we shall see.

THE PRODUCTS OF ELECTROLYSIS

Electrolysis of Molten Salts

The simplest electrolysis processes are those that occur when an electrolytic cell contains a molten salt, such as NaCl (Figure 19-1). In this situation the Na^+

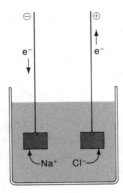

FIGURE 19-1
Electrolysis of molten NaCl.

ions migrate to the negative electrode and, if the applied voltage is sufficiently high, accept electrons from the electrode according to the half-reaction

$$Na^+ + e^- \rightarrow Na$$

At the positive electrode to which the Cl^- ions have migrated, Cl^- ions give up electrons to the electrode according to the half-reaction

$$Cl^- \rightarrow e^- + Cl$$

and the Cl atoms combine with one another to form Cl_2 gas,

$$Cl + Cl \rightarrow Cl_2\uparrow$$

If proper mechanical arrangements are provided, the Na is collected (as a vapor at the temperature of molten NaCl) in the absence of air at the negative electrode, and Cl_2 gas is collected at the positive electrode. Different mechanical provisions must be made if the metal is produced as a liquid or a solid, but the principle is the same in every case. If we look at the completed circuit, we see that electrons have come *from* the power supply to the negative electrode and have gone *to* the power supply from the positive electrode, with a *bi*directional flow of ions within the cell.

Electrolysis of Aqueous Solutions (The Negative Electrode)

The situation is more complicated when the electrolytic cell contains aqueous solutions because, in addition to the ions of the solute, there are present also the H^+ and OH^- ions of water. The basic principle is still the same: positive ions go to the negative electrode, and negative ions go to the positive electrode. But now, at the negative electrode, we must decide whether it is the plus ions of the solute or the H^+ ions of water that accept the electrons. Our natural reaction would be to say that we will first form the atom that has the least tendency to give off electrons. By reference to the table of standard electrode potentials (Table 17-1), we can see that all ions lying above H^+ will accept electrons *less* readily than H^+ *at the same concentration*. In water, the H^+ concentration is $10^{-7}M$, a concentration we can use in the Nernst equation (p 275) to show that the hydrogen electrode potential in water is -0.41 volt. With this knowledge in hand, we would be disposed to say that all those ions lying above -0.41 volt in Table 17-1 will accept electrons less readily than H^+ in aqueous solution. This would imply that metal ions lying below H^+ would plate out as metals from aqueous solution, but only H_2 gas would be obtained from the electrolysis of aqueous solutions of metal ions lying above H^+.

In actual fact this is not true, because a second problem arises. For reasons not yet clearly understood (despite a great many man-years of research), H^+ ions accept electrons at metal surfaces with much greater difficulty than expected, and this difficulty varies with the kind of metal surface. Only on a platinum surface covered with finely divided platinum does this difficulty disappear. But even if you start with platinum and begin to plate out another metal on it, the electrode surface then becomes the "other metal," and it is difficult to produce H_2. Extra voltage is required to produce H_2. This extra voltage is called "hydrogen overvoltage," and it varies with different metal surfaces. The net result of "hydrogen overvoltage" combined with a low H^+ ion concentration ($10^{-7}M$) is that most metal ions can be plated out from aqueous solution, but no H_2 gas is produced; the alkali and alkaline earth metals are the major exceptions. Two common metals that *cannot* be plated out from aqueous solution under any circumstances are iron and aluminum.

The following points sum up our discussion of the negative electrode.

1. Metal is plated out in the electrolysis of solutions of metal salts, other than those of the alkali and alkaline earth metals (and iron and aluminum). For example, in the electrolysis of a $ZnCl_2$ solution, we would obtain metallic zinc,

$$Zn^{2+} + 2e^- \rightarrow Zn$$

because the H_2 overvoltage on metallic zinc changes the H-electrode potential to some value more negative than the Zn-electrode potential of -0.76 volt.

2. H_2 gas is evolved in the electrolysis of solutions of alkali and alkaline earth metal salts. For example, in the electrolysis of an NaCl solution we would obtain H_2 gas,

$$2H^+ + 2e^- \rightarrow H_2\uparrow$$

because the electrode potential (-2.71 volts) for Na (an alkali metal) is more negative than the H-electrode potential, including the H_2 overvoltage.

Electrolysis of Aqueous Solutions (The Positive Electrode)

The electrolytic changes that occur at the positive electrode fall into three categories.

1. At an *inert* electrode (Pt, Au, stainless steel, C), the OH^- ion will decompose more easily than any other oxygen-containing negative ion (even at the low concentration that exists in water) to give

$$4OH^- \rightarrow O_2 + 2H_2O + 4e^-$$

Oxygen-containing ions such as SO_4^{2-}, NO_3^-, or ClO^- remain unchanged.

2. At an *inert* electrode (Pt, Au, stainless steel, C), the Cl^-, Br^-, and I^- ions will be converted to the corresponding halogens. For example,

$$2Cl^- \rightarrow Cl_2 + 2e^-$$

3. At an *active* electrode (almost any metal other than Pt, Au, or stainless steel), the electrode metal goes into solution, and the negative ions in solution remain unchanged. Thus at a positive Cu electrode, a solution of NaCl or $NaNO_3$ will yield neither Cl_2 nor O_2, but a solution of Cu^{2+} ions:

$$Cu \rightarrow Cu^{2+} + 2e^-$$

As a matter of fact, this is the basis for the electrorefining of metals. The impure metal electrode (say, Cu) is used as the positive electrode, and a piece of pure Cu metal is used as the negative electrode; the solution is $CuSO_4$. The composition of the solution is so designed, and the applied voltage so chosen, that the impurities stay in solution or precipitate out, and only copper plates out at the negative electrode:

at the positive electrode,

$$Cu \rightarrow Cu^{2+} + 2e^-$$

at the negative electrode,

$$Cu^{2+} + 2e^- \rightarrow Cu$$

The statements in this summary are, of necessity, oversimplified. Factors such as current density and the presence of strong oxidizing agents may greatly alter the surfaces of some electrodes and make them passive (inert) and incapable of acting as described in category 3. The presence of strong complexing agents such as CN^- can make Au and Pt become active and go into solution as in category 3.

QUANTITATIVE RELATIONS IN ELECTROLYSIS

Electron-Transfer Equivalents (Chemical)

In writing electron-transfer reactions, we express the quantity of electricity in terms of moles of electrons. One mole of electrons is equivalent to 96,487 coulombs of electrical charge. This quantity of electricity is called the Faraday constant (F), in honor of Michael Faraday, the first pioneer in quantitative electrochemistry. The value of F can be expressed either as 96,487 coulombs or as 1 faraday.

It is easy to find the quantity of material that is oxidized or reduced in an electron-transfer reaction when one faraday of electricity is passed. Simply take the half-reaction from Table 17-1 and divide the coefficient of each component by the number (n) of moles of electrons in the half-reaction, thus obtaining an equation involving one mole of electrons (one faraday of electricity). For example,

$$e^- + \tfrac{1}{2}H_2O + \tfrac{1}{4}O_2 \rightleftarrows OH^-$$

We see that one faraday of electricity oxidizes 1 mole of OH^- or reduces $\tfrac{1}{4}$ mole of O_2. In the half-reaction

$$e^- + \tfrac{1}{3}Al^{3+} \rightleftarrows \tfrac{1}{3}Al$$

one faraday oxidizes $\tfrac{1}{3}$ mole of Al or reduces $\tfrac{1}{3}$ mole of Al^{3+}. In the half-reaction

$$e^- + \tfrac{1}{2}Zn^{2+} \rightleftarrows \tfrac{1}{2}Zn$$

one faraday oxidizes $\tfrac{1}{2}$ mole of Zn or reduces $\tfrac{1}{2}$ mole of Zn^{2+}. And in the half-reaction

$$e^- + Fe^{3+} \rightleftarrows Fe^{2+}$$

one faraday oxidizes 1 mole of Fe^{2+} or reduces 1 mole of Fe^{3+}. The weight of material oxidized or reduced by one faraday is called an *electron-transfer equivalent weight* (as distinguished from an acid–base equivalent weight). That is,

$$\text{E-T equiv wt} = \frac{\text{grams}}{F} = \frac{\text{grams}}{\text{mole of electrons}}$$

For aluminum,

$$\text{E-T equiv wt} = \left(27.0\ \frac{g}{mole}\right)\left(\frac{1}{3}\ \frac{mole}{F}\right)$$

$$= 9.0\ \frac{g}{F} = 9.0\ \frac{g}{mole\ of\ electrons}$$

Some substances have more than one electron-transfer equivalent weight. For example, consider iron in the following reactions:

$$2Fe^{3+} + Zn \rightarrow 2Fe^{2+} + Zn^{2+}$$

$$Fe + 2H^+ \rightarrow Fe^{2+} + H_2 \uparrow$$

$$2Fe + 3Cl_2 \rightarrow 2FeCl_3$$

One mole of Fe is involved in the transfer of 1 mole of electrons in the first reaction, 2 moles of electrons in the second reaction, and 3 moles in the third reaction. This, in turn, means that the equivalent weight of Fe in the first reaction is 55.85 g/F, in the second is 27.92 g/F, and in the third is 18.62 g/F.

Electrical Equivalents (Faradays)

The quantity of electricity involved in an electrolysis reaction is determined by the current (amperes) and the length of time the current is passed. By definition,

$$\text{ampere} = \frac{coulomb}{sec}$$

or,

$$\text{coulombs} = (\text{amperes})(\text{secs})$$

A number of practical problems can be solved by applying these simple principles. The following problems are typical.

PROBLEM:
Calculate the weight of copper and the volume of O_2 (at 25.0°C and 730 torr, dry) that would be produced by passing a current of 0.500 amp through a $CuSO_4$ solution between Pt electrodes for a period of 2.00 hr.

SOLUTION:
The number of faradays of electricity passed is

$$\text{faradays} = \frac{(\text{amp})(\text{sec})}{96{,}487 \dfrac{\text{coulombs}}{\text{faraday}}} = \frac{\left(\dfrac{\text{coulombs}}{\text{sec}}\right)(\text{sec})}{96{,}487 \dfrac{\text{coulombs}}{\text{faraday}}}$$

$$= \frac{\left(0.500 \dfrac{\text{coulomb}}{\text{sec}}\right)(2.00 \text{ hr})\left(60 \dfrac{\text{min}}{\text{hr}}\right)\left(60 \dfrac{\text{sec}}{\text{min}}\right)}{96{,}487 \dfrac{\text{coulombs}}{\text{faraday}}} = 0.0373 \text{ faraday}$$

We know that 0.0373 faraday will liberate 0.0373 E-T equiv. The half-reaction ($e^- + \frac{1}{2}Cu^{2+} \rightarrow \frac{1}{2}Cu$) tells us that copper has 2 E-T equiv per mole. Therefore,

$$\text{wt of Cu} = (0.0373 \text{ equiv})\left(\frac{1 \text{ mole}}{2 \text{ equiv}}\right)\left(63.5 \frac{\text{g}}{\text{mole}}\right)$$

$$= 1.18 \text{ g Cu plated out}$$

The half-reaction ($OH^- \rightarrow \frac{1}{4}O_2 + \frac{1}{2}H_2O + e^-$) tells us that O_2 has 4 E-T equiv per mole. Therefore,

$$\text{vol of } O_2 = \frac{nRT}{P} = \frac{(0.0373 \text{ equiv})\left(\dfrac{1 \text{ mole}}{4 \text{ equiv}}\right)\left(62.4 \dfrac{\text{torr liter}}{\text{mole K}}\right)(298 \text{ K})}{730 \text{ torr}}$$

$$= 0.238 \text{ liters} = 238 \text{ ml at } 25.0°C \text{ and } 730 \text{ torr}$$

PROBLEM:

What length of time is required to plate out 0.1000 g Ag from an $AgNO_3$ solution using a current of 0.200 amp?

SOLUTION:

The number of electron-transfer equivalents of Ag plated out is

$$\text{E-T equiv of Ag} = \frac{0.1000 \text{ g}}{\left(107.9 \dfrac{\text{g}}{\text{mole}}\right)\left(1 \dfrac{\text{mole}}{\text{equiv}}\right)} = 9.27 \times 10^{-4}$$

Because 9.27×10^{-4} equiv of Ag requires 9.27×10^{-4} faraday of electricity for plating out,

$$\text{seconds} = \frac{\text{coulombs}}{\text{amperes}} = \frac{\left(96{,}487 \dfrac{\text{coulombs}}{\text{faraday}}\right)(9.27 \times 10^{-4} \text{ faraday})}{\left(0.200 \dfrac{\text{coulomb}}{\text{sec}}\right)}$$

$$= 447 \text{ sec}$$

$$\frac{447 \text{ sec}}{60 \dfrac{\text{sec}}{\text{min}}} = 7.45 \text{ min required for plating}$$

PROBLEM:
What current is required to plate out 0.0200 mole of gold from a $AuCl_3$ solution in 3.00 hr?

SOLUTION:
From the half-cell reaction

$$\tfrac{1}{3} Au^{3+} + e^- \rightarrow \tfrac{1}{3} Au$$

we can see that a mole of Au^{3+} contains 3 E-T equivalents. Thus,

$$\text{E-T equiv of Au} = (0.0200 \text{ mole Au}) \left(3 \frac{\text{equiv}}{\text{mole}} \right) = 0.0600 \text{ equiv Au}$$

Because 0.0600 equiv Au requires 0.0600 faraday of electricity for plating out,

$$\text{amperes} = \frac{\text{coulombs}}{\text{seconds}} = \frac{(0.0600 \text{ faraday}) \left(96{,}487 \dfrac{\text{coulombs}}{\text{faraday}} \right)}{(3 \text{ hr}) \left(3600 \dfrac{\text{sec}}{\text{hr}} \right)}$$

$$= 0.536 \text{ amp required}$$

PROBLEMS A

1. One-half faraday of electricity is passed through aqueous solutions of the compounds listed in the following table. Both electrodes are made of the material indicated. In the appropriate places in the table, give the formula of the substance that is *produced* at each electrode, and the amount that is produced. If a solid is produced, express the amount in moles; if a gas is produced, express the amount in liters (dry) at standard conditions.

		Negative electrode		Positive electrode	
Compound	Electrode material	Substance produced	Amount produced	Substance produced	Amount produced
$NiSO_4$	Ni				
CaI_2	Pt				
$Fe(NO_3)_3$	stainless steel				
$Au(ClO_4)_3$	Au				
$LiHCO_3$	Cu				

2. What weight of each of the following substances will be liberated by the passage of 0.200 faraday of electricity?
   (a) O_2 from Na_2SO_4 (d) Co from $CoSO_4$
   (b) Cl_2 from $AlCl_3$ (e) PbO_2 from $Pb(NO_3)_2$
   (c) Mg from $MgCl_2$

3. What current will be needed to deposit 6.00 g Ag in 30.0 min?

4. How long will it take to deposit the cadmium from 350 ml of a 0.300 M $CdSO_4$ solution, using a current of 1.75 amp?

5. What will be the concentration of $Cd(NO_3)_2$ in solution after 2.50 amp has passed for 5.00 hr through 900 ml of a solution that was originally 0.300 M? (Platinum electrodes are used.)

6. A current of 3.70 amp is passed for 6.00 hr between nickel electrodes in a 2.30 M $NiCl_2$ solution. What will be the concentration of the $NiCl_2$ at the end of the 6 hr? Assume that no other reactions occur.

7. Using a current of 5.50 amp, how long will it take to produce 47.0 liters of O_2 (measured over water at 735 torr and 35.0°C) by the electrolysis of a $CuSO_4$ solution?

8. How many ampere-hours of electricity will be needed to refine electrolytically one-half ton of silver by removing it from an impure positive electrode and depositing it in a pure form on the negative electrode?

9. During the discharge of a lead storage battery, the density of the sulfuric acid falls from 1.294 g/ml to 1.139 g/ml. Calculate the number of ampere-hours the battery must have been used. The overall reaction in the battery is

$$Pb + PbO_2 + 2H_2SO_4 \rightarrow 2PbSO_4 + 2H_2O$$

The more dense sulfuric acid is 39.0% H_2SO_4 by weight; the less dense is 20.0% H_2SO_4 by weight. The battery holds 3.50 liters of acid. Assume that the volume remains unchanged during discharge.

PROBLEMS B

10. One and one-half faradays of electricity are passed through aqueous solutions of the compounds listed in the following table. Both electrodes are made of the material indicated. In the appropriate places in the table give the formula of the substance that is *produced* at each electrode, and the amount that is produced. If a solid or a liquid is produced, express the amount in moles; if a gas is produced, express the amount in liters (dry) at standard conditions.

		Negative electrode		Positive electrode	
Compound	Electrode material	Substance produced	Amount produced	Substance produced	Amount produced
$BaBr_2$	Ag				
$Al_2(SO_4)_3$	Pt				
$SnCl_4$	C				
$Cd(NO_3)_2$	Cd				
K_3PO_4	Zn				

11. What weight of each of the following will be liberated by the passage of 0.650 faraday of electricity?

    (a) Pt from $PtCl_4$ (d) H_2 from H_3PO_4
    (b) O_2 from $Al(NO_3)_3$ (e) Zn from $ZnSO_4$
    (c) Ba from $BaCl_2$

12. How long will it take a current of 2.50 amp to deposit the silver from 650 ml of a 0.200 M $AgNO_3$ solution?

13. What current will be needed to deposit 2.50 g Cu in 15.0 min?

14. Platinum electrodes are placed in 400 ml of a 0.350 M $NiSO_4$ solution. If a current of 1.75 amp is passed for 3.00 hr, what will be the $NiSO_4$ concentration at the end of that time?

15. A current of 4.50 amp is passed between copper electrodes in a 1.85 M $CuSO_4$ solution. What will be the concentration of the $CuSO_4$ at the end of 5.00 hr?

16. Using a current of 3.50 amp, how long will it take to produce 25.0 liters of H_2 (measured over water at 730 torr and 18.0°C) by the electrolysis of an H_2SO_4 solution?

17. How many ampere-hours of electricity will be needed to refine electrolytically 1 ton of Cu by removing it from an impure positive electrode and depositing it in a pure form on the negative electrode?

18. During the discharge of a lead storage battery, the density of the sulfuric acid falls from 1.277 g/ml to 1.155 g/ml. Calculate the number of ampere-hours the battery must have been used. The overall reaction in the battery is

$$Pb + PbO_2 + 2H_2SO_4 \rightarrow 2PbSO_4 + 2H_2O$$

The more dense sulfuric acid is 37.0% H_2SO_4 by weight, and the less dense is 22.0% H_2SO_4 by weight. The battery holds 4.00 liters of acid. Assume that the volume remains unchanged during discharge.

Stoichiometry IV: Equivalent Weight and Normality

The stoichiometric calculations of Chapters 12 and 13 are based on the mole as the fundamental chemical unit in reactions. An alternative method of calculation utilizes the *equivalent* as a fundamental chemical unit. There are two kinds of equivalents, the type depending on the reaction in question; we shall refer to them as acid–base equivalents (or simply as equivalents) and electron-transfer equivalents (or E-T equivalents). The concept of an equivalent is particularly useful when dealing with complex or unknown mixtures, or when working out the structure and properties of unknown compounds. In addition, it emphasizes a basic characteristic of all chemical reactions that is directly applicable to all types of titration analyses.

ACID–BASE EQUIVALENTS

The *acid–base equivalent* is defined as the weight (in grams) of a substance that will provide, react with, or be equivalent to 1 mole of H^+. Thus, 1 mole (98.1 g/mole) of H_2SO_4 contains 2 moles of H^+, and its acid–base equivalent weight is 49.05 g/equiv. One mole (171.4 g/mole) of $Ba(OH)_2$ contains 2 moles of OH^- and will react with 2 moles of H^+; its acid–base equivalent weight therefore is 85.70 g/equiv. For $La_2(SO_4)_3$, there are 6 equivalents/mole, because the 2 La^{3+} ions provide the same total positive charge as 6 H^+; the equivalent weight is $\frac{1}{6}$ of the mole weight.

Care must be taken to avoid the ambiguity that may arise when a substance may have more than one equivalent weight. For example, Na_2CO_3 might react with HCl in either of two ways:

$$2HCl + Na_2CO_3 \rightarrow 2NaCl + CO_2 + H_2O$$

$$HCl + Na_2CO_3 \rightarrow NaHCO_3 + NaCl$$

In the first reaction, Na_2CO_3 acts as though it has two equivalents/mole; in the second it acts as though it has one. The ambiguity is avoided by stating which reaction is involved.

When we prepare a solution that contains 1 equivalent weight per liter of solution, its concentration is said to be 1 normal, designated as 1 N. A 0.200 N $Ba(OH)_2$ solution contains

$$\left(0.200 \ \frac{equiv}{liter}\right)\left(85.7 \ \frac{g}{equiv}\right) = 17.14 \ g/liter$$

The *normality* (N) of the solution is said to be 0.200 equivalents/liter.

Calculations involving equivalents, milliequivalents, normalities, and volumes of solutions are made in just the same way as those involving molarities of solutions. The unique and useful feature about the use of equivalents is that, for *any* chemical reaction, when reactant A has just exactly consumed reactant B, we can say

$$equivalents \ of \ A = equivalents \ of \ B$$

regardless of the number of moles of A required to react with a mole of B. For example, at the endpoint in a titration that uses V_a liters of acid with normality N_a to neutralize V_b liters of base with normality N_b, we can say that

$$endpoint \ equivalents = \left(N_a \ \frac{equiv}{liter}\right)(V_a \ liters) = \left(N_b \ \frac{equiv}{liter}\right)(V_b \ liters)$$

$$N_a V_a = N_b V_b$$

regardless of the base that is used. Or, similarly, if a weighed sample (W grams) of an acid with equivalent weight E is titrated with a standard base solution,

$$endpoint \ equivalents = \frac{W \ g}{E \ \frac{g}{equiv}} = \left(N_b \ \frac{equiv}{liter}\right)(V_b \ liters)$$

$$\frac{W}{E} = N_b B_b$$

When dealing with normality and equivalents, it is not necessary to have balanced chemical equations on which to base your calculations.

PROBLEM:
An HCl solution is standardized by titration of a pure Na_2CO_3 sample. Calculate the normality of the HCl solution if 41.30 ml are required to titrate 0.2153 g of Na_2CO_3.

SOLUTION:
Because there are two equivalents/mole of Na_2CO_3, the equivalent weight is

$$\left(106.00 \ \frac{g}{mole}\right)\left(\frac{1 \ mole}{2 \ equiv}\right) = 53.00 \ \frac{g}{equiv}$$

The number of equivalents of Na_2CO_3 in the weighed sample is

$$\frac{0.2153 \ g}{53.00 \ \dfrac{g}{equiv}} = 4.062 \times 10^{-3} \ equiv \ Na_2CO_3$$

At the endpoint,

$$equiv \ of \ Na_2CO_3 = equiv \ of \ HCl$$

$$4.062 \times 10^{-3} \ equiv = (0.04130 \ liter) \left(N \ \frac{equiv}{liter}\right)$$

$$N = 0.09835 \ \frac{equiv}{liter} = 0.09835 \ N$$

PROBLEM:
A 0.5280 g sample of impure $CaCO_3$ is dissolved in 50.00 ml of 0.09835 N HCl. After the reaction is complete and the CO_2 completely removed by warming, the excess HCl is titrated with 6.30 ml of 0.1052 N NaOH. Calculate the percentage of $CaCO_3$ in the original sample.

SOLUTION:
In this problem, the HCl is used partly by $CaCO_3$ and partly by NaOH, but at the endpoint

$$equiv \ of \ HCl = equiv \ of \ CaCO_3 + equiv \ of \ NaOH$$

Substituting the given data into the equation, we obtain

$$\left(0.09835 \ \frac{equiv}{liter}\right)(0.05000 \ liter) = (equiv \ CaCO_3) + (0.00630 \ liter)\left(0.1052 \ \frac{equiv}{liter}\right)$$

$$equiv \ CaCO_3 = 0.004918 - 0.000663 = 0.004255 \ equiv$$

There are 2 equiv/mole of $CaCO_3$; therefore, the weight of $CaCO_3$ in the original sample is

$$\text{wt of } CaCO_3 = (0.004255 \text{ equiv})\left(\frac{1 \text{ mole}}{2 \text{ equiv}}\right)\left(100.1 \frac{g}{\text{mole}}\right)$$

$$= 0.2277 \text{ g } CaCO_3$$

$$\text{Percentage of } CaCO_3 = \frac{0.2277 \text{ g}}{0.5280 \text{ g}} \times 100 = 43.13\% \ CaCO_3$$

PROBLEM:

An organic chemist synthesizes a new compound X with acidic properties. A 0.7200 g sample requires 30.00 ml of 0.2000 M $Ba(OH)_2$ for titration. What is the equivalent weight of X?

SOLUTION:

$$\text{Equivalents of } Ba(OH)_2 = \left(0.2000 \frac{\text{mole}}{\text{liter}}\right)\left(2 \frac{\text{equiv}}{\text{mole}}\right)(0.03000 \text{ liter})$$

$$= 0.01200 \text{ equiv of X}$$

$$\text{equiv wt of X} = \frac{0.7200 \text{ g}}{0.01200 \text{ equiv}} = 60.00 \frac{g}{\text{equiv}}$$

This knowledge, together with the empirical formula and the molecular weight of X, will help the chemist elucidate the structure of the compound and determine how many acidic groups there are in the molecule.

ELECTRON-TRANSFER EQUIVALENTS

The electron-transfer equivalent weight is defined in Chapter 19 as the weight of material oxidized or reduced by one mole of electrons. This quantity is easily calculated by dividing the coefficient of each component of the half-reaction that is involved in the reaction in question by the number of moles of electrons (n) in the half-reaction. For example,

$$e^- + \tfrac{1}{2} Zn^{2+} \rightleftarrows \tfrac{1}{2} Zn$$

$$e^- + Fe^{3+} \rightleftarrows Fe^{2+}$$

For the first half-reaction, 0.5 mole of Zn (or Zn^{2+}) is oxidized (or reduced) per mole of electrons, and in the second half-reaction one mole of Fe^{2+} (or

Fe^{3+}) is oxidized (or reduced) per mole of electrons. With our definition for E-T equivalent weight as

$$\text{E-T equiv wt} = \frac{\text{grams}}{\text{mole of electrons}}$$

the E-T equivalent weight for zinc can be found from its molecular weight as follows:

$$\text{E-T equiv wt} = \left(65.4 \ \frac{\text{g Zn}}{\text{mole Zn}} \right)\left(0.5 \ \frac{\text{mole Zn}}{\text{mole of electrons}} \right)$$

$$= 32.7 \ \frac{\text{g Zn}}{\text{mole of electrons}} = 32.7 \ \frac{\text{g Zn}}{\text{equiv}}$$

As pointed out on p. 313, many substances have more than one E-T equivalent weight; thus one must state the reaction involved when specifying the E-T equivalent weight for a given substance.

The terms normal and normality are defined and applied as they were for acid–base equivalents. Again, the unique property of normality is that, for any electron-transfer reaction, when the reducing agent has just exactly consumed the oxidizing agent,

$$\text{reducing equivalents} = \text{oxidizing equivalents}$$

regardless of the number of moles of each involved. Moreover, all the statements made about the endpoint in acid–base titrations also apply to the endpoint in electron-transfer titrations.

PROBLEM:
In the standardization of a $KMnO_4$ solution, a 0.2814 g sample of pure $Na_2C_2O_4$ requires 40.45 ml of the $KMnO_4$ solution. Calculate the normality and molarity of the $KMnO_4$ solution.

SOLUTION:
From Table 17-1 or your previous experience, you know that $C_2O_4^{2-}$ is oxidized to CO_2 (losing $2e^-$/mole in the process), and that MnO_4^- is reduced to Mn^{2+} (gaining $5e^-$/mole in the process). Therefore,

$$\text{E-T equiv wt of } Na_2C_2O_4 = \left(134.0 \ \frac{\text{g}}{\text{mole}} \right)\left(\frac{1 \text{ mole}}{2 \text{ equiv}} \right) = 67.0 \ \frac{\text{g}}{\text{equiv}}$$

At the endpoint of the titration,

$$\text{E-T equivalents of } KMnO_4 = \text{E-T equivalents of } Na_2C_2O_4$$

$$(0.04045 \text{ liter}) \left(N \ \frac{\text{equiv}}{\text{liter}} \right) = \frac{0.2814 \text{ g}}{67.00 \ \dfrac{\text{g}}{\text{equiv}}} = 0.004200 \text{ equiv}$$

$$N = \frac{0.004200 \text{ equiv}}{0.04045 \text{ liter}} = 0.1038 \text{ N } KMnO_4$$

Because there are 5 equiv/mole in $KMnO_4$,

$$\text{molarity} = \frac{0.1038 \ \dfrac{\text{equiv}}{\text{liter}}}{5 \ \dfrac{\text{equiv}}{\text{mole}}} = 0.02077 \text{ M } KMnO_4$$

Note that it is not necessary to have a balanced chemical equation to work this problem. You need know only the valence changes that are involved.

PROBLEM:
A 45.34 ml sample of $FeSO_4$ requires 35.76 ml of 0.1047 N $Na_2Cr_2O_7$ in acid solution for titration. Calculate the normality of the $FeSO_4$ solution.

SOLUTION:
Using equiv to mean E-T equiv, we see that (at the endpoint)

$$\text{equiv of } Cr_2O_7^{2-} = \text{equiv of } Fe^{2+}$$

$$\left(0.1047 \ \frac{\text{equiv}}{\text{liter}} \right) (0.03576 \text{ liter}) = (0.04534 \text{ liter}) \left(N \ \frac{\text{equiv}}{\text{liter}} \right)$$

$$N = 0.008258 \text{ N } FeSO_4$$

Note that, in a problem like this, you need not know the equation for the reaction *or* the valence changes involved with the reactants. If you wanted to know the molarities of the two solutions, you would have to know that $Cr_2O_7^{2-}$ has 6 equiv/mole (because $Cr_2O_7^{2-}$ goes to $2Cr^{3+}$) and that Fe^{2+} has 1 equiv/mole (because Fe^{2+} goes to Fe^{3+}). Thus,

$$\text{molarity of } Na_2Cr_2O_7 = \left(0.1047 \ \frac{\text{equiv}}{\text{liter}} \right) \left(\frac{1 \text{ mole}}{6 \text{ equiv}} \right)$$

$$= 0.01745 \text{ M } Na_2Cr_2O_7$$

$$\text{molarity of } FeSO_4 = \left(0.08258 \ \frac{\text{equiv}}{\text{liter}} \right) \left(\frac{1 \text{ mole}}{1 \text{ equiv}} \right)$$

$$= 0.08258 \text{ M } FeSO_4$$

PROBLEM:

A 0.4462 g sample of iron ore is dissolved in H_2SO_4, and the iron is reduced to Fe^{2+} with metallic zinc. The solution requires 38.65 ml of 0.1038 N $KMnO_4$ for titration. Calculate the percentage of iron as Fe_2O_3 in the sample.

SOLUTION:

Again, we do not need the chemical equation for the reaction—only the knowledge that there is 1 equiv/mole for Fe^{2+} as it is oxidized to Fe^{3+} by the $KMnO_4$ and that there are 2 moles of Fe/mole of Fe_2O_3. At the endpoint,

$$\text{equiv of } Fe_2O_3 = \text{equiv of } KMnO_4$$

$$= \left(0.1038 \ \frac{\text{equiv}}{\text{liter}} \right) (0.03865 \text{ liter})$$

$$= 0.004012 \text{ equiv } Fe_2O_3$$

From the number of equivalents, we find

$$\text{wt of } Fe_2O_3 = (0.004012 \text{ equiv}) \left(\frac{1 \text{ mole}}{2 \text{ equiv}} \right) \left(159.6 \ \frac{g}{\text{mole}} \right)$$

$$= 0.3202 \text{ g } Fe_2O_3$$

$$\text{Percentage of } Fe_2O_3 = \frac{0.3202 \text{ g}}{0.4462 \text{ g}} \times 100 = 71.76\% \ Fe_2O_3$$

PROBLEM:

A radiochemist isolates 8.6 μg of a chloride of neptunium (atomic weight = 237 g/mole), which she proves has the formula $NpCl_3$. In trying to find the possible valence states of Np, she finds that titration of the 8.6 μg sample requires 37.5 μl of 0.00200 N $KMnO_4$ solution. To what electrical charge must the Np have been oxidized?

SOLUTION:

The number of μmoles of $NpCl_3$ in the sample is

$$\frac{8.6 \ \mu g}{343.5 \ \dfrac{\mu g}{\mu \text{mole}}} = 0.0250 \ \mu \text{mole } NpCl_3$$

The number of μequiv of $KMnO_4$ required for titration is

$$\left(0.00200 \ \frac{\text{equiv}}{\text{liter}} \right) \left(10^{-6} \ \frac{\text{liter}}{\mu l} \right) \left(10^6 \ \frac{\mu \text{equiv}}{\text{equiv}} \right) (37.5 \ \mu l) = 0.0750 \ \mu \text{equiv}$$

Therefore,

$$\frac{0.0750 \ \mu \text{equiv}}{0.0250 \ \mu \text{mole}} = 3 \ \frac{\text{equiv}}{\text{mole}}$$

With Np^{3+} showing a *change* in electrical charge of 3 on oxidation by $KMnO_4$, its new charge must be $+6$.

PROBLEMS A

1. Tell how you would prepare each of the following solutions.
   (a) 250 ml of 0.100 N sulfamic acid from solid $H(NH_2)SO_3$
   (b) 500 ml of 0.200 N $Ba(OH)_2$ from solid $Ba(OH)_2 \cdot 8H_2O$
   (c) 750 ml of 2.50 N $Al(NO_3)_3$ from solid $Al(NO_3)_3 \cdot 9H_2O$

2. Tell how you would prepare each of the following solutions.
   (a) 2.00 liters of 0.0600 N $Ba(OH)_2$ from 1.86 M $Ba(OH)_2$
   (b) 75.0 ml of 0.300 N $AlCl_3$ from 4.30 M $AlCl_3$
   (c) 150 ml of 0.0320 M $CdSO_4$ from 0.100 N $CdSO_4$

3. What volume of 6.00 M H_2SO_4 should be added to 10.0 liters of 2.00 N H_2SO_4 in order to get 20.0 liters of 3.00 M H_2SO_4 on dilution with water?

4. How much of each of the following substances is needed to react completely with 40.00 ml of 0.1000 N H_2SO_4?
   (a) ml of 0.1500 N KOH   (d) mg of Zn
   (b) mg of KOH   (e) ml of 0.6500 M $Pb(NO_3)_2$
   (c) mg of Al_2O_3 (f) ml of 0.05000 M Na_2CO_3

5. How many ml of 0.1506 N $Ba(OH)_2$ are needed for the titration of 40.00 ml of 0.1000 M H_2SO_4?

6. The NH_3 liberated from a 0.4234 g sample of impure NH_4Cl by the addition of excess concentrated NaOH is collected in 25.00 ml of 0.1527 M H_2SO_4. After removal and rinsing of the delivery tube, the excess H_2SO_4 requires 4.65 ml of 0.1103 M NaOH for titration. Calculate the percentage of NH_4Cl in the original sample.

7. What is the molarity of a Na_2SO_3 solution if 30.00 ml of it require 45.00 ml of 0.04120 M $KMnO_4$ for titration?

8. (a) What is the normality of a ceric sulfate solution if 46.35 ml are required to titrate a 0.2351 g sample of $Na_2C_2O_4$ that is 99.60% pure?
   (b) What is the molarity of the solution?

9. A 0.2586 g sample of $H_2C_2O_4 \cdot 2H_2O$ requires 28.23 ml of KOH for titration, whereas a 0.1875 g sample requires 35.08 ml of $KMnO_4$ for titration. Calculate the molarity and normality of each of these solutions.

10. A 0.3216 g sample of 99.70% pure iron wire is dissolved in acid and converted to the Fe^{2+} state, and then immediately titrated with 36.71 ml of $K_2Cr_2O_7$ solution. What is the normality of the $K_2Cr_2O_7$ solution?

11. A 0.5000 g sample of $KHC_8H_4O_4$ requires 41.66 ml of a $Ba(OH)_2$ solution for titration. A 27.64 ml sample of this same solution requires 25.06 ml of an HNO_3 solution for titration. Calculate the molarity of the $Ba(OH)_2$ and HNO_3 solutions.

12. A soda-lime sample is 85.0% NaOH and 15.0% CaO.
    (a) If 3.50 g are dissolved in a volume of 300 ml, what is the normality of this solution?
    (b) How many ml of 0.5000 N H_2SO_4 would be required to titrate 1.000 ml of the solution?

13. What is the normality of a mixture of H_2SO_4 and HNO_3 if 28.76 ml are required to titrate a 0.4517 g sample of pure Na_2CO_3?

14. What volume of 5.00 N HCl is needed to prepare 12.7 liters of CO_2 at 735 torr and 35.0°C? ($CaCO_3$ is used as the source of CO_2.)

15. (a) What weight of MnO_2 and (b) what volume of 12.0 N HCl are needed for the preparation of 750 ml of 2.00 M $MnCl_2$? (c) What volume of Cl_2 at 745 torr and 23.0°C will also be formed?

16. The amount of $CaCO_3$ in a sample may be determined by dissolving the sample and precipitating the Ca as CaC_2O_4. This precipitate may be washed and dissolved in H_2SO_4, and the $C_2O_4^{2-}$ titrated with standard $KMnO_4$ solution. Following this procedure, the CaC_2O_4 from a 0.4526 g sample of limestone is dissolved in H_2SO_4. The resulting solution requires 41.60 ml of 0.1000 N $KMnO_4$ for titration. What is the percentage of $CaCO_3$ in the original sample?

17. Any of Problems 12 through 23 in Chapter 13 would be suitable for solution by the principles involving equivalents.

PROBLEMS B

18. Tell how you would prepare each of the following solutions.
    (a) 5.50 liters of 0.600 N LiOH from solid LiOH
    (b) 2.80 liters of 0.0100 N $Sr(OH)_2$ from solid $Sr(OH)_2$
    (c) 350 ml of 3.50 N $FeCl_3$ from solid $FeCl_3 \cdot 6H_2O$

19. Tell how you would prepare each of the following solutions.
    (a) 5.32 liters of 1.20 M $CdCl_2$ from 6.85 N $CdCl_2$
    (b) 600 ml of 2.50 N H_3PO_4 from 12.0 M H_3PO_4
    (c) 6.30 liters of 0.00300 M $Ba(OH)_2$ from 0.100 N $Ba(OH)_2$
    (d) 15.8 liters of 0.320 N NH_3 from 8.00 M NH_3

20. Tell how you would prepare each of the following solutions (weight percentage given).
    (a) 500 ml of 0.200 N $HClO_4$ from a 50.0% solution whose density is 1.410 g/ml
    (b) 2.50 liters of 1.50 N H_3PO_4 from an 85.0% solution whose density is 1.689 g/ml
    (c) 750 ml of 0.250 N $Cr_2(SO_4)_3$ from a 35.0% solution whose density is 1.412 g/ml.

21. What volume of 0.0800 M NaOH is needed for the titration of 30.00 ml of 0.1100 M H_2SO_4?

22. 20.00 ml of an H_3AsO_4 solution require 40.00 ml of a 0.1000 N NaOH solution for titration. Calculate (a) the normality and (b) the molarity of the H_3AsO_4 solution.

23. 35.45 ml of an NaOH solution are required to titrate a 2.0813 g sample of benzoic acid, $HC_7H_5O_2$. What is the normality of the NaOH solution?

24. How much of each of the following substances is needed to react completely with 45.00 ml of 0.1000 N H_2SO_4?
    (a) ml of 0.2500 N NH_3 (d) mg of Al
    (b) mg of NaOH   (e) mg of BaO
    (c) mg of $CaCO_3$

25. (a) What is the normality of a ceric sulfate solution if 39.65 ml is required to titrate a 0.3215 g sample of As_2O_3 that is 99.70% pure?
    (b) What is the molarity?

26. A 0.3217 g sample of KHC_2O_4 requires 31.09 ml of NaOH for titration, whereas a 0.2135 g sample requires 29.63 ml of $KMnO_4$ for titration. Calculate the molarity and normality of each of these solutions.

27. A 0.0713 g sample of KIO_3 is treated with an excess of KI in acid solution, and the I_2 that is formed requires 40.00 ml of a $Na_2S_2O_3$ solution for titration. This titration converts the I_2 to I^-, so in essence the titration is equivalent to converting the IO_3^- to I^-, while the $S_2O_3^{2-}$ is being converted to $S_4O_6^{2-}$. Calculate the normality of the $Na_2S_2O_3$ solution.

28. A 0.1056 g sample of $KH(IO_3)_2$ is treated as in Problem 27, and 35.75 ml of $Na_2S_2O_3$ solution is required for the titration. A 0.5965 g sample of $KH(IO_3)_2$ requires 26.43 ml NaOH solution for titration. Calculate the normality and molarity of these solutions.

29. What is the normality of a mixture of HCl and $HClO_4$ if a 25.00 ml sample requires 35.35 ml of 0.2567 M $Ba(OH)_2$ for titration?

30. What volume of 12.0 N HCl must be added to $KMnO_4$ in order to prepare 3.00 liters of Cl_2 at 730 torr and 25.0°C?

31. (a) What volume of 6.00 M HNO_3 and what weight of copper are needed for the production of 1.50 liters of a 0.500 M $Cu(NO_3)_2$ solution?
    (b) What volume of NO, collected over water at 745 torr and 18.0°C, will also be produced?

32. The amount of calcium in blood samples may be determined by precipitating it as CaC_2O_4. This precipitate may then be dissolved in H_2SO_4 and the $C_2O_4^{2-}$ titrated with standard $KMnO_4$ solution. Following this procedure, 10.00 ml of blood are diluted to 100.00 ml, and then the Ca^{2+} is precipitated as CaC_2O_4 from a 10.00 ml portion of this diluted sample. This precipitate, when washed and dissolved in H_2SO_4, requires 1.53 ml of 0.00750 N $KMnO_4$ for titration. How many milligrams of Ca^{2+} were in the 10 ml sample of blood?

33. Any of Problems 39 through 54 in Chapter 13 would be suitable for solution by the principles involving equivalents.

Colligative Properties

The colligative properties of solutions are those properties that depend upon the number of dissolved molecules or ions, irrespective of their kind. They are the lowering of the vapor pressure, the depression of the freezing point, the elevation of the boiling point, and the osmotic pressure. These properties may be used in determining molecular weights of dissolved substances.

VAPOR PRESSURE

When a nonvolatile nonelectrolyte is dissolved in a liquid, the vapor pressure is lowered. *Raoult's law* gives the mathematical relation

$$P = \chi P_0 \tag{21-1}$$

where P is the vapor pressure of the solution, P_0 is the vapor pressure of the pure solvent, and χ is the mole fraction of solvent in the solution. The lowering of vapor pressure is not as widely used as some other colligative properties for experimental determinations of molecular weights, because it is difficult to make the measurements precisely.

PROBLEM:
The vapor pressure of liquid A (mole weight 120) is 70.0 torr at 25.0°C. What is the vapor pressure of a solution containing 10.0 g $C_6H_4Cl_2$ in 30.0 g of A?

SOLUTION:

To use Raoult's law, we must compute the mole fraction of A in the solution (see p 191).

$$\text{Moles of A} = \frac{30.0 \text{ g}}{120 \dfrac{\text{g}}{\text{mole}}} = 0.250 \text{ mole}$$

$$\text{Moles of } C_6H_4Cl_2 = \frac{10.0 \text{ g}}{147 \dfrac{\text{g}}{\text{mole}}} = 0.0680 \text{ mole}$$

$$\text{Mole fraction of A} = \frac{0.250 \text{ mole}}{(0.250 + 0.0680) \text{ mole}} = 0.786$$

$$P = (0.786)(70.0 \text{ torr}) = 55.0 \text{ torr}$$

PROBLEM:

When 6.00 g of substance Z are dissolved in 20.0 g of $C_2H_4Br_2$, the solution has a vapor pressure of 9.00 torr at 22.0°C. Pure $C_2H_4Br_2$ has a vapor pressure of 12.70 torr at 22.0°C. What is the molecular weight of Z?

SOLUTION:

Let n = the number of moles of Z dissolved in

$$\frac{20.0 \text{ g}}{187.8 \dfrac{\text{g}}{\text{mole}}} = 0.106 \text{ moles of } C_2H_4Br_2$$

Substituting the known quantities into Raoult's law,

$$P = \chi P_0$$

we obtain

$$9.00 \text{ torr} = \chi(12.70 \text{ torr})$$

$$\chi = \frac{9.00 \text{ torr}}{12.70 \text{ torr}} = 0.709$$

By definition,

$$\chi = \frac{\text{moles of } C_2H_4Br_2}{(\text{moles of Z}) + (\text{moles of } C_2H_4Br_2)}$$

$$0.709 = \frac{0.106}{n + 0.106}$$

$$n = \frac{(0.106)(1 - 0.709)}{0.709} = 0.0435 \text{ mole Z}$$

Because 0.0435 moles of Z weigh 6.00 g, the mole weight (M) is

$$M = \frac{6.00 \text{ g}}{0.0435 \text{ mole}} = 138 \frac{\text{g}}{\text{mole}}$$

FREEZING AND BOILING POINTS OF SOLUTIONS

The lowering of the vapor pressure by a solute also brings about other changes: the freezing point is lowered, and the boiling point is raised. The amount of the change, ΔT_F or ΔT_B, is determined by the molality (m) of the solution. The relationships are

$$\Delta T_F = K_F m \tag{21-2}$$
$$\Delta T_B = K_B m \tag{21-3}$$

Each solvent has its own characteristic freezing-point constant K_F and boiling-point constant K_B, the changes caused by 1 mole of solute in 1 kilogram of solvent. Selected constants are given in Table 21-1.

TABLE 21-1
Boiling-Point and Freezing-Point Constants for Selected Compounds

Liquid	K_B	Normal BP (°C)	Liquid	K_F	Normal FP (°C)
Benzene	2.53	80.15	Barium chloride	108	962
Bromobenzene	6.26	156.0	Benzene	5.12	5.48
Camphor	5.95	208.25	Camphor	37.7	178.4
Carbon disulfide	2.34	46.13	1,2-Dibromoethane	11.8	9.79
Carbon tetrachloride	5.03	76.50	Diphenyl	8.0	70.0
Chloroform	3.63	60.19	Stannic bromide	28	31
Iodobenzene	8.53	188.47	Urethane	5.14	49.7
Water	0.51	100.00	Water	1.86	0.00

PROBLEM:
Calculate the freezing point of a solution that contains 2.00 g of $C_6H_4Br_2$ (abbreviated here as D) in 25.0 g of benzene (abbreviated B).

SOLUTION:
To find the molality of this solution we need

$$\text{the moles of D} = \frac{2.00 \text{ g D}}{236 \frac{\text{g D}}{\text{mole D}}} = 8.47 \times 10^{-3} \text{ moles D}$$

and the kg of B, which is 0.025 kg B. Therefore

$$\text{molality} = m = \frac{8.47 \times 10^{-3} \text{ moles D}}{0.025 \text{ kg B}} = 0.339 \frac{\text{mole D}}{\text{kg B}}$$

The freezing-point *depression* will be

$$\Delta T_F = mK_F = (0.339)(5.12)$$
$$= 1.74°C$$

the value of $K_F = 5.12$ being taken from Table 21-1. This table also gives the freezing point of pure B (5.48°C). The freezing point of the solution will be 5.48° − 1.74° = 3.74°C.

PROBLEM:
The freezing point of a solution that contains 1.00 g of a compound (Y) in 10.0 g of benzene (B) is found to be 2.07°C. Calculate the mole weight of Y.

SOLUTION:
We find in Table 21-1 that B freezes at 5.48°C and that the value of K_F is 5.12. The freezing-point depression of this solution is

$$\Delta T_F = 5.48° − 2.07° = 3.41°C$$

The molality of this solution is

$$m = \frac{\Delta T_F}{K_F} = \frac{3.41}{5.12} = 0.666 \frac{\text{mole Y}}{\text{kg B}}$$

This solution contains 0.0100 kg B. Therefore,

$$\text{moles of Y} = \left(0.666 \frac{\text{mole Y}}{\text{kg B}} \right) (0.0100 \text{ kg B}) = 6.66 \times 10^{-3} \text{ moles Y}$$

Because the 6.66×10^{-3} moles of Y are contained in 1.00 g,

$$\text{mole weight of Y} = \frac{1.00 \text{ g Y}}{6.66 \times 10^{-3} \text{ moles Y}} = 150 \frac{\text{g}}{\text{mole}}$$

The use of boiling-point elevation to determine molecular weights is based upon the same type of calculation, using K_B instead of K_F.

We recall from Chapter 10 that the percentages of the elements in a compound can be used to compute the simplest formula for the compound. When the substance is soluble in some suitable liquid, we can combine the empirical formula with a molecular-weight determination by freezing-point depression to get the true formula.

PROBLEM:
Compound Y of the preceding problem is found by analysis to have the composition

$$C = 49.0\%, \qquad H = 2.7\%, \qquad Cl = 48.3\%$$

Find the formula and the exact molecular weight.

SOLUTION:
Using the methods of Chapter 10, we find the empirical formula to be C_3H_2Cl. If this were the true formula, the molecular weight would be 73.5. The freezing point depression gives a molecular weight of approximately 150. This is not an accurate value, for the experimental measurement is subject to some error, but it indicates that the true molecular weight is near 150.

Knowing the empirical formula to be C_3H_2Cl, we know that the real formula is $(C_3H_2Cl)_n$, where n is some small integer. If n is 1, a mole is 73.5 g; if n is 2, a mole is 147 g; if n is 3, a mole is 221 g. Our experimental molecular weight of 150 enables us to decide that the true value of n is 2, because 147 is the molecular weight nearest our experimental value. Thus the formula is $(C_3H_2Cl)_2$, or $C_6H_4Cl_2$.

OSMOTIC PRESSURE

When solvent and solution are separated by a semipermeable membrane that permits solvent molecules to pass, an osmotic pressure is developed in the solution. This pressure, π, is defined as the mechanical pressure that must be applied to the solution to prevent solvent molecules from diffusing into it. For water solutions the relationship between π and the molal concentration m is given by the equation

$$\pi = (0.0821T)m \qquad (21\text{-}4)$$

where π is in atmospheres, and T is the absolute temperature. Note that Equation 21-4 is just like Equations 21-2 and 21-3, except that the constant multiplied by m is temperature-dependent. The only common practical solvent for osmosis is water, so the only constant we shall consider is 0.0821. A one molal solution at 0.0°C would have an osmotic pressure of

$$\pi = (0.0821)(273)(1.00) = 22.4 \text{ atm}$$

In comparison with the relatively small vapor pressure lowering caused by relatively concentrated solutions, the osmostic effect is gigantic. For example, a 1.00×10^{-4} molal solution at 25.0°C would have an osmotic pressure of

$$\pi = (0.0821)(298)(1.00 \times 10^{-4}) = 2.45 \times 10^{-3} \text{ atm}$$

$$= (2.45 \times 10^{-3} \text{ atm}) \left(760 \ \frac{\text{torr}}{\text{atm}} \right) = 1.86 \text{ torr}$$

Because of this great sensitivity of osmosis to small changes in concentration, the measurement of osmotic pressure is particularly suitable for the determination of molecular weights of biological materials with extremely high molecular weights. Such solutions will always be of very low molality, partly because of the high molecular weight of the solute and partly because of the generally low solubility of these compounds.

PROBLEM:
The osmotic pressure at 25.0°C of a solution containing 1.35 g of a protein (P) per 100 g of water is found to be 9.12 torr. Estimate the mole weight of the protein.

SOLUTION:
The osmotic pressure (in atm) is

$$\pi = \frac{9.12 \text{ torr}}{760 \ \dfrac{\text{torr}}{\text{atm}}} = 0.0120 \text{ atm}$$

The molality of this solution is

$$m = \frac{\pi}{(0.0821)(T)} = \frac{0.0120}{(0.0821)(298)} = 4.90 \times 10^{-4} \ \frac{\text{mole P}}{\text{kg H}_2\text{O}}$$

The 1.35 g of P per 100 g of water is equivalent to 13.5 g P per kg of water. This 13.5 g P contains 4.90×10^{-4} moles, therefore the mole weight must be

$$M = \frac{13.5 \text{ g P}}{4.90 \times 10^{-4} \text{ moles P}} = 27{,}600 \ \frac{\text{g}}{\text{mole}}$$

COLLIGATIVE PROPERTIES OF ELECTROLYTES

If a mole of NaCl is dissolved in 1 kg of water, the freezing point is not $-1.86°C$, as it would be for a mole of sugar on other nonelectrolyte. Rather, the freezing point is $-3.50°C$, a depression almost twice as great as we should expect. The theory of ionization provides an explanation for this discrepancy. When NaCl is dissolved, it breaks up into Na^+ and Cl^- ions, so that there are twice as many particles in solution as there would be if the dissociation did not occur. The water does not "know" whether the particles are molecules or ions, insofar as the colligative properties are concerned. For every mole of NaCl dissolved we have 2 moles in solution: a mole of Na^+ ions, and a mole of Cl^- ions. Thus we get an abnormal freezing-point depression.

According to modern theory, many strong electrolytes are completely dissociated in dilute solutions. The freezing-point lowering, however, does not indicate complete dissociation. For NaCl, the depression is not quite twice the amount calculated on the basis of the number of moles of NaCl added. In the solution, the ions attract one another to some extent; therefore they do not behave as completely independent particles, as they would if they were nonelectrolytes. From the colligative properties, therefore, we can compute only the "apparent degree of dissociation" of a strong electrolyte in solution.

To illustrate, let us consider the freezing-point depression that occurs when we put 1 mole of NaCl into 1 kg of water. NaCl dissociates according to the equation

$$NaCl \rightarrow Na^+ + Cl^-$$

Let us assume that α is the fraction of NaCl molecules that appear to dissociate, and that $1 - \alpha$ is the fraction that act as if they were still combined as NaCl molecules. Remember that we are talking about our *apparent* degree of dissociation, as measured by the colligative properties. Then we have, if we start with n moles of NaCl,

$$n(1 - \alpha) \text{ mole of undissociated molecules,}$$

$$n\alpha \text{ mole of } Na^+ \text{ ions, and}$$

$$n\alpha \text{ mole of } Cl^- \text{ ions.}$$

Adding, we get

$$\text{total moles in solution} = n(1 - \alpha + 2\alpha) = n(1 + \alpha)$$

This we can use to compute the value of α from the freezing-point lowering. As mentioned previously, we find that a solution of 1 mole of NaCl in 1 kg of H_2O freezes at $-3.50°C$. Here $n = 1$, so we have $m = 1 + \alpha$ moles/kg of water. Applying Equation 21-2, we obtain

$$\Delta T_F = mK_F$$

$$3.50 = (1 + \alpha)(1.86)$$

$$1 + \alpha = 1.88$$

$$\alpha = 0.88$$

Thus, according to these measurements, the apparent degree of dissociation for one-molal NaCl is 0.88, or 88.0%.

If we have an electrolyte such as $CaCl_2$, we get a somewhat different expression. Suppose we add 0.50 moles to 1 kg of water, and it behaves as if α is the fraction of the molecules that dissociate. In solution we would have

$$0.50(1 - \alpha) \text{ moles of } CaCl_2$$
$$0.50\alpha \text{ moles of } Ca^{2+}$$
$$\underline{2 \times 0.50\alpha \text{ moles of } Cl^-}$$
$$m = 0.50(1 + 2\alpha) \text{ total moles/kg of water}$$

From this, and the measured freezing-point depression (2.63°C), we can evaluate α as we did for NaCl and find it to be 0.914.

When you determine the value of α for a weak electrolyte, such as acetic acid or ammonia, the result must be interpreted in a different manner. In such a situation α really does correspond to the degree of dissociation, for most of the molecules actually are present in the undissociated form. The problems are worked in exactly the same way as for strong electrolytes; it is only the interpretation that is different.

PROBLEMS A

1. A solution is prepared by dissolving 1.28 g of naphthalene, $C_{10}H_8$, in 10.0 g of benzene.
   (a) What is the lowering of the freezing point of benzene?
   (b) What is the freezing point of the solution?
   (c) What is the mole fraction of benzene in the solution?
   (d) The vapor pressure of pure benzene is 100 torr at room temperature. What is the vapor pressure of this solution at the same temperature?
   (e) What is the boiling point of the solution?

2. A student uses a thermometer on which she can read temperatures to the nearest 0.1°. In a laboratory experiment, she observes a freezing point of 5.4°C for pure benzene. She then dissolves 0.75 g of an unknown in 15.0 g of benzene and finds the freezing point of the solution to be 2.8°C. What is the molecular weight of the unknown?

3. What will be the freezing point of a 10.0% solution of sucrose, $C_{12}H_{22}O_{11}$, in water?

4. A 10.0 g sample of naphthalene ($C_{10}H_8$) mothballs is added to 50.0 ml of benzene (density = 0.879 g/ml). What will be the boiling point of this solution?

5. Expensive special thermometers are usually needed for determinations of freezing-point depression and boiling-point elevation but, when camphor is used, a common lab thermometer may be employed. A student mixes 0.1032 g

of camphor and 7.32 mg of an unknown compound, and finds the melting point to be 159.3°C. The melting point of the camphor before mixing was 175.1°C. What is the mole weight of the compound?

6. When 2.848 g of sulfur is dissolved in 50.0 ml of CS_2 (density = 1.263 g/ml), the solution boils at a temperature 0.411°C higher than the pure CS_2. What is the molecular formula of sulfur?

7. When 0.532 g of a certain solid organic compound is dissolved in 16.8 g of urethane, whose freezing point is 49.50°C, the freezing point of the solution is lowered to 48.32°C. Chemical analysis shows this compound to be 69.5% C, 7.25% H, and 23.25% O. Determine the true formula of this compound.

8. A 0.356 g sample of a solid organic compound, when dissolved in 9.15 ml of carbon tetrachloride (density = 1.595 g/ml), raises the boiling point of the carbon tetrachloride by 0.560°C. When analyzed, this compound is found to be 55.0% C, 2.75% H, 12.8% N, and 29.4% O. What is the true formula of this compound?

9. Old United States silver coins, which are 10.0% Cu in Ag, melt completely at 875.0°C; pure silver melts at 960.0°C. What is the molal freezing-point constant for Ag?

10. Calculate the mole fraction and the molality of the first-named component in each of the following solutions.
    (a) 10.0% solution of glycerine, $C_3H_8O_3$, in water
    (b) 5.0% solution of water in acetic acid
    (c) 5.0 g CCl_4 and 25 g of C_2H_5OH
    (d) 0.10 g CN_2H_4O in 5.0 g of water
    (e) 8.00 g CH_3OH in 57.0 g C_6H_6

11. How many quarts of ethylene glycol, $C_2H_6O_2$ (density = 1.116 g/ml), will have to be added to 5.0 gal of water (density = 1.00 g/ml) to protect an automobile radiator down to a temperature of −10.0°F? (Ethylene glycol is a "permanent" antifreeze agent.)

12. A beaker containing 100 g of 10.0% sucrose ($C_{12}H_{22}O_{11}$) solution and another containing 150 g of a 30.0% sucrose solution are put under a bell jar at 25.0°C and allowed to stand until equilibrium is attained. What weight of solution will each beaker contain when equilibrium has been reached?

13. By how many torr will the vapor pressure of water at 27.0°C be lowered if 50.0 g of urea, CH_4N_2O, are dissolved in 50.0 g of water?

14. A carbon–hydrogen–oxygen analysis of benzoic acid gives 68.8% C, 5.0% H, 26.2% O. A 0.1506 g sample of benzoic acid dissolved in 100 g of water gives a solution whose freezing point is −0.023°C. A 2.145 g sample of benzoic acid in 50.0 g of benzene, C_6H_6, gives a solution whose freezing point is 4.58°C. The freezing point of the pure benzene is 5.48°C. Determine the molecular formula for benzoic acid in these two solutions, and explain any difference observed.

15. A 75.0 g sample of glucose, $C_6H_{12}O_6$, is dissolved in 250 g of water at 27.0°C. What will be the osmotic pressure of the solution?

16. A solution containing 1.346 g of a certain protein per 100 g of water is found to have an osmotic pressure of 9.69 cm *of water* at 25.0°C. Calculate the mole weight of the protein. The density of water is 1.00 g/ml; that of Hg is 13.6 g/ml.

17. A 0.50% solution of a certain plant polysaccharide (complex plant sugar) has an osmotic pressure of 5.40 torr at 27.0°C. Of how many simple sucrose sugar units, $C_{12}H_{22}O_{11}$, must this polysaccharide be composed?

18. If an aqueous sucrose ($C_{12}H_{22}O_{11}$) solution has an osmotic pressure of 12.5 atm at 23.0°C, what will be the vapor pressure of this solution at 23.0°C?

19. A 0.100 molal solution of acetic acid in water freezes at $-0.190°C$. Calculate the percentage of ionization of acetic acid at this temperature.

20. A 0.500 molal KBr solution freezes at $-1.665°C$. What is its apparent percentage of ionization at this temperature?

21. A 0.0100 molal solution of NH_3 is 4.15% ionized. What will be the freezing point of this solution?

22. A 0.200 molal solution of $Mg(NO_3)_2$ in water freezes at $-0.956°C$. Calculate the apparent percentage of ionization of $Mg(NO_3)_2$ at this temperature.

23. A solution prepared by dissolving 1.00 g of $Ba(OH)_2 \cdot 8H_2O$ in 200 g of water has a freezing point of $-0.0833°C$. Calculate the apparent percentage of ionization of $Ba(OH)_2$ at this temperature.

PROBLEMS B

24. Calculate the mole fraction and the molality of the first-named component in each of the following solutions.
    (a) 15.0% solution of ethylene glycol, $C_2H_6O_2$, in water
    (b) 7.0% solution of water in acetic acid
    (c) 7.00 g of chloroform, $CHCl_3$, and 30.0 g of methyl alcohol, CH_3OH
    (d) 40.0 g of ethyl alcohol, C_2H_5OH, and 60.0 g of acetone, C_3H_6O
    (e) 50.0 g of formamide, $HCONH_2$, and 50.0 g of water

25. What will be the freezing point of a 20.0% solution of glucose, $C_6H_{12}O_6$, in water?

26. A 20.0 g sample of p-dichlorobenzene ($C_6H_4Cl_2$) mothballs is added to 65.0 ml of benzene (density = 0.879 g/ml). What will be the boiling point of this solution?

27. A brass sample composed of 20.0% Zn and 80.0% Cu melts completely at 995.0°C; pure Cu melts at 1084.0°C. What is the molal freezing-point constant for copper?

28. What is the boiling point of a solution of 10.0 g of diphenyl, $C_{12}H_{10}$, and 30.0 g of naphthalene, $C_{10}H_8$, in 60.0 g of benzene?

29. A common way to check the purity of a compound is to take its melting point, because any soluble impurity always lowers the melting point. What is the molal concentration of impurity in urethane if its melting point is 47.7°C?

30. When 1.645 g of phosphorus is dissolved in 60.0 ml of CS_2 (density = 1.263 g/ml), the solution boils at 46.709°C; CS_2 alone boils at 46.300°C. What is the molecular formula of phosphorus?

31. By analysis, a certain solid organic compound is found to be 40.0% C, 6.7% H, and 53.3% O. When a 0.650 g sample of this compound is dissolved in 27.80 g of diphenyl, the freezing point is lowered by 1.56°C. Determine the true formula of this compound.

32. A certain solid organic compound is analyzed as 18.3% C, 0.51% H, and 81.2% Br. When 0.793 g of this compound is dissolved in 14.80 ml of chloroform (density = 1.485 g/ml), whose boiling point is 60.30°C, the solution is found to boil at 60.63°C. What is the true formula of this compound?

33. How many quarts of ethylene glycol, $C_2H_6O_2$ (density = 1.115 g/ml), must be added to 5.0 gal of water (density = 1.00 g/ml) to protect an automobile radiator down to a temperature of −15.0°F? (Ethylene glycol is a "permanent" antifreeze agent.)

34. What is the vapor pressure at 25.0°C of an aqueous solution containing 100 g glycerine, $C_3H_8O_3$, in 150 g of water?

35. What is the osmotic pressure of a solution in which 50.0 g of sucrose, $C_{12}H_{22}O_{11}$, is dissolved in 150 g of water at 21.0°C?

36. If an aqueous glucose ($C_6H_{12}O_6$) solution has an osmotic pressure of 50.0 atm at 35.0°C, what will be its vapor pressure at the same temperature?

37. A solution containing 1.259 g of a certain protein fraction per 100 g of water is found to have an osmotic pressure of 8.32 cm *of water* at 28.0°C. Calculate the mole weight of the protein fraction. The density of water is 1.00 g/ml; that of Hg is 13.6 g/ml.

38. A 0.70% solution of a certain plant polysaccharide (complex plant sugar) has an osmotic pressure of 6.48 torr at 21.0°C. Of how many simple sucrose sugar units, $C_{12}H_{22}O_{11}$, must this polysaccharide be composed?

39. A 0.0200 molal aqueous solution of picric acid, $HC_6H_2O_7N_3$, freezes at −0.0656°C. Picric acid ionizes to a certain extent in water, as

$$HC_6H_2O_7N_3 \rightarrow H^+ + C_6H_2O_7N_3^-$$

Calculate the percentage of ionization of picric acid at this temperature.

40. A 0.200 molal $NaNO_3$ solution freezes at −0.665°C. What is its apparent percentage of ionization?

41. A 0.100 molal solution of HNO_2 in water is 6.50% ionized. What will be the freezing point of this solution?

42. A 0.200 molal solution of $Ni(NO_3)_2$ freezes at $-0.982°C$. Calculate the apparent percentage of ionization of $Ni(NO_3)_2$ at this temperature.

43. If a 0.500 molal Na_2SO_4 solution has an apparent ionization of 72.0%, at what temperature will the solution freeze?

44. o-Chlorobenzoic acid has a composition of 53.8% C, 3.2% H, 20.4% O, and 22.6% Cl. A 0.1236 g sample of this acid dissolved in 100 g of water gives a solution whose freezing point is $-0.0147°C$. A 3.265 g sample of o-chlorobenzoic acid dissolved in 60.0 g of benzene gives a solution whose freezing point is 4.59°C. Determine the molecular formula of o-chlorobenzoic acid in these two solutions, and explain any difference in results.

Hydrogen-Ion Concentration and pH

Chemical reactions in aqueous solutions (including the chemistry of life processes) very often depend on the concentration of hydrogen ion in the solution. As we shall see, we may deal with hydrogen-ion concentrations varying from greater than 1 M to less than 10^{-14} M. Consequently, it is convenient to express these concentrations on a logarithmic basis; for this purpose the terms "pH" and "pOH," which we discuss in this chapter, have been introduced.

HYDROGEN AND HYDROXIDE CONCENTRATIONS IN WATER

Water is a *weak* electrolyte, ionizing slightly and reversibly as

$$H_2O \rightleftarrows H^+ + OH^-$$

Actually the H^+ is hydrated, forming chiefly H_3O^+ ion. Just as we ignore the hydration of all the metal ions (for convenience in writing equations), we also ignore the hydration of H^+. You must always remember, however, that a bare proton (an H^+) can never exist in solution by itself.

This dissociation reaction is always at equilibrium, with extremely rapid formation and recombination of H^+ and OH^-. Because it is always at equilibrium, the principles of chemical equilibrium discussed in Chapter 16 apply, and we can write the equilibrium-constant expression

$$K_w = [H^+][OH^-]$$

This equilibrium is of such importance that K bears the special subscript w. In pure water at 25.0°C, the concentration of H^+ is 1.0×10^{-7} M—that is, $[H^+] = 1.0 \times 10^{-7}$ M. Because the dissociation provides equal numbers of H^+ and OH^- ions, it follows that in pure water $[OH^-] = 1.0 \times 10^{-7}$ M also. Knowing the equilibrium concentrations, we can evaluate K_w numerically:

$$K_w = \left(1.0 \times 10^{-7} \frac{\text{moles}}{\text{liter}}\right)\left(1.0 \times 10^{-7} \frac{\text{moles}}{\text{liter}}\right) = 1.0 \times 10^{-14} \text{ M}^2$$

This constant applies to *all* water solutions. It follows that, if we add acid to water, thereby increasing the $[H^+]$, there must be a corresponding decrease in $[OH^-]$, and vice versa. HCl is a strong acid completely dissociated in water. This means that, in a 0.10 M HCl solution, $[H^+] = 0.10$ M. Because $[H^+][OH^-] = 10^{-14}$ M, it follows that $[OH^-] = 1.0 \times 10^{-13}$ M, one millionth of the concentration in pure water. NaOH is a strong base, also completely dissociated in water. A 0.10 M NaOH solution will have $[OH^-] = 0.10$ M, and an associated $[H^+]$ that is 1.0×10^{-13} M.

DEFINITION OF pH AND pOH

The wide range in the hydrogen-ion concentrations of aqueous solution makes it difficult to plot these values on a linear scale. As a convenience, we use a logarithmic scale introduced many years ago. Hydrogen-ion concentrations are represented by "pH" and hydroxide-ion concentrations by "pOH", defined by the relations

$$pH = -\log [H^+]$$
$$pOH = -\log [OH^-]$$

In keeping with this usage, we also use

$$pK_w = -\log K_w$$

You recall from p 14 that $\log AB = \log A + \log B$. Therefore, because $[H^+][OH^-] = K_w$,

$$pH + pOH = pK_w = 14$$

TABLE 22-1
Relations Among Concentrations of Strong Acid or Strong Base and Solution pH and pOH

Concentration of HCl (M)	10^{-1}	10^{-3}	10^{-5}	0			
pH	1	3	5	7	9	11	13
pOH	13	11	9	7	5	3	1
Concentration of NaOH (M)				0	10^{-5}	10^{-3}	10^{-1}

Relations between pH and pOH for solutions of HCl and NaOH are shown in Table 22-1. Note that in each solution the sum of pH and pOH values is 14. In the neutral solution, containing neither acid nor base, the pH is 7.

CALCULATION OF pH FROM [H⁺]

We need to know how to calculate the pH of any hydrogen-ion or hydroxide-ion concentration. This is done by means of the relation

$$pH = -\log [H^+]$$

A few examples will show the procedure. Limits on instrumentation, control of temperature, and protection of solutions against the effects of CO_2 from the air are such that most calculations of pH to more than two decimals are unwarranted. We shall work all problems on that basis.

PROBLEM:
What is the pH corresponding to a hydrogen-ion concentration of 5.00×10^{-4} M?

SOLUTION:
(a) If you have a hand calculator, all you must do (see pp 14–17) is enter 5×10^{-4} through the keyboard, press the log key(s), and then change the sign to give pH = 3.30.
(b) If you use a log table, then you can proceed as follows:

$$\log [H^+] = \log (5.00 \times 10^{-4}) = \log 5 + \log 10^{-4}$$
$$= 0.70 + (-4.00) = -3.30$$
$$pH = -\log [H^+] = -(-3.30) = 3.30$$

If you are asked to compute the pH of a strong acid solution, remember that the acid is almost completely ionized, and that [H⁺] is the same as the concentration of acid in the solution.

PROBLEM:
What is the pH of a 0.0200 M HCl solution?

SOLUTION:
Because the HCl is considered to be completely ionized, we have $[H^+] =$ 0.0200 M $= 2.00 \times 10^{-2}$ M.
(a) If you have a hand calculator, enter 0.02 through the keyboard, press the log key(s), and then change the sign to give pH = 1.70.
(b) If you use a log table, then

$$\log [H^+] = \log (2.00 \times 10^{-2}) = \log 2 + \log 10^{-2}$$

$$= 0.30 + (-2.00) = -1.70$$

$$pH = -\log [H^+] = -(-1.70) = 1.70$$

PROBLEM:
What is the pH of a 0.0400 M NaOH solution?

SOLUTION
Because NaOH is considered to be completely ionized, $[OH^-] = 0.0400$ M. The simplest of various alternatives for calculation is to first find pOH, then subtract that value from 14 to obtain pH.
(a) If you have a hand calculator, enter 0.04 through the keyboard, press the log key(s), and then change the sign to give pOH = 1.40. Then,

$$pH = 14.00 - pOH = 14.00 - 1.40 = 12.60$$

(b) If you use a log table, then

$$\log [OH^-] = \log (4.00 \times 10^{-2}) = \log 4 + \log 10^{-2}$$

$$= 0.60 + (-2.00) = -1.40$$

$$pOH = -(-1.40)$$

$$pH = 14.00 - pOH = 14.00 - 1.40 = 12.60$$

PROBLEM:
If 25.0 ml of 0.160 M NaOH are added to 50.0 ml of 0.100 M HCl, what is the pH of the resulting solution?

SOLUTION:
The pH of the solution is determined by whether an excess of acid or base is used, or whether they are used in exactly equivalent amounts. The first step is to determine this.

$$\text{Moles of HCl} = (0.0500 \text{ liter}) \left(0.100 \frac{\text{mole}}{\text{liter}}\right) = 0.00500 \text{ mole}$$

$$\text{Moles of NaOH} = (0.0250 \text{ liter}) \left(0.160 \frac{\text{mole}}{\text{liter}}\right) = 0.00400 \text{ mole}$$

The solution is acidic because there is $0.00500 - 0.00400 = 0.00100$ mole more acid than base. The 0.00400 mole of NaCl that is formed has no effect on the pH. The 0.00100 mole of HCl is completely dissociated and is in a total of 75.0 ml of solution. As a result, the concentration of H^+ is

$$[H^+] = \frac{1.00 \times 10^{-3} \text{ moles}}{0.0750 \text{ liter}} = 1.33 \times 10^{-2} \text{ M}$$

The pH is 1.88, as determined directly from your calculator, or as follows with a log table:

$$pH = -\log (1.33 \times 10^{-2}) = -\log 1.33 - \log (10^{-2})$$
$$= -0.12 + 2.00 = 1.88$$

PROBLEM:

What is the pH after 25.0 ml of 0.200 M NaOH are added to 50.0 ml of 0.100 M HCl?

SOLUTION:

As in the previous problem, first find the number of moles of acid and base used.

$$\text{Moles of HCl} = (0.0500 \text{ liter}) \left(0.100 \frac{\text{mole}}{\text{liter}}\right) = 0.00500 \text{ mole}$$

$$\text{Moles of NaOH} = (0.0250 \text{ liter}) \left(0.200 \frac{\text{mole}}{\text{liter}}\right) = 0.00500 \text{ mole}$$

Because the same number of moles of acid and base are used, the solution is the same as that of pure water, with pH = 7.0. It must be emphasized that this conclusion will be correct only if a strong acid reacts with a strong base. The pH of solutions of salts of weak acids or bases is considered in the next chapter.

PROBLEM:

Find the pH of the solution that results from the addition of 26.00 ml of 0.200 M NaOH to 50.0 ml of 0.100 M HCl.

SOLUTION:

As before,

$$\text{Moles of HCl} = (0.0500 \text{ liter}) \left(0.100 \frac{\text{mole}}{\text{liter}}\right) = 0.00500 \text{ mole}$$

$$\text{Moles of NaOH} = (0.0260 \text{ liter}) \left(0.200 \frac{\text{mole}}{\text{liter}}\right) = 0.00520 \text{ mole}$$

The amount of excess NaOH is $0.00520 \text{ mole} - 0.00500 \text{ mole} = 0.00020 \text{ mole}$. This excess of NaOH is present in 76.0 ml, and it will be completely dissociated. Therefore,

$$[OH^-] = \frac{2.00 \times 10^{-4} \text{ moles}}{0.0760 \text{ liter}} = 2.63 \times 10^{-3} \text{ M}$$

$$pOH = -\log (2.63 \times 10^{-3}) = -\log 2.63 - \log (10^{-3})$$
$$= -0.42 + 3.00 = 2.58$$

$$pH = 14.00 - 2.58 = 11.42$$

CALCULATION OF [H⁺] FROM pH

The preceding problems illustrate the computation of the pH for acidic, basic, and neutral solutions. We also need to understand the reverse calculation—how to go from the pH or pOH to the actual concentration of acid or base in a solution.

PROBLEM:
What is the hydrogen-ion concentration in a solution whose pH is 4.30?

SOLUTION:
From the definition of pH,

$$[H^+] = \text{antilog}\,(-pH) = 10^{-4.30}$$

If you simplify this with a hand calculator, enter -4.3 through the keyboard (enter 4.3 followed by the change-sign key), then proceed as on p 16. This gives $[H^+] = 5.01 \times 10^{-5}$ M.

If you simplify by using a log table, then the decimal part of this exponent (-4.30) must first be changed to a positive number. To do this, rewrite it as $10^{0.70} \times 10^{-5}$. The antilog of 0.70 (from the log table) is 5.01. This gives

$$[H^+] = 5.01 \times 10^{-5} \text{ M}$$

PROBLEM:
What is the hydroxide-ion concentration in a solution whose pH is 8.40?

SOLUTION:
Because we want [OH⁻], we first convert to pOH:

$$pOH = 14.00 - pH = 14.00 - 8.40 = 5.60$$
$$[OH^-] = \text{antilog}\,(-pOH) = 10^{-5.60}$$

If you simplify this with a hand calculator, enter -5.6 through the keyboard (enter 5.6 followed by the change-sign key), then proceed as on p 16. This gives $[OH^-] = 2.51 \times 10^{-6}$ M.

If you simplify by using a log table, then the decimal part of this exponent

(-5.60) must first be changed to a positive number. To do this, rewrite it as $10^{0.40} \times 10^{-6}$. The antilog of 0.40 (from the log table) is 2.51. This gives

$$[OH^-] = 2.51 \times 10^{-6} \text{ M}$$

PROBLEM:

How many grams of NaOH must be added to 200 ml of water to give a solution of pH 11.5?

SOLUTION:

From the pH we find the pOH, and from that the $[OH^-]$. Since NaOH is completely ionized, this $[OH^-]$ will also be the NaOH concentration needed.

$$pOH = 14.00 - pH = 14.00 - 11.5 = 2.50$$

Proceeding as in the previous problem, $[OH^-] = 3.16 \times 10^{-3}$ M. We need 200 ml of 3.16×10^{-3} M NaOH, so

$$\text{wt of NaOH needed} = (0.200 \text{ liter}) \left(3.16 \times 10^{-3} \frac{\text{moles}}{\text{liter}}\right) \left(40.0 \frac{\text{g}}{\text{mole}}\right)$$

$$= 0.0253 \text{ g NaOH}$$

PROBLEMS A

1. Calculate the pH of solutions with the following H^+ concentrations (in moles/ liter).
   (a) 10^{-4} (f) 8.9×10^{-2}
   (b) 10^{-6} (g) 3.7×10^{-5}
   (c) 10^{-8} (h) 6.5×10^{-8}
   (d) 10   (i) 3.5
   (e) 0.012   (j) 0.5

2. Calculate the pH of solutions with the following OH^- concentrations (in moles/liter).
   (a) 10^{-4} (f) 7.91×10^{-2}
   (b) 10^{-6} (g) 4.65×10^{-5}
   (c) 10^{-8} (h) 2.56×10^{-8}
   (d) 10   (i) 6.5
   (e) 0.025   (j) 0.72

3. Calculate the H^+ concentration for each of the solutions with the following values for pH.
   (a) 3.61   (f) 8.96
   (b) 7.52   (g) 0
   (c) 13.43   (h) 2.80
   (d) 0.77   (i) -0.6
   (e) 6.45   (j) 14.8

4. Calculate the pOH for each of the solutions in Problem 1.

5. Calculate the pH of a solution made by dissolving 1.00 g of KOH in 250 ml of distilled water.

6. Calculate the pH of a solution made by diluting 25.0 ml of commercial concentrated HBr solution (48.0% by weight, and density $= 1.49$ g/ml) to 2.00 liters.

7. A student pipets 50.0 ml of 0.1000 M HCl into a flask and then adds increments of 0.1000 M NaOH solution from a buret. Calculate the pH after the addition of each of the following volumes of NaOH solution.
   (a) 0.00 ml   (f) 49.9 ml
   (b) 10.0 ml   (g) 50.0 ml
   (c) 25.0 ml   (h) 50.1 ml
   (d) 45.0 ml   (i) 51.0 ml
   (e) 49.0 ml   (j) 55.0 ml
   Plot the results, showing pH on the y axis and volume of NaOH added on the x axis. This graph is called a titration curve.

PROBLEMS B

8. Calculate the pH of solutions with the following H^+ concentrations (in moles/liter).
   (a) 10^{-3} (f) 7.6×10^{-4}
   (b) 10^{-5} (g) 4.3×10^{-6}
   (c) 10^{-9} (h) 8.3×10^{-9}
   (d) 0.0056   (i) 5.6
   (e) 10   (j) 0.35

9. Calculate the pH of solutions with the following OH^- concentrations (in moles/liter).
   (a) 10^{-2} (f) 8.5×10^{-3}
   (b) 10^{-4} (g) 1.67×10^{-6}
   (c) 10^{-10} (h) 4.73×10^{-10}
   (d) 10   (i) 7.65
   (e) 0.077   (j) 0.22

10. Calculate the H^+ concentration for each of the solutions with the following values for pH.
    (a) 6.35   (f) 7.32
    (b) 2.78   (g) 15.21
    (c) 12.91   (h) 0.76
    (d) 0.55   (i) 0
    (e) 10.47   (j) -0.36

11. Calculate the pOH for each of the solutions in Problem 8.

12. Calculate the pH of a solution made by dissolving 20.0 g of sulfamic acid, $H(NH_2)SO_3$, and diluting to 200 ml with water. (Sulfamic acid is a strong acid, like HCl.)

13. Calculate the pH of a solution made by diluting 50.0 ml of a HNO_3 solution (56.0% by weight, and density $= 1.350$ g/ml) to 2.75 liters.

14. The following volumes of 0.1000 M HCl are added from a buret to a flask that initially contains 50.0 ml of 0.1000 M NaOH solution:

(a) 0.00 ml	(f) 49.9 ml
(b) 10.0 ml	(g) 50.0 ml
(c) 25.0 ml	(h) 50.1 ml
(d) 45.0 ml	(i) 51.0 ml
(e) 49.0 ml	(j) 55.0 ml

Calculate the pH of the solution after each addition of HCl, and plot pH on the y axis versus volume of HCl on the x axis. This graph is called a titration curve.

Acid–Base Equilibria

In Chapter 16, we apply the fundamental general equilibrium expression to gaseous equilibrium reactions. In this chapter, we apply the same expression to the equilibria that involve weak acids and bases in aqueous solution, the principal difference being that all concentrations are expressed in moles/liter (rather than in atmospheres as for gases). All the general conclusions given in Chapter 16, and summarized in the principle of Le Châtelier, apply to equilibria in solutions as well as to those in gases.

WEAK ACIDS

When we put a strong electrolyte (such as HCl) into solution, essentially all the molecules dissociate to ions—in this case, H^+ and Cl^-. But when we put into solution a weak electrolyte, such as acetic acid ($HC_2H_3O_2$), only a small fraction of the molecules dissociate. The equation is

$$HC_2H_3O_2 \rightleftarrows H^+ + C_2H_3O_2^-$$

Because this reaction is at equilibrium, we can apply the mathematical expression

$$\frac{[H^+][C_2H_3O_2^-]}{[HC_2H_3O_2]} = K_e$$

The equilibrium constant for the ionization of a weak electrolyte usually is designated as K_i, which we call the ionization constant.

Ionization constants are determined by experimental measurements of equilibrium concentrations. For example, to determine K_i for acetic acid, we prepare a solution of known concentration and by any of several methods measure the H^+ concentration or the pH. The method most widely used today is measuring with a pH meter, which gives a direct dial reading for the pH. We find experimentally that, in a 0.100 M solution of acetic acid, the pH is 2.88. From this we calculate the concentrations in the solution, and we use these to evaluate K_i. Starting with $[H^+]$, we have

$$[H^+] = 10^{-2.88} = 10^{0.12} \times 10^{-3} = 1.31 \times 10^{-3} \text{ mole/liter}$$

Each molecule that ionizes yields a H^+ ion and a $C_2H_3O_2^-$ ion, so the concentration of $C_2H_3O_2^-$ also is 1.31×10^{-3}. We have put 0.100 mole/liter of $HC_2H_3O_2$ in solution. Because 1.31×10^{-3} mole/liter has dissociated, there remains $0.10000 - 0.00131 = 0.09869$ mole/liter of undissociated molecules. Substituting these molar concentrations into the mathematical equation for equilibrium, we obtain

$$\frac{(1.31 \times 10^{-3})(1.31 \times 10^{-3})}{(0.09869)} = 1.74 \times 10^{-5} = K_i$$

Experimental values for selected ionization constants are given in Table 23-1. Although some of these values, such as the constant for acetic acid, are reliable to at least two significant figures, keep in mind that others may be in error, some as much as tenfold. For example, the dissociation constant for the HS^- ion is listed in different tables with values ranging from 10^{-13} to 10^{-15}. This uncertainty is due to the difficulty of determining the concentrations of ions (such as that of S^{2-}) that are present in very low concentration. In general, the second dissociation constant for a diprotic acid is less accurately known than the one for the first stage. In using tables of dissociation constants one must, therefore, keep in mind the limitations of many of the computations based on them. It must also be remembered that, even if a constant is accurately known, computations based on it are really accurate only when used in conjunction with activities, rather than concentrations, of the various ions present.

Uses of ionization constants to compute concentrations of the ions present in solution and the pH of the solution are illustrated in the following problems. First we consider the dissociation of a monoprotic acid, using acetic acid as an example. Later we examine the dissociation of a diprotic acid, H_2S, in connection with precipitation of metal sulfides.

TABLE 23-1
Ionization Constants at 25°C

Name	Reaction	K_i
Weak acids:		
Acetic acid	$HC_2H_3O_2 \rightleftharpoons H^+ + C_2H_3O_2^-$	1.74×10^{-5}
Boric acid	$H_3BO_3 \rightleftharpoons H^+ + H_2BO_3^-$	5.89×10^{-10}
Carbonic acid	$H_2CO_3 \rightleftharpoons H^+ + HCO_3^-$	$K_1 = 4.47 \times 10^{-7}$
	$HCO_3^- \rightleftharpoons H^+ + CO_3^{2-}$	$K_2 = 4.68 \times 10^{-11}$
Cyanic acid	$HCNO \rightleftharpoons H^+ + CNO^-$	2.19×10^{-4}
Formic acid	$HCHO_2 \rightleftharpoons H^+ + CHO_2^-$	1.70×10^{-4}
Hydrazoic acid	$HN_3 \rightleftharpoons H^+ + N_3^-$	1.91×10^{-5}
Hydrocyanic acid	$HCN \rightleftharpoons H^+ + CN^-$	6.31×10^{-10}
Hydrofluoric acid	$HF \rightleftharpoons H^+ + F^-$	6.76×10^{-4}
Hydrogen sulfide	$H_2S \rightleftharpoons H^+ + HS^-$	$K_1 = 1.00 \times 10^{-7}$
	$HS^- \rightleftharpoons H^+ + S^{2-}$	$K_2 = 1.20 \times 10^{-13}$
Nitrous acid	$HNO_2 \rightleftharpoons H^+ + NO_2^-$	5.13×10^{-4}
Oxalic acid	$H_2C_2O_4 \rightleftharpoons H^+ + HC_2O_4^-$	$K_1 = 5.62 \times 10^{-2}$
	$HC_2O_4^- \rightleftharpoons H^+ + C_2O_4^{2-}$	$K_2 = 5.25 \times 10^{-5}$
Phosphoric acid	$H_3PO_4 \rightleftharpoons H^+ + H_2PO_4^-$	$K_1 = 5.89 \times 10^{-3}$
	$H_2PO_4^- \rightleftharpoons H^+ + HPO_4^{2-}$	$K_2 = 6.17 \times 10^{-8}$
	$HPO_4^{2-} \rightleftharpoons H^+ + PO_4^{3-}$	$K_3 = 4.79 \times 10^{-13}$
Phosphorous acid	$H_3PO_3 \rightleftharpoons H^+ + H_2PO_3^-$	$K_1 = 1.00 \times 10^{-2}$
Bisulfate ion	$HSO_4^- \rightleftharpoons H^+ + SO_4^{2-}$	$K_2 = 1.02 \times 10^{-2}$
Sulfurous acid	$H_2SO_3 \rightleftharpoons H^+ + HSO_3^-$	$K_1 = 1.74 \times 10^{-2}$
	$HSO_3^- \rightleftharpoons H^+ + SO_3^{2-}$	$K_2 = 6.17 \times 10^{-8}$
Weak bases:		
Ammonia	$NH_3 + H_2O \rightleftharpoons NH_4^+ + OH^-$	1.74×10^{-5}
Methylamine	$CH_3NH_2 + H_2O \rightleftharpoons CH_3NH_3^+ + OH^-$	4.59×10^{-4}
Ethylamine	$C_2H_5NH_2 + H_2O \rightleftharpoons C_2H_5NH_3^+ + OH^-$	5.00×10^{-4}
Dimethylamine	$(CH_3)_2NH + H_2O \rightleftharpoons (CH_3)_2NH_2^+ + OH^-$	5.81×10^{-4}
Trimethylamine	$(CH_3)_3N + H_2O \rightleftharpoons (CH_3)_3NH^+ + OH^-$	6.11×10^{-5}
Aniline	$C_6H_5NH_2 + H_2O \rightleftharpoons C_6H_5NH_3^+ + OH^-$	4.17×10^{-10}
Pyridine	$C_5H_5N + H_2O \rightleftharpoons C_5H_5NH^+ + OH^-$	1.48×10^{-9}
Water	$H_2O \rightleftharpoons H^+ + OH^-$	1.00×10^{-14}

PROBLEM:
What is the pH of a 0.0500 M acetic acid solution?

SOLUTION:
Write the chemical equation, and above each term of the equation write the equilibrium molar concentration. Because we are not given the H^+ and $C_2H_3O_2^-$

concentrations, and because the two are the same, we represent this value by x. This gives a concentration of $(0.0500 - x)$ mole/liter of undissociated $HC_2H_3O_2$ molecules, because 0.0500 mole of acid is put into solution and x moles dissociate:

$$0.0500 - x \qquad x \qquad\qquad x$$
$$HC_2H_3O_2 \rightleftarrows H^+ + C_2H_3O_2^-$$

Substitute the molar concentrations into the K_i equation:

$$\frac{x^2}{0.0500 - x} = 1.74 \times 10^{-5}$$

To solve an equation of this type, we usually first assume x to be so small that $0.0500 - x$ may be considered as 0.0500 (in other words, subtraction of x from 0.0500 does not appreciably change the value). This gives, as the simplified equation,

$$\frac{x^2}{0.0500} = 1.74 \times 10^{-5}$$

$$x^2 = 8.70 \times 10^{-7}$$

$$x = [H^+] = 9.33 \times 10^{-4} \text{ M}$$

$$pH = -\log (9.33 \times 10^{-4}) = -(0.97 - 4.00) = 3.03$$

You can see that x is much smaller than the original concentration of 0.0500 M, and that we were justified in neglecting it compared to 0.0500. In general, we shall say that, if the calculated value of x is less than 10.0% of the number from which it is subtracted or to which it is added, it is permissible to make the approximation as we did in this problem.

Polyprotic Acids

When a solution contains a weak acid that can lose more than one proton (such as H_2S, H_2SO_3, or H_3AsO_4) the question arises as to whether we should write chemical equations and K_i expressions to show the loss of all these protons. The answer is NO. As you can tell from Table 23-1, each successive proton comes off with very much greater difficulty than the one before, and even the first one doesn't contribute much in the way of $[H^+]$. In addition, the H^+ ions that come from the first dissociation tend to repress second and third dissociations, just as would H^+ ions from some other source. The net result is that second and third dissociations can be neglected, and the equilibrium expressions are always written for the loss of only *one* proton.

PROBLEM:
Calculate the pH of a 0.100 M H_2S solution.

SOLUTION:
For the reasons just given, the proper chemical equilibrium to consider is

$$0.100 - x \qquad x \qquad x$$
$$H_2S \rightleftarrows H^+ + HS^-$$

with equal small unknown concentrations of H^+ and HS^- of x moles/liter, leaving $(0.100 - x)$ moles/liter of undissociated H_2S. Substituting these values into the K_i expression, we obtain

$$K_i = \frac{[H^+][HS^-]}{[H_2S]} = \frac{x^2}{(0.100 - x)} = 1.00 \times 10^{-7}$$

Neglecting x compared to 0.100, we have

$$x_2 = 1.00 \times 10^{-8}$$
$$x = [H^+] = 1.00 \times 10^{-4} \text{ M}$$
$$pH = -\log (1.00 \times 10^{-4}) = 4.00$$

WEAK BASES

When a weak base is dissolved in water, a few of the molecules accept protons from water, leaving OH^- ions in the solution to make it slightly basic. For many years it was said that such solutions contain the *hydrated* form of the base (instead of the base itself), and that the hydrated base then subsequently dissociates to a slight degree. For ammonia, it was said that NH_3 first reacts with water to form NH_4OH, which then dissociates slightly as a weak base. Because most of the dissolved base probably does *not* exist in the hydrated form in solution, it is now more acceptable to write the chemical equilibrium equation as

$$NH_3 + H_2O \rightleftarrows NH_4^+ + OH^-$$

Just as it is customary to consider the concentration of water to be constant (or at unit activity) in dilute solutions of weak acids, so we shall consider that the water concentration remains constant in dilute solutions of weak bases; $[H_2O]$ will not appear in any K_i expression.

PROBLEM:
What is the pH of a 0.100 M NH_3 solution?

SOLUTION:

First, write the chemical equation:

$$NH_3 + H_2O \rightleftarrows NH_4^+ + OH^-$$

Second, write the K_i expression based on the chemical equation, obtaining the needed value of K_i from Table 23-1.

$$K_i = \frac{[NH_4^+][OH^-]}{[NH_3]} = 1.74 \times 10^{-5}$$

Third, write what you know and do not know. You are asked for the pH of a *basic* solution, so you will first have to find the $[OH^-]$. Let $[OH^-] = x$. Because NH_4^+ and OH^- ions are formed in equal amounts, $[NH_4^+]$ also equals x. Of the original 0.100 mole/liter of NH_3, x moles/liter will dissociate and leave $(0.100 - x)$ mole/liter at equilibrium. Associate these concentrations with the chemical equation, and substitute them into the K_i expression.

$$\overset{0.100 - x}{NH_3} + H_2O \rightleftarrows \overset{x}{NH_4^+} + \overset{x}{OH^-}$$

$$K_i = \frac{(x)(x)}{(0.100 - x)} = 1.74 \times 10^{-5}$$

We try to simplify the solution by neglecting x compared to 0.100, obtaining

$$x^2 = (0.100)(1.74 \times 10^{-5})$$
$$x = [OH^-] = 1.32 \times 10^{-3} \text{ M}$$

We see that the assumption about neglecting x was sound.

$$pOH = -\log(1.32 \times 10^{-3}) = -(0.12 - 3.00) = 2.88$$
$$pH = 14.00 - 2.88 = 11.12$$

COMMON-ION EFFECT

If we add some sodium acetate or other source of $C_2H_3O_2^-$ to a solution of $HC_2H_3O_2$, we shift the equilibrium

$$HC_2H_3O_2 \rightleftarrows H^+ + C_2H_3O_2^-$$

to the left, and use up some of the H^+ ions to form more undissociated $HC_2H_3O_2$. We may use the regular K_i expression for acetic acid to compute the pH of such a mixture of acetic acid and sodium acetate.

PROBLEM:
What is the pH after 1.00 g $NaC_2H_3O_2$ is added to 150 ml of 0.0500 M $HC_2H_3O_2$?

SOLUTION:
First, find the concentrations of the substances put into solution. $NaC_2H_3O_2$ is a soluble salt that is completely dissociated in solution; for every mole of $NaC_2H_3O_2$ put into solution, we get one mole of $C_2H_3O_2^-$. So, from $NaC_2H_3O_2$,

$$[C_2H_3O_2^-] = \frac{1.00 \text{ g}}{\left(82.0 \dfrac{\text{g}}{\text{mole}}\right)(0.150 \text{ liter})} = 0.0813 \text{ M}$$

If we let x = moles/liter of $HC_2H_3O_2$ that dissociate, then at equilibrium $[HC_2H_3O_2] = 0.0500 - x$. For every mole/liter of $HC_2H_3O_2$ that dissociates, there will be formed x moles/liter of H^+ and x moles/liter of $C_2H_3O_2^-$. These x moles/liter of $C_2H_3O_2^-$ will be added to the 0.0813 mole/liter of $C_2H_3O_2^-$ that come from $NaC_2H_3O_2$ to give a total equilibrium concentration of $[C_2H_3O_2^-] = 0.0813 + x$. Writing the chemical equation and placing the equilibrium concentration above each substance, we have

$$\begin{array}{ccc} 0.0500 - x & x & 0.0813 + x \\ HC_2H_3O_2 \rightleftharpoons H^+ &+& C_2H_3O_2^- \end{array}$$

The K_i expression is

$$K_i = \frac{[H^+][C_2H_3O_2^-]}{[HC_2H_3O_2]} = \frac{(x)(0.0813 + x)}{(0.0500 - x)} = 1.74 \times 10^{-5}$$

Assuming that x is negligible compared to 0.0500 and 0.0813, we have

$$\frac{(x)(0.0813)}{0.0500} = 1.74 \times 10^{-5}$$

$$x = [H^+] = 1.07 \times 10^{-5} \text{ M}$$

We see that the assumption about neglecting x was sound.

$$pH = -\log(1.07 \times 10^{-5}) = -(0.03 - 5.00) = 4.97$$

When we compare this pH with the pH of 3.03 obtained from the 0.0500 M $HC_2H_3O_2$ solution on p 352, we see that the addition of the salt $NaC_2H_3O_2$ has greatly repressed the ionization of the acid, decreasing the $[H^+]$ about 100-fold from 9.33×10^{-4} M to 1.11×10^{-5} M. The shift in equilibrium caused by adding a substance with an ion in common with that equilibrium is known as the "common-ion effect."

BUFFERS

A solution that contains a weak acid plus a salt of that acid, or a weak base plus a salt of that base, is known as a *buffer*. Such a solution has the capability to buffer against (to resist) changes in pH when small amounts of strong acid or base are added. Only very small changes occur.

To illustrate, consider a buffer containing s moles/liter of $NaC_2H_3O_2$ and a moles/liter of $HC_2H_3O_2$. If x moles/liter of $HC_2H_3O_2$ dissociate, there will be $(a - x)$ moles/liter of $HC_2H_3O_2$ left at equilibrium, and a total of $(s + x)$ moles/liter of $C_2H_3O_2^-$ at equilibrium, along with x moles/liter of H^+. If we substitute these equilibrium concentrations into the K_i expression, we get

$$K_i = \frac{[H^+][C_2H_3O_2^-]}{[HC_2H_3O_2]} = \frac{(x)(s + x)}{(a - x)} = 1.74 \times 10^{-5}$$

From the preceding problems, we can see that x usually will be negligible compared to s and a, so

$$x = [H^+] = K_i \frac{(a)}{(s)} = K_i \frac{[acid]}{[salt]}$$

If the buffer contained s moles/liter of the salt of a weak base and b moles/liter of the weak base, the arguments would be exactly the same as for the acid buffer, and the resulting expression would be for $[OH^-]$:

$$[OH^-] = K_i \frac{(b)}{(s)} = K_i \frac{[base]}{[salt]}$$

We see that, for a given acid, the pH is determined primarily by the *concentration ratio* of the weak acid and its salt. If we add a small amount of strong base to this buffer it is used up by reaction with some of the weak acid and converted to the salt; the ratio of acid to salt is changed, but not by much. Likewise, if we add strong acid, it is used up by reaction with the salt and converted to the weak acid; again, the ratio of acid to salt is changed, but not appreciably. The following problems illustrate this buffer action.

PROBLEM:
A buffer solution contains 1.00 mole/liter each of acetic acid and sodium acetate. Calculate the pH of the buffer.

SOLUTION:

$$[H^+] = K_i \frac{[acid]}{[salt]} = (1.74 \times 10^{-5}) \left(\frac{1.00}{1.00}\right) = 1.74 \times 10^{-5} \text{ M}$$

$$pH = -\log (1.74 \times 10^{-5}) = 4.76$$

PROBLEM:
What is the pH of the solution when 0.10 mole of HCl is added to 1.00 liter
of the buffer in the preceding problem?

SOLUTION:
The 0.10 mole of HCl (strong acid) reacts with the salt,

$$H^+ + C_2H_3O_2^- \rightarrow HC_2H_3O_2$$

to give 0.10 mole *more* $HC_2H_3O_2$ for a total of 1.10 mole/liter, and 0.10 mole *less*
$C_2H_3O_2^-$ for a total of 0.90 mole/liter. If we substitute these new acid and salt
concentrations into our expression for $[H^+]$, we get

$$[H^+] = K_i \frac{[\text{acid}]}{[\text{salt}]} = (1.74 \times 10^{-5}) \left(\frac{1.1}{0.9}\right) = 2.13 \times 10^{-5} \text{ M}$$

$$pH = -\log (2.13 \times 10^{-5}) = 4.67$$

You see that the pH changes only 0.09 units, whereas 0.10 mole HCl added to
one liter of water would have given a solution whose pH = 1.00—an enormous
change in pH without the buffer.

PROBLEM:
If 20.0 ml of 0.200 M NaOH are added to 50.0 ml of 0.100 M $HC_2H_3O_2$, what is
the pH of the resulting solution?

SOLUTION:
Some of the $HC_2H_3O_2$ is converted to $NaC_2H_3O_2$, and all of the NaOH is used
up in the process. The resulting solution is a buffer, and we need to find the
concentrations of $HC_2H_3O_2$ and $NaC_2H_3O_2$ in solution in order to find the pH.

$$\text{Original moles } HC_2H_3O_2 = (0.0500 \text{ liter}) \left(0.100 \frac{\text{mole}}{\text{liter}}\right) = 0.00500 \text{ mole}$$

$$\text{Original moles NaOH} = (0.0200 \text{ liter}) \left(0.200 \frac{\text{mole}}{\text{liter}}\right) = 0.00400 \text{ mole}$$

$$\text{Moles of } HC_2H_3O_2 \text{ left} = 0.00500 - 0.00400 = 0.00100 \text{ mole}$$

$$\text{Moles of } C_2H_3O_2^- \text{ formed} = 0.00400 \text{ mole}$$

These moles are present in 70.0 ml = 0.0700 liter, so

$$[\text{acid}] = \frac{0.00100}{0.0700} \quad \text{and} \quad [\text{salt}] = \frac{0.00400}{0.0700}$$

$$[H^+] = K_i \frac{[\text{acid}]}{[\text{salt}]} = (1.74 \times 10^{-5}) \left(\frac{\frac{0.00100}{0.0700}}{\frac{0.00400}{0.0700}}\right) = 4.35 \times 10^{-6} \text{ M}$$

$$pH = -\log (4.35 \times 10^{-6}) = 5.36$$

This problem shows that one way to make a buffer is to partially neutralize a weak acid with a strong base. Partial neutralization of a weak base with a strong acid also will work.

PROBLEM:

How many grams of sodium acetate must be added to 250 ml of 0.200 M $HC_2H_3O_2$ in order to prepare a buffer with pH $= 5.00$?

SOLUTION:

The pH is given as 5.00; therefore, $[H^+] = 1.00 \times 10^{-5}$ M. We are also told that [acid] $= 0.200$ M. Substitution of these values into the expression for $[H^+]$ gives

$$[H^+] = K_i \frac{[acid]}{[salt]} = (1.74 \times 10^{-5}) \left(\frac{0.200}{[salt]}\right) = 1.00 \times 10^{-5} \text{ M}$$

$$[salt] = \frac{(1.74 \times 10^{-5})(0.200)}{1.00 \times 10^{-5}} = 0.348 \text{ M}$$

We find that the sodium acetate concentration must be 0.348 mole/liter, but we want only 0.250 liter. Therefore,

$$\text{wt of } NaC_2H_3O_2 \text{ needed} = (0.250 \text{ liter}) \left(0.348 \frac{\text{mole}}{\text{liter}}\right) \left(82.0 \frac{\text{g}}{\text{mole}}\right)$$

$$= 7.13 \text{ g}$$

Addition of the 7.13 g $NaC_2H_3O_2$ to the 250 ml will give a solution that is 0.348 M in salt and 0.200 M in acid, and that has a pH of 5.00.

HYDROLYSIS

Pure water has a pH of 7.00. If we add a salt of a strong base and a strong acid (such as NaCl), it does not affect the pH, because neither the Na^+ ion nor the Cl^- ion can react with the H^+ ion or the OH^- ion of water.

If we add to water a salt whose ions come from a weak acid or base, some of the salt reacts with water, or hydrolyzes.* An ion of the salt ties up some of the H^+ or OH^- ions of the water, leaving the other ion in excess. The reaction

---

* The simplified discussion in this chapter does not include hydrolysis effects due to dissociation of hydrated metal ions; see Chapter 25 for such a discussion.

of water with sodium acetate is

$$HOH \rightleftarrows \quad H^+ \quad + OH^-$$
$$+ \qquad +$$
$$NaC_2H_3O_2 \rightleftarrows C_2H_3O_2^- + Na^+$$
$$\updownarrow \qquad \updownarrow$$
$$HC_2H_3O_2 \qquad NR \text{ (no reaction)}$$

Because some of the H^+ ions of water are tied up, the solution is left with an excess of OH^- ions and is basic.

Similarly, NH_4Cl gives by hydrolysis an acid solution, because it removes OH^- ions to form the weak base NH_3, thereby leaving an excess of H^+ ions:

$$HOH \rightleftarrows H^+ + OH^-$$
$$+ \qquad +$$
$$NH_4Cl \rightleftarrows Cl^- + NH_4^+$$
$$\updownarrow \qquad \updownarrow$$
$$NR \qquad NH_3 + H_2O$$

We have used here a form of equation that emphasizes the ions of the salt that tie up ions of water. Ordinarily, we do not write hydrolysis equations in this way. The common forms of the equations above are

$$C_2H_3O_2^- + H_2O \rightleftarrows HC_2H_3O_2 + OH^-$$
$$NH_4^+ + H_2O \rightleftarrows NH_3 + H_3O^+$$

The first solution is basic because of the OH^- ions that have been produced (*not* acidic because of the undissociated acetic acid produced). The second solution is acidic because of the H_3O^+ ions that have been produced (*not* basic because of the undissociated NH_3 produced). If we apply the usual equilibrium expression to the first of these equations, and consider $[H_2O]$ to remain constant at unit activity, we have

$$K_h = \frac{[HC_2H_3O_2][OH^-]}{[C_2H_3O_2^-]}$$

K_h is called the hydrolysis constant. Unlike ionization constants, which are tabulated in extensive tables, values of K_h are never tabulated because they are easily calculated from values of K_i. For example, if we take the product of the expressions for K_i and K_h for acetic acid and the acetate ion, we get

$$K_i \times K_h = \frac{[H^+][C_2H_3O_2^-]}{[HC_2H_3O_2]} \times \frac{[HC_2H_3O_2][OH^-]}{[C_2H_3O_2^-]} = [H^+][OH^-] = K_w = 10^{-14}$$

from which we may solve for K_h:

$$K_h = \frac{K_w}{K_i}$$

This equation is a general expression, in which K_i is the ionization constant for the weak acid or weak base that is *formed* on hydrolysis. For the hydrolysis of the $C_2H_3O_2^-$ ion, we get

$$K_h = \frac{1.00 \times 10^{-14}}{1.74 \times 10^{-5}} = 5.75 \times 10^{-10} = \frac{[HC_2H_3O_2][OH^-]}{[C_2H_3O_2^-]}$$

For the hydrolysis of the NH_4^+ ion, we get

$$K_h = \frac{1.00 \times 10^{-14}}{1.74 \times 10^{-5}} = 5.75 \times 10^{-10} = \frac{[NH_3][H^+]}{[NH_4^+]}$$

If the salt has ions of both a weak acid and a weak base,

$$K_h = \frac{K_w}{K_a K_b}$$

We may use the hydrolysis constant to compute the pH of a salt solution, as illustrated in the following problems.

PROBLEM:
Calculate the pH of a 0.0500 M $NaC_2H_3O_2$ solution.

SOLUTION:
Write the chemical equation, showing the equilibrium concentrations above the symbols.

$$\begin{array}{cccc} 0.0500 - x & & x & x \\ C_2H_3O_2^- & + H_2O \rightleftarrows & HC_2H_3O_2 & + OH^- \end{array}$$

As in previous problems, we represent the small unknown concentration of OH^- by x. Because $HC_2H_3O_2$ is formed simultaneously with OH^- and in equimolar amounts, its concentration also is x. The *un*hydrolyzed concentration of $C_2H_3O_2^-$ is $0.0500 - x$. Substitution of these values into the K_h expression gives

$$K_h = \frac{K_w}{K_i} = \frac{1.00 \times 10^{-14}}{1.74 \times 10^{-5}} = 5.75 \times 10^{-10} = \frac{x^2}{(0.0500 - x)}$$

Neglecting x compared to 0.0500, we have

$$\frac{x^2}{0.0500} = 5.75 \times 10^{-10}$$

$$x^2 = 2.88 \times 10^{-11}$$

$$x = [OH^-] = 5.36 \times 10^{-6} \text{ M}$$

$$pOH = -\log (5.36 \times 10^{-6}) = 5.27$$

$$pH = 14.00 - 5.27 = 8.73$$

PROBLEM:

If 25.0 ml of 0.200 M NaOH are added to 50.0 ml of 0.100 M $HC_2H_3O_2$, what is the pH of the resulting solution?

SOLUTION:

We must first find the number of moles of acid and base used, in order to determine whether there is an excess of either one.

$$\text{Moles of NaOH added} = (0.0250 \text{ liter}) \left(0.200 \ \frac{\text{mole}}{\text{liter}} \right) = 0.00500 \text{ mole}$$

$$\text{Moles of } HC_2H_3O_2 \text{ added} = (0.0500 \text{ liter}) \left(0.100 \ \frac{\text{mole}}{\text{liter}} \right) = 0.00500 \text{ mole}$$

Because equal numbers of moles of acid and base are added, only the salt is present; 0.00500 mole of $NaC_2H_3O_2$ in 0.0750 liter to give

$$[C_2H_3O_2^-] = \frac{0.00500 \text{ mole}}{0.0750 \text{ liter}} = 0.0667 \text{ M}$$

Because only a salt is present in solution, the problem involves the hydrolysis equilibrium

$$\begin{matrix} 0.0667 - x & & x & & x \\ C_2H_3O_2^- & + H_2O \rightleftarrows & HC_2H_3O_2 & + & OH^- \end{matrix}$$

just as in the last problem. Solving in the same manner gives

$$K_h = 5.75 \times 10^{-10} = \frac{x^2}{0.0667 - x} = \frac{x^2}{0.0667}$$

$$x^2 = (0.0667)(5.75 \times 10^{-10}) = 3.84 \times 10^{-11}$$

$$x = [OH^-] = 6.19 \times 10^{-6} \text{ M}$$

$$pOH = -\log (6.19 \times 10^{-6}) = 5.21$$

$$pH = 14.00 - 5.21 = 8.79$$

This problem emphasizes the point that, at the endpoint in a titration, there is only the salt present (no excess acid or base), and the pH of the solution will be determined entirely by the hydrolysis of that salt.

---

pH OF "ACID-SALT" SOLUTIONS

How should the salt of a partially neutralized acid (such as $NaHSO_4$ or $KHCO_3$) be considered? Should the negative ion be treated as a weak acid that ionizes to make the solution acidic, or as an ion that hydrolyzes to make the solution basic? We know that such salts are soluble (they are Na^+ and K^+ salts) and that we can ignore the hydrolysis of the Na^+ and K^+ ions.

Each ion must be considered on its own merits, but the answer is easy to obtain. We compare the equilibrium constants (K_i and K_h) for the two possible reactions of the negative ion, and whichever has the larger constant will be the predominant reaction, so that we can ignore the other. The following problem illustrates the decision-making process.

PROBLEM:
What is the pH of a 0.100 M $KHCO_3$ solution?

SOLUTION:
The two possible reactions of the HCO_3^- ion and the associated equilibrium constants are the following.

$$HCO_3^- \rightleftarrows H^+ + CO_3^{2-}$$

$$K_i = \frac{[H^+][CO_3^{2-}]}{[HCO_3^-]} = 4.68 \times 10^{-11}$$

$$HCO_3^- + H_2O \rightleftarrows H_2CO_3 + OH^-$$

$$K_h = \frac{[H_2CO_3][OH^-]}{[HCO_3^-]} = \frac{1.00 \times 10^{-14}}{4.47 \times 10^{-7}} = 2.24 \times 10^{-8}$$

Because $K_h \gg K_i$, the hydrolysis reaction predominates, and the solution will be basic. Letting $x = [OH^-] = [H_2CO_3]$, and $[HCO_3^-] = 0.100 - x \cong 0.100$ M, we have

$$K_h = 2.24 \times 10^{-8} = \frac{x^2}{0.100}$$

$$x = [OH^-] = 4.73 \times 10^{-5} \text{ M}$$

$$pOH = -\log (4.73 \times 10^{-5}) = 4.33$$

$$pH = 14.00 - 4.33 = 9.67$$

THE pH OF AQUEOUS SOLUTIONS (SUMMARY)

We are now in a position to summarize the various types of aqueous solutions we may deal with in chemistry and the computation of the pH of each type. In these calculations we use the ionization constants of Table 23-1.

1. *Pure water* (see p 341). The equilibrium reaction is

$$H_2O \rightleftarrows H^+ + OH^-$$

$$[H^+] = [OH^-] = 1.00 \times 10^{-7} \text{ M}$$

$$pH = 7$$

2. *Strong acid* (see p 342). Because of complete dissociation of acid in dilute solution,

$$[H^+] = \text{molarity of acid}$$

Thus, for 10^{-3} M HCl,

$$[H^+] = 10^{-3} \text{ M}$$

$$pH = 3$$

3. *Strong base* (see p 343). Because of complete dissociation of base in dilute solution,

$$[OH^-] = \text{molarity of base}$$

Thus, for 10^{-3} M NaOH,

$$[OH^-] = 10^{-3} \text{ M}$$

$$pOH = 3$$

$$pH = 14 - 3 = 11$$

4. *Weak acid* (see p 350). The equilibrium reaction is

$$HA \rightleftarrows H^+ + A^-$$

$$[H^+] = [A^-]$$

$$[H^+][A^-] = [H^+]^2 = [HA]K_i$$

$$[H^+] = \sqrt{[HA]K_i}$$

5. *Weak base* (see p 353). The equilibrium reaction is

$$B + H_2O \rightleftharpoons BH^+ + OH^-$$

$$[BH^+] = [OH^-]$$

$$[OH^-][BH^+] = [OH^-]^2 = [B]K_i$$

$$[OH^-] = \sqrt{[B]K_i}$$

6. *Weak acid and its salt,* or *weak base and its salt* (see p 356). Here the common-ion effect is involved; the solutions are buffers. The equilibria involved are the same as in type 4 or 5 above, *except* that for the acid solution $[H^+] \neq [A^-]$, and for the basic solution $[OH^-] \neq [BH^+]$. After first calculating the salt concentration and the concentration of the corresponding acid or base, we find

$$[H^+] = \frac{[HA]}{[A^-]} K_i = \frac{[acid]}{[salt]} K_i \text{ for the acid solution}$$

$$[OH^-] = \frac{[B]}{[BH^+]} K_i = \frac{[base]}{[salt]} K_i \text{ for the base solution}$$

7. *Salt of a strong acid and a strong base* (see p 358). Salts of this type do not hydrolyze. Consequently, the pH of such solutions is 7.00, the same as that of pure water.

8. *Salt of a weak acid and a strong base* (see p 359). This is a hydrolysis equilibrium reaction:

$$A^- + H_2O \rightleftharpoons HA + OH^-$$

$$[OH^-] = [HA]$$

$$[OH^-]^2 = [A^-]K_h = \frac{[A^-]10^{-14}}{K_i} = \frac{[salt]10^{-14}}{K_i}$$

$$[OH^-] = \sqrt{\frac{[A^-]10^{-14}}{K_i}}, \text{ where } K_i \text{ is for the acid, HA}$$

9. *Salt of a weak base and a strong acid* (see p 359). This is a hydrolysis equilibrium reaction:

$$BH^+ + H_2O \rightleftharpoons B + H_3O^+$$

$$[H^+] = [B]$$

$$[H^+]^2 = [BH^+]K_h = \frac{[BH^+]10^{-14}}{K_i} = \frac{[salt]10^{-14}}{K_i}$$

$$[H^+] = \sqrt{\frac{[BH^+]10^{-14}}{K_i}}, \quad \text{where } K_i \text{ is for the base, B}$$

TITRATION CURVES

When an acid or base solution is titrated, each addition of reagent causes a change in pH. A plot of pH versus volume of reagent added is known as a titration curve. Figure 23-1 shows an example, for titration of 25 ml of 0.100 M $HC_2H_3O_2$ by 0.100 M NaOH. The points used to plot this curve are computed by the methods of the preceding section. Four different types of calculations are involved.

1. Initial point, before any base is added. The solution contains only $HC_2H_3O_2$ (calculate as in type 4 of the preceding section).

2. Intermediate region. The solution contains excess acid plus its salt (calculate as in type 6).

3. Equivalence point. The solution contains the salt of a weak acid and a strong base (calculate as in type 8).

4. Beyond equivalence point. The solution contains an excess of strong base (calculate as in type 3).

Every titration curve involves the same four categories, but exactly which of the nine types of calculation you use will depend on whether (a) the acid is added to the base, or vice versa; (b) the acid is strong or weak; and (c) the base is strong or weak.

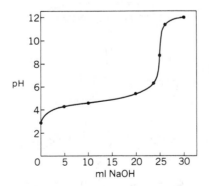

FIGURE 23-1
Titration of acetic acid with sodium hydroxide:
pH versus ml of NaOH.

ENDPOINT INDICATORS

When a pH meter is used to locate the endpoint of a titration, the midpoint of the nearly vertical region of the curve (Figure 23-1) is taken as the endpoint. When an indicator is used to locate the endpoint, you must select one that changes color at the pH corresponding to the pure salt formed in the titration (points 7, 8, and 9 of the pH summary).

Acid–base indicators are themselves weak acids (or bases) that possess an intense color in the acid form (HIn) and a different intense color in the salt form (In⁻). The equilibrium between the two color forms is

$$HIn \rightleftarrows H^+ + In^-$$

and the equilibrium constant expression is

$$K_i = \frac{[H^+][In^-]}{[HIn]}$$

If the equilibrium lies far to either side, the solution will have the color corresponding to the predominant form; the eye will be unable to perceive the small amount of the other form. When the equilibrium is such that there are equal concentrations of the two forms, $[HIn] = [In^-]$, the eye will detect a rapid change in color if the ratio changes slightly. This is the situation where the indicator will be useful, and the $[H^+]$ that corresponds to the point of most sensitive color change is just equal to the K_i of the indicator.

To select a proper indicator, first calculate the $[H^+]$ that will exist at the endpoint (points 7, 8, and 9 of the pH summary), then select an indicator whose K_i is as nearly the same as possible as the calculated $[H^+]$. At the endpoint of the addition of NaOH to $HC_2H_3O_2$ (see p 361), the pH is 8.79, corresponding to $[H^+] = 1.62 \times 10^{-9}$. You can see from Table 23-2 that phenolphthalein ($K_i = 1.6 \times 10^{-9}$) would be an ideal indicator for this titration.

TABLE 23-2
Characteristics of Some Acid–Base Indicators

Indicator	K_i	Colors	
		Acid form	Base form
Methyl yellow	3.5×10^{-4}	Red	Yellow
Bromcresol green	2.5×10^{-5}	Yellow	Blue
Methyl red	5.6×10^{-6}	Red	Yellow
Bromthymol blue	1.6×10^{-7}	Yellow	Blue
Cresol purple	6.3×10^{-9}	Yellow	Purple
Phenolphthalein	1.6×10^{-9}	Colorless	Magenta
Thymolphthalein	1.2×10^{-10}	Colorless	Blue

PROBLEMS A

1. Calculate the ionization constants for each of the weak electrolytes in the following table, using the experimental data given.

Electrolyte	Concentration	[H$^+$]	[OH$^-$]
(a) HCN	1.00 M	2.51×10^{-5} M	——
(b) HC$_2$H$_3$O$_2$	0.00100 M	1.23×10^{-4} M	——
(c) NH$_3$	0.00690 M	——	3.38×10^{-4} M
(d) HC$_2$H$_3$O$_2$	0.200 M	1.86×10^{-3} M	——
(e) HCHO$_2$	0.200 M	5.75×10^{-3} M	——

2. By pH meter or by color comparison with indicators, the solutions listed below are found to have the pH values given. Calculate the ionization constant for each.
   (a) 0.0100 M HC$_2$H$_3$O$_2$ has a pH of 3.39
   (b) 0.0500 M HCN has a pH of 5.25
   (c) 0.0400 M NH$_3$ has a pH of 10.92
   (d) 0.0200 M HCNO has a pH of 2.70
   (e) 0.0100 M CH$_3$NH$_2$ has a pH of 11.28

3. Calculate the pH of each of the following solutions.
   (a) 0.500 M HCNO   (f) 1.50 M NH$_3$
   (b) 0.100 M HN$_3$ (g) 0.500 M C$_2$H$_5$NH$_2$
   (c) 0.0100 M H$_3$BO$_3$ (h) 0.200 M C$_5$H$_5$N
   (d) 0.250 M HC$_2$H$_3$O$_2$ (i) 0.00500 M NH$_3$
   (e) 0.750 M HNO$_2$ (j) 1.25 M (CH$_3$)$_3$N

4. Calculate the percentage of ionization of each of the following in water solution.
   (a) 0.00300 M HCN   (d) 0.0500 M H$_2$S
   (b) 0.600 M HCHO$_2$ (e) 0.00100 M C$_5$H$_5$N
   (c) 1.25 M (CH$_3$)$_2$NH

5. A 5.00 g sample of HC$_2$H$_3$O$_2$ is added to 500 ml of water.
   (a) What is the pH of the solution?
   (b) Now 5.00 g of NaC$_2$H$_3$O$_2$ are added to the solution. What is the pH of the solution now?

6. How many grams of NaC$_2$H$_3$O$_2$ must be added to 250 ml of a 0.100 M HC$_2$H$_3$O$_2$ solution to give a pH of 6.50?

7. A 10.0 ml sample of 2.00 M HC$_2$H$_3$O$_2$ is added to 30.0 ml of 1.00 M NaC$_2$H$_3$O$_2$. What is the pH of the solution?

8. A 5.00 g sample of NH$_4$NO$_3$ is added to 100 ml of 0.100 M NH$_3$. What is the pH of the solution?

9. How many grams of NH$_4$Cl must be added to 250 ml of 0.200 M NH$_3$ in order that the solution shall have a pH of 7.20?

10. A 125 ml sample of 1.00 M NH_4Cl is added to 50.0 ml of 2.00 M NH_3. What is the pH of the resulting solution?

11. What is the pH of the solution that results from the addition of 50.0 ml of 0.100 M NaOH to 50.0 ml of 0.200 M $HC_2H_3O_2$?

12. What is the pH of the solution that results from the addition of 50.0 ml of 0.100 M NH_3 to 15.0 ml of 0.200 M HCl?

13. The pH of each of the following solutions is determined experimentally by pH meter or by color comparison. Calculate the hydrolysis constant for each salt.
    (a) 0.200 M NH_4Cl has a pH of 4.97
    (b) 0.0100 M KCN has a pH of 10.59
    (c) 0.100 M CH_3NH_3Cl has a pH of 5.83
    (d) 0.100 M $KCHO_2$ has a pH of 8.38
    (e) 0.0100 M $LiC_2H_3O_2$ has a pH of 8.38

14. Calculate the pH of each of the following solutions.
    (a) 0.400 M NaN_3 (e) 0.250 M $(CH_3)_2NH_2Cl$
    (b) 1.50 M NH_4NO_3 (f) 0.100 M KNO_3
    (c) 0.100 M KF   (g) 0.500 M $LiCHO_2$
    (d) 0.250 M NaCl   (h) 1.00×10^{-9} M HCl

15. Calculate the pH of the solution resulting from each of the following titration processes.
    (a) 50.0 ml of 0.100 M HCl + 50.0 ml 0.100 M NaOH
    (b) 25.0 ml 0.100 M HCl + 50.0 ml 0.0500 M NH_3
    (c) 35.0 ml 0.200 M HCN + 35.0 ml 0.200 M LiOH
    (d) 30.0 ml 0.100 M $HC_2H_3O_2$ + 10.0 ml 0.300 M NaOH

16. Select a suitable endpoint indicator for each of the titrations listed in Problem 15.

17. If 40.0 ml of 1.00 M $HC_2H_3O_2$ are added to 50.0 ml of 2.00 M $NaC_2H_3O_2$ solution, what will be the pH of the resulting solution?

18. How would you prepare a solution, using NH_4NO_3 and NH_3, so that its pH is 8.0?

19. You have a buffer solution that contains 1.00 mole of NH_4Cl and 1.00 mole of NH_3 per liter.
    (a) Calculate the pH of this solution.
    (b) Calculate the pH of the solution after the addition of 0.100 mole of solid NaOH to a liter.
    (c) Calculate the pH of this solution after the addition of 0.10 mole of HCl gas to a separate one-liter portion of the buffer.
    (d) Calculate the pH of a solution made by adding 0.10 mole of solid NaOH to 100 ml of water.
    (e) Calculate the pH of a solution made by adding 0.10 mole of HCl gas to 100 ml of H_2O.
    (f) Compare the answers to (a) through (e). Comment on their significance.

20. Compute the pH for each of the points shown in Figure 23-1. The titration involves the successive addition of 0.00, 5.0, 10.0, 24.0, 24.9, 25.0, 25.1, 26.0, and 30.0 ml of 0.100 M NaOH to 25.0 ml of 0.100 M $HC_2H_3O_2$.

21. Compute pH values and construct a curve for the titration of HCl by NaOH. Assume that an initial volume of 25.0 ml of 0.100 M HCl is used and is titrated by 0.100 M NaOH. Make calculations for additions of 0.00, 5.0, 10.0, 24.0, 24.9, 25.0, 25.1, 26.0, and 30.0 ml of NaOH solution.

PROBLEMS B

22. Experimental measurement shows that each of the following is ionized in water as shown. Calculate the ionization constant for each.
    (a) 0.100 M HF is 7.89% ionized
    (b) 0.200 M HNO_2 is 4.94% ionized
    (c) 0.100 M $HC_2H_3O_2$ is 1.31% ionized
    (d) 0.250 M NH_3 is 0.83% ionized
    (e) 0.200 M $HAsO_2$ is 0.0055% ionized

23. By pH meter or by color comparison with indicators, the solutions listed below are found to have the pH values given. Calculate the ionization constant for each.
    (a) 0.0100 M HNO_2 has a pH of 2.69
    (b) 0.00100 M HN_3 has a pH of 3.89
    (c) 0.00500 M HCN has a pH of 5.75
    (d) 0.200 M $(CH_3)_2NH$ has a pH of 12.02
    (e) 0.00200 M $C_6H_5NH_2$ has a pH of 7.96

24. Calculate the pH of each of the following solutions.
    (a) 0.00200 M HCN   (f) 2.50 M CH_3NH_2
    (b) 0.750 M $HCHO_2$ (g) 0.500 M NH_3
    (c) 1.20 M HF   (h) 1.75 M $C_6H_5NH_2$
    (d) 0.00200 M H_3BO_3 (i) 0.75 M $(CH_3)_2NH$
    (e) 0.100 M H_2S (j) 0.0200 M NH_3

25. Calculate the percentage of ionization of each of the following in water solution.
    (a) 0.00500 M C_5H_5N (d) 1.25 M NH_3
    (b) 0.800 M HN_3 (e) 2.50 M HF
    (c) 0.650 M HCNO

26. A 10.0 g sample of formic acid is added to 200 ml of water.
    (a) What is the pH of the solution?
    (b) Now 20.0 g of sodium formate are added to the solution. What is the pH of the resulting solution?

27. How many grams of $NaNO_2$ must be added to 300 ml of a 0.200 M HNO_2 solution to give a pH of 4.70?

28. A 20.0 ml sample of 1.00 M HNO_2 is added to 50.0 ml of 0.600 M $NaNO_2$. What is the pH of the solution?

29. A 25.0 g sample of NH_4Cl is added to 200 ml of 0.500 M NH_3. What is the pH of the solution?

30. How many grams of NH_4NO_3 must be added to 100 ml of 3.00 M NH_3 in order that the solution shall have a pH of 8.50?

31. A 75.0 ml sample of 2.00 M NH_4NO_3 is added to 100 ml of 1.00 M NH_3. What is the pH of the resulting solution?

32. What is the pH of the solution that results from the addition of 25.0 ml of 0.200 M KOH to 50.0 ml of 0.150 M HNO_2?

33. What is the pH of the solution that results from the addition of 25.0 ml of 0.200 M NH_3 to 30.0 ml of 0.150 M HCl?

34. The pH of each of the following solutions is determined by pH meter or color comparison. Calculate the hydrolysis constant for each salt.
   (a) 0.100 M NH_4Cl has a pH of 5.12
   (b) 0.0100 M $LiCHO_2$ has a pH of 7.89
   (c) 0.100 M $NaC_2H_3O_2$ has a pH of 8.88
   (d) 0.0100 M $(CH_3)_2NH_2Cl$ has a pH of 6.38
   (e) 0.100 M NaCN has a pH of 11.10

35. Calculate the pH of each of the following solutions.
   (a) 0.100 M NaCNO   (e) 0.800 M $C_2H_5NH_3Cl$
   (b) 0.0500 M NH_4NO_3 (f) 0.0100 M KCl
   (c) 1.50 M KNO_2 (g) 1.50 M $KC_2H_3O_2$
   (d) 0.350 M $NaNO_3$ (h) 1.00×10^{-9} M NaOH

36. Calculate the pH of the solution resulting from each of the following titration processes.
   (a) 25.0 ml of 0.200 M HNO_3 + 25.0 ml of 0.200 M KOH
   (b) 40.0 ml of 0.100 M $HCHO_2$ + 20.0 ml of 0.200 M NaOH
   (c) 35.0 ml of 0.100 M HNO_3 + 35.0 ml of 0.100 M C_5H_5N
   (d) 20.0 ml of 0.250 M H_3BO_3 + 10.0 ml of 0.500 M KOH

37. Select a suitable endpoint indicator for each of the titrations listed in Problem 36.

38. If 10.0 ml of 1.00 M HNO_2 are added to 20.0 ml of 2.00 M $NaNO_2$, what will be the pH of the resulting solution?

39. How would you prepare a solution, using $HC_2H_3O_2$ and $NaC_2H_3O_2$, so that the pH of the resulting solution is 5.5?

40. You have a buffer solution that contains 1.00 mole of $NaC_2H_3O_2$ and 1.00 mole of $HC_2H_3O_2$ per liter.
   (a) Calculate the pH of this solution.
   (b) Calculate the pH of the solution after the addition of 0.10 mole of solid NaOH to a liter.

   (c) Calculate the pH of this solution after the addition of 0.10 mole of HCl gas to a separate one-liter portion of the buffer.

   (d) Calculate the pH of a solution made by adding 0.10 mole of solid NaOH to 100 ml of water.

   (e) Calculate the pH of a solution made by adding 0.10 mole of HCl gas to 100 ml of H_2O.

   (f) Compare the answers for (a) through (e). Comment on their significance.

41. Construct a titration curve for the titration of 40.0 ml of 0.200 M NaOH with 0.200 M HCl. Make calculations for the addition of 0.00, 5.0, 20.0, 30.0, 39.0, 39.9, 40.0, 40.1, 41.0, and 45.0 ml of HCl.

42. Calculate pH values and construct a titration curve for the titration of 25.0 ml of 0.100 M NH_3 with 0.100 M HCl. Make calculations for the addition of 0.0, 5.0, 10.0, 20.0, 24.0, 24.9, 25.0, 25.1, 26.0, and 30.0 ml of HCl.

Solubility Product and Precipitation

In Chapter 23, we discuss applications of equilibrium relations to the pH of aqueous solutions. Another important application of these principles is to precipitation reactions.

We tend to think of the precipitates we separate in chemical analysis as insoluble. Actually, any precipitate is somewhat soluble. When we have a precipitate of AgCl, for example, in contact with supernatant liquid, the liquid is saturated with the ions of the precipitate, Ag^+ and Cl^-. The equation is

$$AgCl_{(s)} \rightleftarrows Ag^+ + Cl^-$$

The symbol (s) designates a solid. At the surface of the precipitate, Ag^+ and Cl^- ions are constantly going into solution and redepositing from the solution. Because this is an equilibrium process, we can apply the mathematical relation

$$\frac{[Ag^+][Cl^-]}{[AgCl_{(s)}]} = K_e$$

The concentration of solid AgCl crystals is constant and invariable, so we may simplify the equation by combining the two constants:

$$[Ag^+][Cl^-] = K_e \times [AgCl_{(s)}] = K_{sp}$$

When written in this form, to involve only the product of the concentrations of the ions, the constant K_{sp} is called the "solubility-product constant."

EVALUATION OF K_{sp}

It is found experimentally that, in a saturated solution of AgCl at room temperature, the concentration of dissolved salt is 0.00191 g/liter, or 1.33×10^{-5} mole/liter. When a molecule of AgCl dissolves, it dissociates to give an Ag^+ ion and a Cl^- ion. Therefore, the concentrations of Ag^+ and Cl^- in the saturated solution are each 1.33×10^{-5} M. Substituting these concentrations, we have

$$[Ag^+][Cl^-] = (1.33 \times 10^{-5})^2 = 1.78 \times 10^{-10} = K_{sp}$$

When the precipitate has different numbers of positive and negative ions, the calculation of K_{sp} is a little more complicated, as illustrated by the next problem.

PROBLEM:
The solubility of $Mg(OH)_2$ is 8.34×10^{-4} g/100 ml at 25.0°C. Calculate the K_{sp} value.

SOLUTION:
The chemical equilibrium involved is

$$Mg(OH)_{2(s)} \rightleftarrows Mg^{2+} + 2OH^-$$

Each mole of $Mg(OH)_2$ that dissolves produces 1 mole of Mg^{2+} and 2 moles of OH^-.

$$\text{Solubility of } Mg(OH)_2 = \left(\frac{8.34 \times 10^{-4} \text{ g}}{100 \text{ ml}}\right)\left(10^3 \frac{ml}{liter}\right)\left(\frac{1 \text{ mole}}{58.3 \text{ g}}\right)$$

$$= 1.43 \times 10^{-4} \text{ M}$$

$$[Mg^{2+}] = 1.43 \times 10^{-4} \text{ M}$$

$$[OH^-] = 2.86 \times 10^{-4} \text{ M}$$

Based on the chemical equation, the K_{sp} expression is

$$K_{sp} = [Mg^{2+}][OH^-]^2$$

$$= (1.43 \times 10^{-4})(2.86 \times 10^{-4})^2$$

$$= 1.16 \times 10^{-11}$$

Table 24-1 lists experimentally-determined values of K_{sp} for some common insoluble compounds.

TABLE 24-1
Solubility Products at 25°C

Salt	K_{sp}	Salt	K_{sp}	Salt	K_{sp}
Carbonates:		Iodates:		Sulfates:	
Ag_2CO_3	7.94×10^{-12}	$AgIO_3$	3.09×10^{-8}	Ag_2SO_4	1.58×10^{-5}
$BaCO_3$	5.50×10^{-10}	$Ce(IO_3)_3$	3.16×10^{-10}	$BaSO_4$	1.07×10^{-10}
$CaCO_3$	4.79×10^{-9}	$Pb(IO_3)_2$	2.51×10^{-13}	$CaSO_4$	2.40×10^{-5}
$CuCO_3$	2.34×10^{-10}	Oxalates:		Hg_2SO_4	7.41×10^{-7}
$FeCO_3$	3.16×10^{-11}	BaC_2O_4	1.58×10^{-7}	$PbSO_4$	1.66×10^{-8}
$MgCO_3$	1.00×10^{-5}	CaC_2O_4	2.29×10^{-9}	$SrSO_4$	2.51×10^{-7}
$MnCO_3$	5.00×10^{-10}	MgC_2O_4	7.94×10^{-5}	Sulfides:	
$PbCO_3$	6.31×10^{-14}	Hydroxides:		Ag_2S	6×10^{-50}
$SrCO_3$	1.10×10^{-10}	$Al(OH)_3$	6.31×10^{-32}	Bi_2S_3	1×10^{-97}
Chromates:		$Ca(OH)_2$	3.72×10^{-6}	CdS	8×10^{-27}
Ag_2CrO_4	1.12×10^{-12}	$Cd(OH)_2$	4.00×10^{-15}	CoS	2×10^{-25}
$BaCrO_4$	1.17×10^{-10}	$Cr(OH)_3$	1.00×10^{-30}	CuS	6×10^{-36}
$PbCrO_4$	1.78×10^{-14}	$Cu(OH)_2$	1.58×10^{-19}	FeS	6×10^{-18}
$SrCrO_4$	2.24×10^{-5}	$Fe(OH)_2$	7.94×10^{-16}	HgS	1.6×10^{-52}
Halides:		$Fe(OH)_3$	7.94×10^{-40}	MnS	3×10^{-10}
$AgCl$	1.78×10^{-10}	$Mg(OH)_2$	1.16×10^{-11}	NiS	1×10^{-24}
$AgBr$	4.90×10^{-13}	$Mn(OH)_2$	1.58×10^{-13}	PbS	2.5×10^{-27}
AgI	8.32×10^{-17}	$Ni(OH)_2$	6.50×10^{-18}	SnS	1×10^{-25}
CaF_2	2.69×10^{-11}	$Pb(OH)_2$	5.00×10^{-16}	Tl_2S	5×10^{-21}
Hg_2Cl_2	1.32×10^{-18}	$Sn(OH)_2$	6.31×10^{-27}	ZnS	1.6×10^{-24}
$PbCl_2$	1.62×10^{-5}	$Zn(OH)_2$	3.16×10^{-16}		
PbI_2	6.46×10^{-9}				
SrF_2	2.45×10^{-9}				

SOLUBILITY IN WATER

Values of K_{sp} make it possible to calculate the solubility of compounds under a wide variety of conditions. The simplest of these calculations is the solubility in water.

PROBLEM:
What is the solubility of $PbSO_4$ in grams/100 ml of water?

SOLUTION:
The chemical equilibrium is

$$PbSO_4 \rightleftarrows Pb^{2+} + SO_4^{2-}$$

If we let s = the molar solubility of $PbSO_4$, we see that $[Pb^{2+}] = s$, and $[SO_4^{2-}] = s$ because, for every mole of $PbSO_4$ that dissolves, there is produced one mole each of Pb^{2+} and SO_4^{2-}. Substituting these values in the K_{sp} expression, we obtain

$$K_{sp} = [Pb^{2+}][SO_4^{2-}] = (s)(s) = s^2 = 1.66 \times 10^{-8}$$

$$s = 1.29 \times 10^{-4} \text{ moles/liter of } PbSO_4$$

$$\text{g of } PbSO_4/100 \text{ ml} = \left(1.29 \times 10^{-4} \frac{\text{mole}}{\text{liter}}\right)(0.100 \text{ liter})\left(303.3 \frac{\text{mole}}{\text{mole}}\right)$$

$$= 3.91 \times 10^{-3} \text{ g}$$

PROBLEM:

What is the solubility of Ag_2CrO_4 in grams/100 ml of water?

SOLUTION:

The chemical reaction is

$$Ag_2CrO_4 \rightleftarrows 2Ag^+ + CrO_4^{2-}$$

If we let s = the molar solubility of Ag_2CrO_4, we see that $[Ag^+] = 2s$, and $[CrO_4^{2-}] = s$ because, for every mole of Ag_2CrO_4 that dissolves, there are produced 2 moles of Ag^+ and 1 mole of CrO_4^{2-}. Substituting these values in the K_{sp} expression, we have

$$K_{sp} = [Ag^+]^2[CrO_4^{2-}] = (2s)^2(s) = 4s^3 = 1.12 \times 10^{-12}$$

$$s = \left(\frac{1.12 \times 10^{-12}}{4}\right)^{\frac{1}{3}} = 6.54 \times 10^{-5} \frac{\text{moles}}{\text{liter}} \text{ of } Ag_2CrO_4$$

$$\text{g of } Ag_2CrO_4/100 \text{ ml} = \left(6.54 \times 10^{-5} \frac{\text{moles}}{\text{liter}}\right)(0.100 \text{ liter})\left(331.8 \frac{\text{g}}{\text{mole}}\right)$$

$$= 2.17 \times 10^{-3} \text{ g}$$

SOLUBILITY IN THE PRESENCE OF A COMMON ION

If there is introduced into the solution from some other source an ion that is in common with an ion of the insoluble solid, the chemical equilibrium is shifted to the left, and the solubility of that solid will be greatly decreased from what it is in pure water. This is called the "common-ion effect." This effect is important in gravimetric analysis, where one wishes to precipitate essentially all of the ion being analyzed for, by adding an excess of the "common-ion" precipitating reagent. There is a practical limit to the excess, however, which involves such factors as purity of precipitate and possibility of complex formation. You can calculate the solubility under a variety of conditions, as illustrated in the following problem.

PROBLEM:

What is the solubility of Ag_2CrO_4 in 0.100 M $AgNO_3$?

SOLUTION:

Just as in the preceding problem, the chemical equilibrium is

$$Ag_2CrO_4 \rightleftarrows 2Ag^+ + CrO_4^{2-}$$

Again, we let $s =$ the molar solubility of Ag_2CrO_4, but in this case the Ag^+ concentration is determined primarily by the 0.100 mole/liter of $AgNO_3$, to which there is added $2s$ moles/liter of Ag^+ from the Ag_2CrO_4 to give a total $[Ag^+] = (0.100 + 2s)$ moles/liter. The only source of CrO_4^{2-} is the Ag_2CrO_4, so $[CrO_4^{2-}] = s$ moles/liter. Substituting these values into the K_{sp} expression, we have

$$K_{sp} = [Ag^+][CrO_4^{2-}] = (0.100 + 2s)^2(s) = 1.12 \times 10^{-12}$$

In this problem, s will be even smaller than in the preceding problem, because the solubility is decreased by the common ion; as a consequence, $2s$ can be neglected compared to 0.100. This gives

$$(0.100)^2(s) = 1.12 \times 10^{-12}$$

$$s = 1.12 \times 10^{-10} \ \frac{\text{moles}}{\text{liter}} \ \text{of} \ Ag_2CrO_4$$

You can see that, compared to water solubility, the solubility has been decreased by a factor of more than 100,000.

CRITERION FOR PRECIPITATION

A K_{sp} value represents the product of concentrations that exist in a *saturated* solution. If you try to put into solution a combination of ions whose concentrations exceed the solubility product, you will get a precipitate because you have exceeded the quantities needed to saturate the solution. The *ion product* (Q) is an expression that *looks* like the K_{sp} expression, but it uses the actual concentrations you *propose* to have in solution, not the equilibrium concentrations. If the value of Q is larger than K_{sp}, you will get a precipitate; if it is smaller than K_{sp}, no precipitate will form; and if $Q = K_{sp}$, the solution will be just saturated. The prediction of precipitation is illustrated in the following problems.

PROBLEM:

If 1.00 mg of Na_2CrO_4 is added to 200 ml of 2.00×10^{-3} M $AgNO_3$, will a precipitate form?

SOLUTION:

We must find out whether the proposed concentrations of $[Ag^+]$ and $[CrO_4^{2-}]$ will give a value of Q larger than K_{sp}. The *proposed* concentrations are

$$[Ag^+] = 2.00 \times 10^{-3} \text{ M (given)}$$

$$[CrO_4^{2-}] = \frac{(1.00 \times 10^{-3} \text{ g})}{\left(162.0 \ \frac{\text{g}}{\text{mole}}\right)(0.200 \text{ liter})} = 3.09 \times 10^{-5} \text{ M}$$

$$Q = (2.00 \times 10^{-3})^2(3.09 \times 10^{-5}) = 1.24 \times 10^{-10}$$

$$K_{sp} = 1.12 \times 10^{-12} \text{ (from Table 24-1)}$$

We see that $Q > K_{sp}$. Therefore, a precipitate will form.

PROBLEM:
Will a precipitate of $Mn(OH)_2$ form if you add 1.00 ml of 0.0100 M NH_3 to 250 ml of 2.00×10^{-4} M $Mn(NO_3)_2$?

SOLUTION:
Again, we must find whether the *proposed* concentrations of Mn^{2+} and OH^- will give a value of Q that is larger than K_{sp}. This problem is more complicated than the preceding one, in that NH_3 is a weak base and we will have to find the $[OH^-]$ from the NH_3 dissociation equilibrium

$$NH_3 + H_2O \rightleftarrows NH_4^+ + OH^-$$

The NH_3 concentration before dissociation will be

$$\left(\frac{1 \text{ ml}}{251 \text{ ml}}\right)\left(0.0100 \ \frac{\text{mole}}{\text{liter}}\right) = 4.00 \times 10^{-5} \text{ M}$$

As in the problem on p. 354, we let $x = [OH^-] = [NH_4^+]$, and $[NH_3] = 4.00 \times 10^{-5} - x$. Substituting these values into the K_i expression, we obtain

$$K_i = 1.74 \times 10^{-5} = \frac{x^2}{4.00 \times 10^{-5} - x}$$

We cannot neglect x compared to 4×10^{-5}, because x is greater than 10% of 4×10^{-5}. Solving the quadratic equation, we find

$$x = [OH^-] = 1.91 \times 10^{-5} \text{ M}$$

$$[Mn^{2+}] = \left(\frac{250 \text{ ml}}{251 \text{ ml}}\right)\left(2.00 \times 10^{-4} \ \frac{\text{moles}}{\text{liter}}\right) = 2.00 \times 10^{-4} \text{ M}$$

$$Q = [Mn^{2+}][OH^-]^2 = (2.00 \times 10^{-4})(1.91 \times 10^{-5})^2 = 7.29 \times 10^{-14}$$

$$K_{sp} = 1.58 \times 10^{-13} \text{ (from Table 24-1)}$$

We see that $Q < K_{sp}$, so no precipitate will form. Note that, if you had not taken into account the fact that NH_3 is a *weak* base, you would have concluded that $[OH^-] = 4.00 \times 10^{-5}$ M, obtained a value of $Q = 3.20 \times 10^{-13}$, and reached the *erroneous* conclusion that a precipitate would form.

CONTROL OF PRECIPITATION

Precipitation by Hydrogen Sulfide

For many reasons, H_2S is an important reagent for the separation of metal ions in solution. H_2S is a gas and therefore is easy to obtain from a commerical cylinder; it is easy to saturate a solution with H_2S (the saturation concentration is 0.10 M); the *enormous* range in K_{sp} values makes precipitations and separations amenable to control by pH control; and most metals form sulfides that are insoluble to some degree.

H_2S is a diprotic acid that ionizes in two stages,

$$H_2S \rightleftarrows H^+ + HS^-$$

$$HS^- \rightleftarrows H^+ + S^{2-}$$

with the corresponding equilibrium expressions

$$K_1 = \frac{[H^+][HS^-]}{[H_2S]} = 1.0 \times 10^{-7}$$

$$K_2 = \frac{[H^+][S^{2-}]}{[HS^-]} = 1.2 \times 10^{-13}$$

If we are interested only in the *acidity* of an H_2S solution, we consider only K_1, because the number of H^+ ions produced in the second stage is negligible compared to the number produced in the first (see p 353). In a simple aqueous solution of H_2S with no other source of H^+ or S^{2-}, $[H^+] = [HS^-]$ as a result of the first stage of ionization; $[S^{2-}]$ is calculated from the expression for K_2:

$$[S^{2-}] = \frac{[HS^-](1.2 \times 10^{-13})}{[H^+]} = 1.2 \times 10^{-13} \text{ M}$$

a concentration that is negligible compared to the $[HS^-]$ from which it came.

The variation and control of the $[S^{2-}]$ by control of pH is best calculated from the overall reaction,

$$H_2S \rightleftarrows 2H^+ + S^{2-}$$

whose equilibrium constant is the product of the constants for the two stages:

$$\frac{[H^+][HS^-]}{[H_2S]} \times \frac{[H^+][S^{2-}]}{[HS^-]} = \frac{[H^+]^2[S^{2-}]}{[H_2S]} = K_1 \times K_2 = 1.2 \times 10^{-20}$$

There is an important special form of this equilibrium expression for solutions *saturated* with H_2S, where $[H_2S] = 0.10$ M at 25.0°C. For this common situation, we have

$$[H^+]^2[S^{2-}] = 1.2 \times 10^{-21}$$

By means of buffers, you can maintain the $[S^{2-}]$ at whatever value you wish when you saturate a solution with H_2S. How this capability permits the separation of ions is illustrated in the next problem.

PROBLEM:
To what pH must a solution be buffered in order that the maximum amount of CdS be precipitated, on saturation with H_2S, without precipitating any FeS? The original solution is 0.020 M in both Cd^{2+} and Fe^{2+}.

SOLUTION:
The whole calculation is determined by the concentration of Fe^{2+}, the ion that forms the more soluble sulfide. This determines the maximum tolerable $[S^{2-}]$, which in turn fixes the minimum required acidity of the buffer. From the equilibrium,

$$FeS \rightleftarrows Fe^{2+} + S^{2-}$$

the maximal tolerable $[S^{2-}] = \dfrac{K_{sp}}{[Fe^{2+}]} = \dfrac{6.0 \times 10^{-18}}{0.020} = 3.0 \times 10^{-16}$ M. Any higher $[S^{2-}]$ will exceed K_{sp} and cause a precipitation of FeS. The minimal required $[H^+]$ to produce this small a value of $[S^{2-}]$ is

$$[H^+] = \left(\frac{1.20 \times 10^{-21}}{[S^{2-}]} \right)^{\frac{1}{2}} = \left(\frac{1.20 \times 10^{-21}}{3.0 \times 10^{-16}} \right)^{\frac{1}{2}} = 2.0 \times 10^{-3} \text{ M}$$

$$pH = -\log (2.0 \times 10^{-3}) = 2.70$$

If the $[H^+]$ is lower than 2.0×10^{-3} M (or the pH higher than 2.70), the $[S^{2-}]$ will be greater than 3.0×10^{-16} M, and a precipitate of FeS will form.
    When buffered at a pH of 2.70, which gives a $[S^{2-}]$ of 3.0×10^{-16} M, the concentration of Cd^{2+} remaining in solution will be given by

$$[Cd^{2+}] = \frac{K_{sp}}{[S^{2-}]} = \frac{8.0 \times 10^{-27}}{3.0 \times 10^{-16}} = 2.7 \times 10^{-11} \text{ M}$$

Most of the Cd^{2+} has been precipitated as CdS, but none of the Fe^{2+} has precipitated; a good separation has been achieved.

When a metal ion is precipitated as the sulfide, H^+ ions are produced. In the preceding example, the precipitation of 0.020 M Cd^{2+} by the reaction

$$Cd^{2+} + H_2S \rightleftarrows CdS + 2H^+$$

produces 0.040 M H^+. As precipitation proceeds, therefore, the solution becomes more acidic, and the concentration of the S^{2-} becomes less. This increase in acidity may seriously interfere with metal-ion separations or may cause inadequate precipitation of some of the ions. This problem is overcome by using a buffer that maintains a constant desired pH.

Dissolving a Precipitate

When a precipitate is treated with a reagent that can react with one of its ions, its solubility is increased, because the reduction of the concentration of one ion causes the solubility equilibrium to shift to the right. Two different types of reaction may be used to dissolve precipitates.

1. The negative ion of the precipitate may react with an added ion to form a weak acid or base.
2. The positive ion of the precipitate may react with a substance to form a slightly ionized complex ion (Chapter 25).

The most common example of the first type is the dissolving of insoluble salts of weak acids by strong acids. Many hydroxides, carbonates, sulfides, phosphates, borates, oxalates, and salts of other weak acids may be dissolved by strong acids, even though their solubility in water is extremely low. In the following problems, we consider two common questions: "How much precipitate will dissolve under certain conditions?" and "What conditions are needed to totally dissolve a given amount of precipitate?"

PROBLEM:
How many moles of CaC_2O_4 will dissolve in one liter of 0.100 M HCl?

SOLUTION:
Two equilibria are involved: the solubility of CaC_2O_4 (horizontal equation), and the dissociation of $HC_2O_4^-$ (vertical equation). The concentration of $C_2O_4^{2-}$ is common to both of them.

$$CaC_2O_4 \rightleftarrows Ca^{2+} + C_2O_4^{2-}$$
$$+$$
$$H^+$$
$$\updownarrow$$
$$HC_2O_4^-$$

For the solubility equilibrium,

$$[C_2O_4^{2-}] = \frac{K_{sp}}{[Ca^{2+}]}$$

For the dissociation equilibrium,

$$[C_2O_4^{2-}] = \frac{K_1[HC_2O_4^-]}{[H^+]}$$

Setting these two expressions for $[C_2O_4^{2-}]$ equal to each other, we obtain

$$\frac{K_{sp}}{K_i} = \frac{[Ca^{2+}][HC_2O_4^-]}{[H^+]} = \frac{2.29 \times 10^{-9}}{5.25 \times 10^{-5}} = 4.36 \times 10^{-5}$$

an expression that corresponds to the overall equilibrium

$$CaC_2O_{4(s)} + H^+ \rightleftarrows Ca^{2+} + HC_2O_4^-$$

If s moles of CaC_2O_4 dissolve per liter, there will be produced s moles/liter each of Ca^{2+} and $HC_2O_4^-$, leaving $(0.100 - s)$ moles/liter of H^+. Substitution of these values into the equilibrium expression gives

$$\frac{s^2}{0.100 - s} = 4.36 \times 10^{-5}$$

Neglecting s compared to 0.100, we have

$$s = 2.09 \times 10^{-3} \text{ M} = \text{moles of } CaC_2O_4 \text{ that dissolve per liter}$$

SEPARATION OF IONS

Separation of ions can be accomplished by selectively dissolving a mixture of precipitates, or by selectively precipitating a mixture of ions. The following problems illustrate this.

PROBLEM:
You have a precipitate that consists of 0.100 moles each of $Mg(OH)_2$ and $Mn(OH)_2$. You would like to dissolve all of the $Mg(OH)_2$, and as little as possible of the $Mn(OH)_2$, in 500 ml of 0.100 M NH_3 containing NH_4Cl. How many grams of NH_4Cl must be added?

SOLUTION:
The calculations for problems such as this must be based on the K_{sp} of the *more* soluble compound—in this case, the $Mg(OH)_2$. Two simultaneous equilibria are involved: the solubility of $Mg(OH)_2$ (horizontal equation), and the dissociation of NH_3 (vertical equation). The coefficient of 2 belongs only with the horizontal equation. The concentration of OH^- is common to both equilibria.

$$Mg(OH)_2 \rightleftarrows Mg^{2+} + 2OH^-$$
$$+$$
$$NH_4^+$$
$$\updownarrow$$
$$NH_3 + H_2O$$

For the solubility equilibrium,

$$[OH^-]^2 = \frac{K_{sp}}{[Mg^{2+}]}$$

For the dissociation equilibrium,

$$[OH^-] = \frac{K_i[NH_3]}{[NH_4^+]}$$

$$[OH^-]^2 = \frac{K_i^2[NH_3]^2}{[NH_4^+]^2}$$

Setting these two expressions for $[OH^-]^2$ equal to each other, we obtain

$$\frac{K_{sp}}{K_i^2} = \frac{[Mg^{2+}][NH_3]^2}{[NH_4^+]^2} = \frac{1.16 \times 10^{-11}}{(1.74 \times 10^{-5})^2} = 0.0383$$

an expression that corresponds to the overall equilibrium,

$$Mg(OH)_{2(s)} + 2NH_4^+ \rightleftarrows Mg^{2+} + 2NH_3 + 2H_2O$$

When the $Mg(OH)_2$ has *just* dissolved, the solution will be saturated with $Mg(OH)_2$, and $[Mg^{2+}] = 0.200$ M. Because 2 moles of NH_3 are produced for every Mg^{2+} produced, 0.400 mole/liter of NH_3 will be formed in addition to the 0.100 mole/liter already present, to make the total $[NH_3] = 0.500$ M. $[NH_4^+]$ is the unknown. Substituting these values, we have

$$[NH_4^+]^2 = \frac{[Mg^{2+}][NH_3]^2}{0.0383} = \frac{(0.20)(0.50)^2}{0.0383} = 1.31$$

$$[NH_4^+] = 1.14 \ \frac{moles}{liter} \ NH_4^+ \ \text{required in final solution}$$

The *dissolving* of the precipitate also requires NH_4^+; in fact, the production of 0.200 mole/liter of Mg^{2+} will require 0.400 mole/liter of NH_4^+, an amount that must be supplied *in addition* to the 1.14 moles/liter. The total $[NH_4^+]$ required in the *initial* solution must therefore be 1.54 moles/liter so as to have 1.14 moles/liter remaining in the final solution.

$$\text{Wt of } NH_4Cl \text{ needed} = \left(1.54 \ \frac{moles}{liter}\right)(0.500 \text{ liter})\left(53.5 \ \frac{g}{mole}\right)$$

$$= 41.2 \text{ g}$$

PROBLEM:
A solution is 0.200 M in both Mg^{2+} and Mn^{2+}. How many grams of NH_4Cl must be added to 500 ml of this solution if it is desired to make it 1.00 M in NH_3 and to precipitate the maximum amount of Mn^{2+} as $Mn(OH)_2$ without precipitating any Mg^{2+}?

SOLUTION:
At first sight, this might appear to be essentially the same problem as the preceding one, but it's not. All of the equilibria are the same, and the calculations must be based on $Mg(OH)_2$ just as before. The difference is this: in the last problem the $[NH_3]$ was *increased* as a result of dissolving the $Mg(OH)_2$,

whereas in this problem the $[NH_3]$ is *decreased* as a result of precipitating $Mn(OH)_2$:

$$Mn^{2+} + 2NH_3 + 2H_2O \rightarrow Mn(OH)_2\downarrow + 2NH_4^+$$

Because we are precipitating 0.20 mole of Mn^{2+} per liter, we will use up 0.40 mole of NH_3 per liter, leaving $[NH_3] = 1.00 - 0.40 = 0.60$ M. The $[NH_4^+]$, 0.40 mole/liter of which is produced in the precipitation reaction, is our unknown. Substituting these values as we did in the last problem, we obtain

$$[NH_4^+]^2 = \frac{[Mg^{2+}][NH_3]^2}{0.0383} = \frac{(0.20)(0.60)^2}{0.0383} = 1.88$$

$$[NH_4^+] = 1.37 \text{ M}$$

Thus, 1.37 moles/liter is the final concentration of NH_4^+ that must be in solution if the Mg^{2+} is to remain unprecipitated, but 0.40 mole/liter of this is produced by the precipitation of $Mn(OH)_2$, leaving only 0.97 mole/liter to be added in the form of NH_4Cl.

$$\text{Wt of } NH_4Cl \text{ needed} = \left(0.97 \, \frac{\text{mole}}{\text{liter}} \right) (0.500 \text{ liter}) \left(53.5 \, \frac{\text{g}}{\text{mole}} \right)$$

$$= 25.9 \text{ g}$$

PROBLEM:
To what pH must a solution be buffered if you wish to permit the maximum precipitation of $Mn(OH)_2$ with no precipitation of $Mg(OH)_2$, if the solution is 0.200 M in both Mg^{2+} and Mn^{2+}?

SOLUTION:
This problem presents the *easy* alternative for making the separation of Mg^{2+} and Mn^{2+} in the preceding two problems. The chemical equilibrium is based solely on

$$Mg(OH)_2 \rightleftarrows Mg^{2+} + 2OH^-$$

The maximal tolerable $[OH^-]$ that can exist without precipitating any of the 0.200 M Mg^{2+} is

$$[OH^-] = \left(\frac{K_{sp}}{[Mg^{2+}]} \right)^{\frac{1}{2}} = \left(\frac{1.16 \times 10^{-11}}{0.200} \right)^{\frac{1}{2}} = 7.62 \times 10^{-6}$$

$$pOH = 5.12$$

$$pH = 14.00 - 5.12 = 8.88$$

If you use a buffer of pH = 8.88 that possesses high concentrations of a weak base and its salt, the precipitation of Mn^{2+} will not use up enough base to significantly change the [base]/[salt] ratio (see p 356), the pH will remain essentially unchanged, and no Mg^{2+} will precipitate. The same buffer solution could be used to dissolve the $Mg(OH)_2$ from the mixture of $Mn(OH)_2$ and $Mg(OH)_2$ in the problem on p 381, because the $Mg(OH)_2$ will not use up enough salt from the buffer to significantly change the [base]/[salt] ratio, and the pH will remain essentially unchanged. In the two preceding problems, it is easily shown that the pH of the final solutions is 8.88.

PROBLEMS A

1. The following solubilities have been determined by experiment. From these experimental data, calculate the solubility products for the solids involved. The solubilities are given in grams/100 ml of solution.
   (a) AgI, 2.14×10^{-7} (e) Ag_3PO_4, 2.01×10^{-3}
   (b) $BaSO_4$, 2.41×10^{-4} (f) PbS, 1.20×10^{-12}
   (c) $Cd(OH)_2$, 1.46×10^{-4} (g) $BaCO_3$, 4.63×10^{-4}
   (d) SrF_2, 1.07×10^{-2} (h) Sb_2S_3, 3.63×10^{-18}

2. Which of the following solids will dissolve readily in an excess of 1.00 M HCl?
   (a) AgI   (d) $Zn(OH)_2$
   (b) $MnCO_3$ (e) $SrSO_4$
   (c) $SrCrO_4$ (f) BaC_2O_4

3. In each of the following cases, show whether a precipitate will form under the given conditions.
   (a) 1.00 ml of 0.100 M $AgNO_3$ is added to 1 liter of 0.0100 M Na_2SO_4
   (b) 1.00 g of $Pb(NO_3)_2$ is added to 100 ml of 0.0100 M HCl
   (c) 1.00 ml of 1.00 M NaOH is added to 1 liter of 1.00×10^{-4} M $Mg(NO_3)_2$
   (d) 1.00 ml of 1.00 M NH_3 is added to 1 liter of 1.00×10^{-4} M $Mg(NO_3)_2$
   (e) 1.00 ml of 0.100 M $Sr(NO_3)_2$ is added to 1 liter of 0.0100 M HF
   (f) 1.00 ml of 0.0100 M $Ba(NO_3)_2$ is added to 1 liter of 0.100 M $NaHC_2O_4$
   (g) 1.00 mg of $CaCl_2$ and 1.00 mg of $Na_2C_2O_4$ are added to 100 ml of H_2O

4. Calculate the solubility (in moles/liter) of each of the following compounds in water (neglect hydrolysis effects).
   (a) HgS   (d) Ag_2S
   (b) $SrCrO_4$ (e) PbI_2
   (c) CaC_2O_4 (f) $Zn(OH)_2$

5. Calculate how many grams of CaC_2O_4 will dissolve in 1.00 liter of (a) water, (b) 0.100 M $Na_2C_2O_4$, (c) 0.0100 M $CaCl_2$, (d) 0.100 M $NaNO_3$.

6. Calculate how many grams of PbI_2 will dissolve in 250 ml of (a) water, (b) 0.0100 M $Pb(NO_3)_2$, (c) 0.0100 M CaI_2.

7. What concentration of sulfide ion is needed to commence precipitation of each of the following metals from solution as the sulfide?
   (a) 0.100 M $CuCl_2$ (d) 1.00×10^{-3} M $Bi(NO_3)_3$
   (b) 1.00×10^{-4} M $AgNO_3$ (e) 1.00×10^{-6} M $Hg(NO_3)_2$
   (c) 0.0200 M $TlNO_3$

8. What OH^- concentration is needed to commence precipitation of each of the following metals as the hydroxide, if each is present to the extent of 1.00 mg of metal ion/milliliter?

(a) Cu^{2+} (d) Sn^{2+}
(b) Cr^{3+} (e) Al^{3+}
(c) Zn^{2+}

9. K_2CrO_4 is slowly added to a solution that is 0.0200 M in $Pb(NO_3)_2$ and 0.0200 M in $Ba(NO_3)_2$.
    (a) Which ion precipitates first?
    (b) What will be its concentration when the second ion begins to precipitate?

10. A 1.00 ml sample of 0.100 M $Pb(NO_3)_2$ is added to 100 ml of a solution saturated with $SrCrO_4$. How many milligrams of $PbCrO_4$ will precipitate?

11. What concentration of $C_2O_4^{2-}$ is needed to start precipitation of calcium from a saturated solution of $CaSO_4$?

12. An analyst wishes to determine the amount of calcium by precipitation as the oxalate. If his sample contains 50.0 mg Ca^{2+} in 250 ml of solution, and if he adds $(NH_4)_2C_2O_4$ to give an oxalate ion concentration of 0.500 M, what percentage of his Ca^{2+} will remain unprecipitated?

13. What will be the concentration of Cu^{2+} remaining in solution when Cd^{2+} just begins to precipitate as CdS from a solution that was originally 0.0200 M in both Cu^{2+} and Cd^{2+}?

14. A solution is 0.0200 M in Mg^{2+} and 0.100 M in NH_4NO_3. What concentration of NH_3 must be attained in order to begin precipitation of $Mg(OH)_2$?

15. In order to prevent precipitation of $Mg(OH)_2$, what is the minimal number of grams of NH_4Cl that must be added to 500 ml of a solution containing 3.00 g of $Mg(NO_3)_2$ and 5.00 g of NH_3?

16. How many moles of SrF_2 will dissolve in 250 ml of 0.100 M HNO_3?

17. What is the pH of a water solution saturated with $Fe(OH)_3$?

18. (a) To what final pH must a solution be adjusted in order to precipitate as much CdS as possible without precipitating any ZnS? (The solution is originally 0.0200 M with respect to each metal ion.)
    (b) How much Cd^{2+} will be left in solution when the Zn^{2+} begins to precipitate?

19. What must be the final pH of an H_2SO_4 solution in order to just dissolve 0.0500 mole of ZnS in 1 liter of the solution?

20. What is the lowest pH that will permit the sulfide of each of the following metals to precipitate from a 1.00×10^{-4} M solution saturated with H_2S?
    (a) Pb^{2+} (c) Co^{2+}
    (b) Bi^{3+} (d) Mn^{2+}

21. Can a suspension of AgI that is 5.00 M in HCl be converted to Ag_2S by saturation with H_2S? Explain.

22. How many liters of a solution saturated with HgS would one have to take to find, statistically, one Hg^{2+} ion? (Neglect hydrolysis of the sulfide ion.)

23. When a metal ion precipitates from an unbuffered solution on saturation with H_2S, the acidity of the solution increases as the metal removes the S^{2-}. This in turn increases the solubility of the metal sulfide. What will be the concentration of each of the following metals remaining after saturation of 0.100 M aqueous solutions with H_2S?
    (a) Cu^{2+} (b) Cd^{2+} (c) Zn^{2+}

24. A 0.0500 M $ZnCl_2$ solution is buffered with 2.00 moles of $NaC_2H_3O_2$ and 1.00 mole of $HC_2H_3O_2$ per liter. What will be the Zn^{2+} concentration remaining after saturation with H_2S?

25. A solution saturated with CO_2 at 1 atm and 25.0°C has a concentration of about 0.0340 M.
    (a) Write an equilibrium expression, in terms of H^+ and CO_3^{2-}, that could be usefully applied to the separation of metal ions as carbonates by saturation with CO_2.
    (b) At what pH must a solution be adjusted so as not to precipitate $BaCO_3$ from a 0.0200 M $Ba(NO_3)_2$ solution on saturation with CO_2?
    (c) What would be the concentration of Pb^{2+} remaining in a solution adjusted to the pH in (b)?
    (d) Why can't saturation with CO_2 be applied in a practical way to the separation of metal ions in the same way that saturation with H_2S is?

26. Which will precipitate first when solid Na_2S is slowly added to a 0.0100 M $MnSO_4$ solution: MnS or $Mn(OH)_2$?

27. Calculate the solubility in water of (a) PbS and (b) Tl_2S. (Do not neglect hydrolysis of the sulfide ion.)

PROBLEMS B

28. The following solubilities in water have been determined by experiment. From these experimental data, calculate the solubility products for the solids involved. The solubilities are given in grams/100 ml of solution.
    (a) $CaCO_3$, 6.93×10^{-4} (e) CaF_2, 1.47×10^{-3}
    (b) AgCN, 1.50×10^{-7} (f) $Ce(IO_3)_3$, 0.123
    (c) PbI_2, 5.41×10^{-2} (g) $MgNH_4PO_4$, 8.66×10^{-3}
    (d) Ag_2S, 6.11×10^{-16} (h) Hg_2Br_2, 1.36×10^{-6}

29. Which of the following solids will dissolve readily in an excess of 1.00 M HCl?
    (a) CaF_2 (e) $Cd(OH)_2$
    (b) AgBr   (f) $BaSO_4$
    (c) $MgCO_3$ (g) MgC_2O_4
    (d) $BaCrO_4$

30. In each of the following cases show whether a precipitate will form under the given conditions.
    (a) 1.00 g of $Sr(NO_3)_2$ is added to 1 liter of 0.00100 M K_2CrO_4
    (b) 1.00 ml of 0.0100 M $Pb(NO_3)_2$ is added to 1 liter of 0.0100 M H_2SO_4

(c) 1.00 ml of 1.00 M NaOH is added to 1 liter of 1.00×10^{-5} M $MnSO_4$
(d) 1.00 ml of 1.00 M NH_3 is added to 1 liter of 1.00×10^{-5} M $MnSO_4$
(e) 1.00 ml of 0.00100 M $Ca(NO_3)_2$ is added to 1 liter of 0.0100 M HF
(f) 1.00 ml of 0.0100 M $Ca(NO_3)_2$ is added to 1 liter of 0.100 M $NaHC_2O_4$
(g) 1.00 mg of $Ba(NO_3)_2$ is added to 1 liter of 0.0100 M $K_2C_2O_4$

31. Calculate the solubility (in moles/liter) of each of the following compounds in water (neglect hydrolysis effects).
    (a) BaC_2O_4 (d) Ag_2SO_4
    (b) $PbSO_4$ (e) $Ca(OH)_2$
    (c) AgBr

32. Calculate how many grams of $BaSO_4$ will dissolve in a liter of (a) water, (b) 0.100 M Na_2SO_4, (c) 0.100 M $Ba(NO_3)_2$, (d) 0.0100 M KCl.

33. Calculate how many grams of $Cd(OH)_2$ will dissolve in 250 ml of (a) water, (b) 0.0100 M KOH, (c) 0.0100 M $CdCl_2$.

34. What concentration of Ag^+ is needed to initiate precipitation of the negative ions from each of the following solutions? Assume in each case that the negative ion has not reacted with the water.
    (a) 0.0500 M NaBr   (d) 2.00 M H_2SO_4
    (b) 0.100 M K_2CO_3 (e) 0.0200 M $KH(IO_3)_2$
    (c) 0.00100 M $(NH_4)_2CrO_4$

35. What OH^- concentration is needed to commence precipitation of each of the following metals as the hydroxide if each is present in the amount of 1.00 mg/ml?
    (a) Cd^{2+} (d) Mn^{2+}
    (b) Fe^{2+} (e) Ca^{2+}
    (c) Fe^{3+}

36. KI is slowly added to a solution that is 0.0200 M in $Pb(NO_3)_2$ and 0.0200 M in $AgNO_3$.
    (a) Which ion precipitates first?
    (b) What will be its concentration when the second ion begins to precipitate?

37. What is the pH of a water solution saturated with $Sn(OH)_2$?

38. A 1.00 ml sample of 0.500 M K_2CrO_4 is added to 100 ml of a solution saturated with $PbCl_2$. How many milligrams of $PbCrO_4$ will precipitate?

39. What concentration of I^- is needed to start precipitation of silver from a saturated solution of AgCl?

40. An analyst wishes to determine the amount of Pb by precipitation as the sulfate. If her sample contains 40.0 mg Pb^{2+} in 300 ml of solution, and if she adds H_2SO_4 to give a sulfate concentration of 0.300 M, what percentage of the Pb^{2+} will remain unprecipitated?

41. What will be the concentration of Hg^{2+} remaining in solution when Tl^+ first begins to precipitate as Tl_2S from a solution that was originally 0.100 M in Hg^{2+} and Tl^+?

42. A 100 ml sample of 0.250 M NH_3 is added to 100 ml of 0.0200 M $Mn(NO_3)_2$ solution. What is the minimal number of grams of NH_4Cl that must be added to prevent the precipitation of $Mn(OH)_2$?

43. What volume of 3.00 M NH_3 must be added to 100 ml of a solution containing 1.00 g of $Mn(NO_3)_2$ and 75.0 g of NH_4Cl in order to commence the precipitation of $Mn(OH)_2$?

44. How many grams of NaCN must be added to 1.00 liter of 0.0100 M $Mg(NO_3)_2$ in order to commence precipitation of $Mg(OH)_2$?

45. What will be the molar concentration of the following ions remaining in solutions saturated with H_2S and adjusted to a final pH of 2.00?
    (a) Ag^+ (d) Mn^{2+}
    (b) Bi^{3+} (e) Tl^+
    (c) Ni^{2+}
    Initial concentrations are 0.0100 M.

46. Is it possible to precipitate PbS from a suspension of PbI_2 in 2.00 M HCl by saturation with H_2S? Explain.

47. (a) To what final pH must a solution saturated with H_2S be adjusted in order to precipitate SnS without precipitating any FeS? (This solution is originally 0.100 M with respect to each metal ion.)
    (b) How much Sn^{2+} will be left in solution when the Fe^{2+} begins to precipitate?

48. What must be the final pH of an HCl solution in order to just dissolve 0.200 mole FeS in 1.00 liter of solution?

49. What is the lowest pH that will permit the sulfide of each of the following metals to precipitate from a 1.00×10^{-4} M solution saturated with H_2S?
    (a) Tl^+ (c) Ni^{2+}
    (b) Cd^{2+} (d) Sn^{2+}

50. What volume of saturated Ag_2S solution would you have to take in order to obtain, statistically, one Ag^+ ion? (Neglect hydrolysis of the sulfide ion.)

51. A 100 ml sample of a solution contains 10.0 g of $NaC_2H_3O_2$, 5.00 ml of 6.00 M $HC_2H_3O_2$, and 50.0 mg of Fe^{2+}. When the solution is saturated with H_2S, how many milligrams of Fe^{2+} remain in solution?

52. A solution that is 0.0500 M in Cd^{2+}, 0.0500 M in Zn^{2+}, and 0.100 M in H^+ is saturated with H_2S. Calculate the concentration of Zn^{2+} and Cd^{2+} remaining in solution after saturation. (Do not neglect the H^+ formed on precipitation.)

53. Calculate the solubility in water of (a) MnS and (b) Ag_2S. (Do not neglect hydrolysis of the sulfide ion.)

54. Which will precipitate first when solid Na_2S is slowly added to a 0.0200 M $CdCl_2$ solution: CdS or $Cd(OH)_2$? Explain.

Complex Ions

Chapter 24 deals with the control of precipitation by controlling the way in which H^+ ions react with the *negative* ion of the solid. In this chapter, we turn our attention to control by using the way reagents react with the *positive* ion to form a complex ion. We also see how the pH of many salt solutions is related to the properties of complex ions.

THE NATURE OF COMPLEX IONS

All metal ions are hydrated in aqueous solution and exist in the form of aquo complex ions, such as hexaaquozinc(II), $Zn(H_2O)_6^{2+}$. For convenience, these water molecules normally are not shown when writing chemical equations. Some hydrated metal ions have the ability to exchange their water molecules for other molecules or ions to form, for example, $Zn(NH_3)_4^{2+}$ and $Zn(CN)_4^{2-}$. The molecules and ions that attach themselves to a metal ion to form a slightly ionized complex ion are called *ligands*. The structure of these complexes is considered in Chapter 9.

Like weak acids, complex ions dissociate to some extent. Unlike polyprotic weak acids that normally lose protons one after the other with enormously increasing difficulty, complex ions lose ligands one after the other with only slightly increasing difficulty. One consequence of this difference is that, unless there is an excess of ligand present in solution, there will be a whole collection of related complex ions in equilibrium with each other—such as $Zn(NH_3)^{2+}$, $Zn(NH_3)_2^{2+}$, $Zn(NH_3)_3^{2+}$, and $Zn(NH_3)_4^{2+}$, each with its characteristic and somewhat different dissociation constant. Problems that deal with solutions not having an excess of ligand are too messy for consideration at the elementary level but, in solutions with an excess of ligand, all of the intermediate stages of complexation can be ignored, and we can deal solely with the equilibrium between the fully complexed ion and its completely dissociated components. For example, with an excess of NH_3, we need to consider only the equilibrium

$$Zn(NH_3)_4^{2+} \rightleftarrows Zn^{2+} + 4NH_3$$

The corresponding equilibrium constant

$$K_{inst} = \frac{[Zn^{2+}][NH_3]^4}{[Zn(NH_3)_4^{2+}]}$$

is designated as the instability constant. By convention, it always refers to the dissociation of the complex ion, just as K_i refers to the dissociation of a weak acid.

THE INSTABILITY CONSTANT

Normally the equilibrium concentration of the uncomplexed ion in the presence of an excess of ligand is very small, and its determination may be difficult. One straightforward way is to place the equilibrium solution in a half-cell that uses the metal in question as the electrode. When this half-cell is used with another (as in Figure 17-1), the voltage of the cell can be used to determine the equilibrium concentration of the uncomplexed metal ion. The rest of the calculation is simple, as shown in the following problem.

PROBLEM:
When 0.100 mole of $ZnSO_4$ is added to one liter of 6.00 M NH_3, voltage measurements show that the $[Zn^{2+}]$ in this solution is 8.13×10^{-14} M. Calculate the value of K_{inst} for $Zn(NH_3)_4^{2+}$.

SOLUTION:
The equilibrium equation

$$Zn(NH_3)_4^{2+} \rightleftarrows Zn^{2+} + 4NH_3$$

shows that 4 moles of NH_3 are used up for every 1 mole of Zn^{2+} complexed. The fact that only 8.13×10^{-14} M Zn^{2+} remains uncomplexed means that, from a practical standpoint, $[Zn(NH_3)_4^{2+}] = 0.100$ M, and that $4 \times 0.100 = 0.400$ mole/liter of NH_3 was used up, leaving $[NH_3] = 5.60$ M. Substituting these values into the K_{inst} expression, we have

$$K_{inst} = \frac{[Zn^{2+}][NH_3]^4}{[Zn(NH_3)_4^{2+}]} = \frac{(8.13 \times 10^{-14})(5.60)^4}{0.100}$$

$$= 8.00 \times 10^{-10}$$

The K_{inst} values for a few common complex ions are listed in Table 25-1. One of the uses of these constants is the calculation of the concentration of the un-complexed metal in solution, as illustrated by the following problem.

TABLE 25-1
Instability Constants of Complex Ions at 25°C

Name*	Equilibrium	K_{inst}
Tetraamminecadmium(II)	$Cd(NH_3)_4^{2+} \rightleftarrows Cd^{2+} + 4NH_3$	1.0×10^{-7}
Tetraamminecopper(II)	$Cu(NH_3)_4^{2+} \rightleftarrows Cu^{2+} + 4NH_3$	2.5×10^{-13}
Diamminesilver(I)	$Ag(NH_3)_2^{+} \rightleftarrows Ag^{+} + 2NH_3$	4.0×10^{-8}
Tetraamminezinc(II)	$Zn(NH_3)_4^{2+} \rightleftarrows Zn^{2+} + 4NH_3$	8.0×10^{-10}
Tetrachloromercurate(II)	$HgCl_4^{2-} \rightleftarrows Hg^{2+} + 4Cl^{-}$	8.0×10^{-16}
Dichloroargentate(I)	$AgCl_2^{-} \rightleftarrows Ag^{+} + 2Cl^{-}$	3.2×10^{-5}
Tetracyanocadmate(II)	$Cd(CN)_4^{2-} \rightleftarrows Cd^{2+} + 4CN^{-}$	1.3×10^{-19}
Tetracyanocuprate(I)	$Cu(CN)_4^{3-} \rightleftarrows Cu^{+} + 4CN^{-}$	2.8×10^{-28}
Tetracyanomercurate(II)	$Hg(CN)_4^{2-} \rightleftarrows Hg^{2+} + 4CN^{-}$	3.2×10^{-42}
Dicyanoargentate(I)	$Ag(CN)_2^{-} \rightleftarrows Ag^{+} + 2CN^{-}$	8.0×10^{-22}
Tetrahydroxoaluminate(III)	$Al(OH)_4^{-} \rightleftarrows Al(OH)_{3(s)} + OH^{-}$	0.050
Tetrahydroxochromate(III)	$Cr(OH)_4^{-} \rightleftarrows Cr(OH)_{3(s)} + OH^{-}$	2.5
Trihydroxoplumbate(II)	$Pb(OH)_3^{-} \rightleftarrows Pb(OH)_{2(s)} + OH^{-}$	25.0
Trihydroxostannate(II)	$Sn(OH)_3^{-} \rightleftarrows Sn(OH)_{2(s)} + OH^{-}$	8.0
Tetrahydroxozincate(II)	$Zn(OH)_4^{2-} \rightleftarrows Zn(OH)_{2(s)} + 2OH^{-}$	1.6

*In the names of complex ions, the Roman numeral in parentheses after the name indicates the valence of the metal ion involved. The prefixes *di-, tri-, tetra-*, and so on are used to tell how many coordination partners are attached to the central metal ion. If the complex ion has a positive valence, the central metal ion uses only the name of the element, with no suffix. If the complex ion has a negative valence, the name of the central metal ion is always a common root of the element with the suffix *-ate*. The names of coordination partners usually end in *-o*; the common exception is NH_3 whose name here is *ammine*.

PROBLEM:
What is the concentration of Ag^+ in a solution made by adding 0.0500 moles of $AgNO_3$ to 500 ml of 0.500 M NH_3:

SOLUTION:
The equation for the equilibrium involved is

$$Ag(NH_3)_2^{+} \rightleftarrows Ag^{+} + 2NH_3$$

Because NH_3 is in excess and K_{inst} is small (4.0×10^{-8}), we know that almost all of the silver ion is tied up in the complex, and that it will be smart to let $x = [Ag^+]$. The *total* silver concentration is 0.0500 mole/0.500 liter = 0.100 M, with x moles/liter as Ag^+, and $(0.100 - x)$ moles/liter as $Ag(NH_3)_2^+$. Two moles of NH_3 per liter are used for every mole of $Ag(NH_3)_2^+$ formed per liter, leaving $0.500 - 2(0.100 - x)$ $= (0.300 + 2x)$ moles NH_3 per liter for the equilibrium concentration. Substituting these equilibrium concentrations into the K_{inst} expression, we obtain

$$K_{inst} = \frac{[Ag^+][NH_3]^2}{[Ag(NH_3)_2^+]} = \frac{(x)(0.300 + 2x)^2}{(0.100 - x)} = 4.00 \times 10^{-8}$$

Because x should be negligible compared to 0.10 and 0.30, we can simplify the equation to

$$x = [Ag^+] = \frac{(0.100)(4.00 \times 10^{-8})}{(0.300)^2} = 4.44 \times 10^{-8} \text{ M}$$

AQUO IONS AS WEAK ACIDS

Aqueous solutions of the +2 and +3 metal salts of strong acids are always slightly acid. This would probably seem reasonable to you if you reasoned from our general statement about the hydrolysis of salts on p 358, assuming that the metal ions come from weak bases (though they are actually "insoluble"). A more satisfactory explanation is that hydrated aquo metal ions can act as weak acids. For example, hexaaquo iron(III) could dissociate to give

$$Fe(H_2O)_6^{3+} \rightleftarrows Fe(H_2O)_5(OH)^{2+} + H^+$$

That the dissociation goes even farther is evident from the small bit of colloidal suspension of ferric hydroxide that is obtained by boiling a solution of $FeCl_3$ (you can see the beam of scattered light even though the solution is perfectly clear). This results from the reaction

$$Fe(H_2O)_6^{3+} \rightleftarrows Fe(H_2O)_3(OH)_3 + 3H^+$$

The acidity, though slight, is easily detected by litmus paper. A similar experiment with $FeCl_2$ would show an appreciably less acid solution. A little thought easily rationalizes this difference: the $Fe(H_2O)_6^{3+}$, which has a much higher positive charge density than $Fe(H_2O)_6^{2+}$, will exert more repulsion toward the H^+ on the ligand water molecules, the net result being more dissociation (and greater acidity). If we compare several aquo ions with the same charge, we usually find that the smallest ion has the most dissociation (because it has the highest charge density), and the largest ion has the least dissociation. We also find that strong-acid salts of singly-charged ions, such as $Na(H_2O)_6^+$ in NaCl solution, have negligible dissociation. Table 25-2 lists values of K_i for a few

TABLE 25-2
Ionization Constants of Aquo Metal Ions at 25°C

Dissociation Reaction	K_i
$Al(H_2O)_6^{3+} \rightleftarrows Al(H_2O)_5(OH)^{2+} + H^+$	1.00×10^{-5}
$Cu(H_2O)_6^{2+} \rightleftarrows Cu(H_2O)_5(OH)^+ + H^+$	4.57×10^{-8}
$Fe(H_2O)_6^{2+} \rightleftarrows Fe(H_2O)_5(OH)^+ + H^+$	1.20×10^{-6}
$Fe(H_2O)_6^{3+} \rightleftarrows Fe(H_2O)_5(OH)^{2+} + H^+$	7.41×10^{-3}
$Mg(H_2O)_6^{2+} \rightleftarrows Mg(H_2O)_5(OH)^+ + H^+$	2.00×10^{-12}
$Hg(H_2O)_6^{2+} \rightleftarrows Hg(H_2O)_5(OH)^+ + H^+$	3.24×10^{-3}
$Zn(H_2O)_6^{2+} \rightleftarrows Zn(H_2O)_5(OH)^+ + H^+$	3.55×10^{-10}

aquo metal ions. Note that the aquo Al^{3+} ion has about the same strength as acetic acid, and that the aquo Hg^{2+} and Fe^{3+} ions are appreciably stronger.

PROBLEM:
What is the pH of a 0.100 M $Al(NO_3)_3$ solution?

SOLUTION:
$Al(NO_3)_3$ is the salt of a strong acid, so we do not have to consider the effect of the NO_3^- ion; the pH will be determined by the dissociation of the hexaaquo-aluminum(III) ion:

$$Al(H_2O)_6^{3+} \rightleftarrows Al(H_2O)_5(OH)^{2+} + H^+$$

Applying the usual equilibrium expression, using the needed value of K_i from Table 25-2, we obtain

$$K_i = 1.00 \times 10^{-5} = \frac{[Al(H_2O)_5(OH)^{2+}][H^+]}{[Al(H_2O)_6^{3+}]}$$

As on p 350, we know that

$$[Al(H_2O)_5(OH)^{2+}] = [H^+] = x$$

and

$$[Al(H_2O)_6^{3+}] = 0.100 - x \cong 0.100 \text{ M}$$

Substitution gives

$$\frac{x^2}{0.100} = 1.00 \times 10^{-5}$$

$$x^2 = 1.00 \times 10^{-6}$$

$$[H^+] = x = 1.00 \times 10^{-3} \text{ M}$$

$$pH = 3.00$$

AMPHOTERISM (HYDROXO COMPLEXES)

Some insoluble hydroxides dissolve not only in acid but also in excess of strong base. Those that do so are said to be *amphoteric*. We illustrate this by $Al(OH)_3$:

$$Al(OH)_{3(s)} + 3H^+ \rightleftarrows Al^{3+} + 3H_2O \qquad \text{(in acid)}$$

$$Al(OH)_{3(s)} + OH^- \rightleftarrows Al(OH)_4^- \qquad \text{(in base)}$$

The hydroxo complex ions formed in this way have instability constants, just as ammine or other complexes do. These instability constants are somewhat special, in that one of the products of the equilibrium is the insoluble amphoteric hydroxide. Thus, for aluminum hydroxide,

$$Al(OH)_4^- \rightleftarrows Al(OH)_{3(s)} + OH^-$$

and the instability constant is

$$K_{inst} = \frac{[OH^-]}{[Al(OH)_4^-]} = 0.050$$

The $[Al(OH)_{3(s)}]$ is omitted, as usual, because it is a solid. Because the solids are part of the equilibrium, the instability constants of these hydroxyl complex ions can be applied *only* to solutions that are saturated with respect to the solid.

PROBLEM:
Solid $Pb(OH)_2$ is added to 250 ml of 1.00 M NaOH solution until no more dissolves. What weight of the solid goes into solution?

SOLUTION:
The equilibrium reaction is

$$Pb(OH)_3^- \rightleftarrows Pb(OH)_{2(s)} + OH^-$$

You can see that, for every mole of $Pb(OH)_2$ that dissolves, one mole of OH^- is used up, and one mole of $Pb(OH)_3^-$ is formed. If we let s = moles of $Pb(OH)_2$ that dissolve per liter, then s moles of $Pb(OH)_3^-$ will be formed per liter, and there will remain $(1.00 - s)$ moles of OH^- per liter at equilibrium. Substitution of these equilibrium concentrations into the K_{inst} expression gives

$$K_{inst} = \frac{[OH^-]}{[Pb(OH)_3^-]} = \frac{1.00 - s}{s} = 25.0$$

$s = 0.0385$ mole of $Pb(OH)_2$ that dissolves per liter

$$\text{Wt of Pb(OH)}_2 \text{ that dissolves} = \left(0.0385 \; \frac{\text{mole}}{\text{liter}}\right)(0.250 \; \text{liter})\left(241.2 \; \frac{\text{g}}{\text{mole}}\right)$$
$$= 2.32 \; \text{g in } 250 \; \text{ml}$$

CONTROL OF PRECIPITATION

The insoluble salts of metals that are able to form complex ions often can be dissolved by adding an excess of ligand, just as the insoluble salts of weak acids can be dissolved by adding an excess of strong acid. The high ligand concentration uses up the metal ions to form complex ions, causing the solubility equilibrium to shift to the right. The following problems illustrate this effect.

PROBLEM:
What weight of AgCl will dissolve in one liter of 1.00 M NH_3?

SOLUTION:
Two simultaneous equilibria are involved: the solubility equilibrium (horizontal equation), and the dissociation of the complex ion (vertical equation). The Ag^+ is shared in common with both equilibria.

$$AgCl_{(s)} \rightleftarrows Ag^+ + Cl^-$$
$$+$$
$$2NH_3$$
$$\updownarrow$$
$$Ag(NH_3)_2^+$$

For the solubility equilibrium

$$[Ag^+] = \frac{K_{sp}}{[Cl^-]}$$

For the complex-ion equilibrium

$$[Ag^+] = \frac{K_{inst}[Ag(NH_3)_2^+]}{[NH_3]^2}$$

Setting these two expressions for $[Ag^+]$ equal to each other, we obtain

$$\frac{K_{sp}}{K_{inst}} = \frac{1.78 \times 10^{-10}}{4.00 \times 10^{-8}} = 4.45 \times 10^{-3} = \frac{[Ag(NH_3)_2^+][Cl^-]}{[NH_3]^2}$$

an expression that corresponds to the equilibrium

$$AgCl_{(s)} + 2NH_3 \rightleftarrows Ag(NH_3)_2^+ + Cl^-$$

For every mole of AgCl that dissolves, 1 mole each of $Ag(NH_3)_2^+$ and Cl^- are formed, and 2 moles of NH_3 are consumed. If s = the moles of AgCl that dissolve per liter, then $[Ag(NH_3)_2^+] = [Cl^-] = s$, and the remaining $[NH_3]$ = $(1.00 - 2s)$ moles/liter. Substituting these equilibrium concentrations into the equilibrium expression, we have

$$4.45 \times 10^{-3} = \frac{s^2}{(1.00 - 2s)^2}$$

Taking the square root of both sides, we get

$$6.67 \times 10^{-2} = \frac{s}{1.00 - 2s}$$

$s = 0.0589$ mole of AgCl dissolves per liter

$$\text{Wt of AgCl that dissolves} = \left(0.0589 \ \frac{\text{mole}}{\text{liter}}\right)\left(143.4 \ \frac{\text{g}}{\text{mole}}\right) = 8.45 \ \frac{\text{liter}}{\text{liter}}$$

PROBLEM:
An excess of $AgNO_3$ is used to precipitate 0.0500 moles each of Cl^- and I^- from solution. What must be the concentration of an NH_3 solution in order to dissolve in one liter all of the AgCl and a minimal amount of AgI? In other words, how can you separate the Cl^- and I^- ions?

SOLUTION:
The equilibria and the equilibrium expressions involving AgCl are identical to those in the last problem. In this problem, however, we know that we shall dissolve 0.0500 mole of AgCl per liter to give $[Ag(NH_3)_2^+] = [Cl^-] = 0.0500$ M. This will require

$$(2) \left(0.0500 \ \frac{\text{mole}}{\text{liter}}\right) = 0.100 \text{ mole of } NH_3 \text{ per liter}$$

leaving an equilibrium concentration of $[NH_3] = (x - 0.100)$ moles/liter. Substituting these equilibrium values into the equilibrium expression, we have

$$4.45 \times 10^{-3} = \frac{(0.0500)^2}{(x - 0.100)^2}$$

$$x = \frac{0.05667}{0.0667}$$

$$= 0.850 \text{ M } NH_3 \text{ needed to dissolve 0.0500 mole AgCl}$$

To determine how much AgI will dissolve at the same time, we need an equilibrium constant similar to that for AgCl, but involving $[I^-]$ and the K_{sp} for AgI:

$$\frac{K_{sp}}{K_{inst}} = \frac{8.32 \times 10^{-17}}{4.00 \times 10^{-8}} = 2.08 \times 10^{-9} = \frac{[Ag(NH_3)_2^+][I^-]}{[NH_3]^2}$$

Once the AgCl is dissolved, $[Ag(NH_3)_2^+] = 0.500 + s$, where $s =$ moles of AgI that dissolve. $[I^-] = s$ because, for every mole of AgI that dissolves, one mole of I^- is formed. The $[NH_3]$ will be $0.750 - 2s$. Substituting these values into the equilibrium expression, we obtain

$$2.08 \times 10^{-9} = \frac{(0.0500 + s)(s)}{(0.750 - 2s)^2}$$

The equilibrium constant is very small, so s will be negligible compared to 0.0500 and 0.375, giving

$$s = \frac{(0.750)^2(2.08 \times 10^{-9})}{(0.0500)} = 2.34 \times 10^{-8} \text{ moles of AgI dissolved per liter}$$

As you can see, almost no AgI dissolves while *all* of the AgCl dissolves in 0.85 M NH_3. A good separation is achieved.

PROBLEMS A

1. Give the proper name for each of the following complexes:
   (a) $Mg(H_2O)_6^{2+}$
   (b) $Sb(OH)_6^{3-}$
   (c) $Ag(S_2O_3)_2^{3-}$
   (d) $Ni(C_2O_4)_2^{2-}$
   (e) $Co(NH_3)_6^{3+}$
   (f) $[Ni(NH_3)_2(NO_2)_2]$
   (g) HgI_4^{2-}

2. What is the concentration of Cu^{2+} in a solution made by diluting 0.100 mole of $CuSO_4$ and 2.00 moles of NH_3 to 500 ml?

3. K_{sp} for MX (mol wt = 80) is 1.00×10^{-10}; K_{inst} for $M(NH_3)_2^+$ is 1.00×10^{-8}. If 0.100 mole of MNO_3 is added to 1.00 liter of 2.00 M NH_3, what are the concentrations of (a) $M(NH_3)_2^+$; (b) NH_3; (c) M^+; (d) NO_3^-; (e) H^+; (f) OH^-?

4. If solid MX (see Problem 3) is added to 1.00 liter of 0.500 M NH_3 solution, how many grams dissolve?

5. How many grams of AgBr will dissolve in 500 ml of 2.00 M NH_3?

6. What must be the final NH_3 concentration to dissolve 2.00 g of AgCl in 250 ml of solution?

7. What must be the concentration of an HCl solution to dissolve 1.00 mg of AgCl in 100 ml of solution?

8. How many grams of HgS will dissolve in 100 ml of 1.00 M NaCN? (Neglect hydrolysis of S^{2-} and CN^-.)

9. What must be the NH_3 concentration in a 0.100 M $AgNO_3$ solution to prevent the precipitation of AgCl when enough NaCl has been added to raise its concentration to 0.500 M?

10. A 100 ml sample of a solution containing 5.00 mg of Cd^{2+} is treated with 1.00 ml of 15.0 M NH_3. Calculate the Cd^{2+} concentration in the resulting solution.

11. What is the minimum concentration of CN^- that will prevent the precipitation of AgI from a solution that is 0.100 M in I^- and 0.0100 M in Ag^+?

12. What must be the concentration of an NH_3 solution if 100 ml of it must dissolve 3.00 g of $AgCl$ together with 3.00 g of $AgNO_3$?

13. Calculate the pH of (a) a 0.100 M $Cu(NO_3)_2$ solution, and (b) a 0.100 M $Fe(NO_3)_3$ solution.

14. The following questions involve the typical amphoteric hydroxides, $Pb(OH)_2$ and $Sn(OH)_2$.
    (a) Calculate the concentrations of Pb^{2+}, $Pb(OH)_3^-$, and OH^- in a water solution saturated with $Pb(OH)_2$.
    (b) Calculate the Pb^{2+} and $Pb(OH)_3^-$ concentrations in a solution that is saturated with $Pb(OH)_2$ and whose equilibrium $NaOH$ concentration is 1.00 M.
    (c) Calculate the Sn^{2+} and $Sn(OH)_3^-$ concentrations that result from the addition of excess solid $Sn(OH)_2$ to 500 ml of 0.500 M $NaOH$.
    (d) What is the minimum volume of 0.500 M $NaOH$ needed to dissolve 0.100 mole of $Pb(OH)_2$?

PROBLEMS B

15. Give the proper names for each of the following complexes.
    (a) $Hg(H_2O)_6^{2+}$
    (b) $Fe(SCN)_6^{3-}$
    (c) $Co(NH_3)_6^{2+}$
    (d) $Al(C_2O_4)_3^{3-}$
    (e) BF_4^-
    (f) $Cr(OH)_4^-$
    (g) $[Co(NH_3)_3Cl_3]$

16. What is the concentration of Cd^{2+} in a solution made by diluting 0.100 mole of $CdSO_4$ and 2.00 moles of NH_3 to 250 ml?

17. K_{sp} for MX (mol wt = 125) is 1.00×10^{-12}; K_{inst} for $M(NH_3)_2^+$ is 1.00×10^{-10}. If 0.100 mole of MNO_3 is added to 500 ml of 3.00 M NH_3 solution, what are the concentrations of (a) $M(NH_3)_2^+$; (b) NH_3; (c) M^+; (d) NO_3^-; (e) H^+; (f) OH^-?

18. If solid MX (see Problem 17) is added to 500 ml of 2.00 M NH_3 solution, how many grams dissolve?

19. How many grams of $AgBr$ will dissolve in 250 ml of a 6.00 M NH_3 solution?

20. What must be the final NH_3 concentration to dissolve 5.00 g of $AgCl$ in 500 ml of solution?

21. What must be the concentration of $CaCl_2$ solution to dissolve 3.00 g of $AgCl$ in 500 ml of solution?

22. How many grams of $Cu(OH)_2$ will dissolve in 100 ml of 6.00 M NH_3?

23. What must be the NH_3 concentration in order to prevent the precipitation of $AgCl$ if the solution is 0.250 M in $AgNO_3$ and 2.00 M in $NaCl$?

24. A 100 ml sample of a solution containing 10.0 mg Cu^{2+} is treated with 50.0 ml of 3.00 M NH_3. Calculate the Cu^{2+} concentration in the resulting solution.

25. What is the minimum concentration of CN^- that will prevent the precipitation of $Hg(OH)_2$ from a solution that is 0.500 M in OH^- and 0.100 M in Hg^{2+}? [K_{sp} for $Hg(OH)_2$ is 2.00×10^{-22}.]

26. What volume of 6.00 M NH_3 must be added to 250 ml of 0.100 M $AgNO_3$ to prevent the precipitation of AgCl when 4.00 g of NaCl are added?

27. Calculate the pH of (a) a 0.100 M $Mg(NO_3)_2$ solution, and (b) a 0.100 M $Hg(NO_3)_2$ solution.

28. Answer the following questions involving the typical amphoteric hydroxides, $Zn(OH)_2$ and $Cr(OH)_3$.
    (a) Calculate the Zn^{2+}, $Zn(OH)_4^{2-}$, and OH^- concentrations in a water solution saturated with $Zn(OH)_2$.
    (b) Calculate the Cr^{3+} and $Cr(OH)_4^-$ concentrations that result when an excess of solid $Cr(OH)_3$ is added to 500 ml of 1.50 M NaOH.
    (c) A saturated $Zn(OH)_2$ solution is shown to be 0.250 M in NaOH. Calculate the concentrations of Zn^{2+} and $Zn(OH)_4^{2-}$ ions.
    (d) An excess of solid $Zn(OH)_2$ is added to 100 ml of 1.00 M NaOH. How many moles of $Zn(OH)_2$ are dissolved?
    (e) Compare (i) the minimal volume of 1.00 M NH_3 required to dissolve 0.100 mole of $Zn(OH)_2$ to give $Zn(NH_3)_4^{2+}$ with (ii) the minimal volume of 1.00 M NaOH required to dissolve 0.100 mole of $Zn(OH)_2$ to give $Zn(OH)_4^{2-}$.

29. It is proposed, by treatment with the minimal amount of NH_3 necessary to dissolve the AgCl, to separate a solid mixture of 0.500 g of AgCl and 0.750 g of AgBr.
    (a) What must be the NH_3 concentration in the final reaction mixture if all the AgCl is dissolved in a total volume of 200 ml?
    (b) What volume of 1.00 M NH_3 should be added to the solids to just dissolve the AgCl?
    (c) How many grams of AgBr would also dissolve if the separation were performed as in (b)?

30. A 2.0000 g mixture of solids containing $NaNO_3$, NaCl, and NaBr is dissolved in water, and the Cl^- and Br^- are precipitated with an excess of $AgNO_3$. This silver halide precipitate, which weighs 2.5000 g, is treated with 200 ml of 0.900 M NH_3 that dissolves the AgCl. The remaining residue of AgBr weighs 0.9000 g.
    (a) Calculate the percentages of NaCl and NaBr in the original mixture.
    (b) By what percentage are the analyses in (a) in error as a result of the solution of some AgBr by the NH_3?

Nuclear Chemistry

All the chemical changes and many of the physical changes that we have studied so far involve alterations in the electronic structures of atoms. Electron-transfer reactions, emission and absorption spectra, and X rays result from the movement of electrons from one energy level to another. In all of these, the nuclei of the atoms remain unchanged, and different isotopes of the same element have the same chemical activity. Nuclear chemistry, or radioactivity, differs from other branches of chemistry in that the important changes occur in the nucleus. These nuclear changes also are represented by chemical equations. However, because the isotopes of the same element may, from a *nuclear* standpoint, be very different in reactivity, it is necessary that the equations show which isotopes are involved.

NUCLEAR EQUATIONS

Nuclear symbolism disregards valence and electrons and always includes the mass number of the isotope (the whole number nearest the atomic weight of the isotope), the atomic number of the element, and the symbol of the element. The mass number is shown as a superscript preceding the symbol, and the atomic number is shown as a subscript preceding the symbol. This notation is illustrated by the following examples:

$^1_1$H is a proton, or a hydrogen atom with mass number 1 and atomic number 1;

$^2_1$H is a deuteron, or a hydrogen atom with mass number 2 and atomic number 1;

$^1_0$n is a neutron, or a particle with mass number 1 and no electrical charge (atomic number 0).

Four other particles that occur in nuclear reactions, but that are not nuclear particles, also must be described by similar symbols if we are to write nuclear equations:

$_{-1}^{0}$e is an electron, with mass number 0 and atomic number -1;

$^0_1$e is a positron, with mass number 0 and atomic number $+1$;

$^0_0\nu$ is a neutrino, with mass number 0 and atomic number 0;

$^0_0\bar{\nu}$ is an antineutrino, with mass number 0 and atomic number 0.

The actual masses of $^0_1$e and $_{-1}^{0}$e are about 1/1845 that of a proton, and the masses of $^0_0\nu$ and $^0_0\bar{\nu}$ are virtually negligible.

In the early days of nuclear chemistry, before an established system of symbols was used, three other terms were invented: alpha particle, α; beta particle, β; and gamma ray, γ. We still use these terms, but for writing nuclear equations we must know that

α is $^4_2$He (a He^{2+} ion);

β is $_{-1}^{0}$e (an electron);

γ is radiation similar to X rays, which have no mass.

Two rules must be followed in balancing nuclear equations: (1) the sum of the mass numbers on the left side of the equation must equal the sum of the mass numbers on the right side; and (2) the sum of the atomic numbers on the left side must equal the sum of the atomic numbers on the right side. These rules are illustrated in the following nuclear reactions:

$$^{226}_{88}\text{Ra} \rightarrow {}^{222}_{86}\text{Rn} + {}^4_2\text{He}$$

$$^{232}_{90}\text{Th} \rightarrow {}^{228}_{88}\text{Ra} + {}^4_2\text{He}$$

$$^{239}_{93}\text{Np} \rightarrow {}^{239}_{94}\text{Pu} + {}_{-1}^{0}\text{e}$$

$$^{14}_{6}\text{C} \rightarrow {}^{14}_{7}\text{N} + {}_{-1}^{0}\text{e}$$

These equations also illustrate the following general statements.

1. Whenever a radioactive element emits an α particle, its product (daughter) has a mass number that is 4 less than that of the parent and an atomic number that is 2 less than that of the parent.

2. Whenever a radioactive element emits a β particle, its daughter has a mass number that is the same as that of the parent and an atomic number that is 1 greater than that of the parent. Because electrons as such are not present in the nuclei of atoms, it is not obvious at first why the loss of a β particle should cause an increase in atomic number. What actually happens is that a neutron disintegrates:

$$\ _0^1n \rightarrow \ _1^1H + \ _{-1}^0e + \ _0^0\bar{\nu}$$

The $_{-1}^0e$ is ejected as a β particle, and the proton, $_1^1H$, stays in the nucleus in the place of the neutron. The additional positive charge causes the atomic number to be increased by 1.

Alpha particles from radioactive samples, or He^{2+} ions accelerated in a cyclotron, may be used to bring about other nuclear reactions. For example, they may bombard a beryllium metal target:

$$\ _4^9Be + \ _2^4He \rightarrow \ _6^{12}C + \ _0^1n$$

The neutrons produced by such bombardments may be used to cause additional nuclear reactions. For example, they may bombard lithium:

$$\ _3^6Li + \ _0^1n \rightarrow \ _2^4He + \ _1^3H$$

Protons ($_1^1H$) and deuterons ($_1^2H$) are the other common particles that may be accelerated and used to bring about nuclear reactions:

$$\ _7^{14}N + \ _1^1H \rightarrow \ _6^{11}C + \ _2^4He$$
$$\ _{26}^{54}Fe + \ _1^2H \rightarrow \ _{27}^{55}Co + \ _0^1n$$

If a product of one of these man-made reactions is unstable and spontaneously disintegrates further, it is said to be "artificially radioactive." In the preceding equation, for example, $_{27}^{55}Co$ is artificially radioactive, disintegrating as

$$\ _{27}^{55}Co \rightarrow \ _{26}^{55}Fe + \ _1^0e$$

NUCLEAR ENERGY

In all the nuclear equations we have just written, we have shown that there is no change in the sum of the mass numbers. However, there actually is a small change in mass, and this change is one of the most important properties of nuclear reactions. Consider the nuclear fusion of hydrogen and tritium:

$$\ce{^{1}_{1}H + ^{3}_{1}H \rightarrow ^{4}_{2}He}$$

The sum of the weight (per mole) of the products is 0.0246 g less than the sum of the weights of the reactants. In the Bethe cycle (the series of reactions in the sun by which solar energy is produced), 4 moles of protons weighing 4.03228 g are converted to 1 mole of helium weighing 4.00336 g, a loss of 0.02892 g. In the nuclear fission of $^{235}_{92}U$,

$$^{235}_{92}U + ^{1}_{0}n \rightarrow \text{fission products} + 2 \text{ or } 3\ ^{1}_{0}n$$

there is a loss of 0.205 g/mole. The weight that is lost is converted to energy, according to the Einstein equation:

$$E = mc^2$$

where E is the energy (in ergs) liberated by converting mass to energy, m is the mass (in grams) converted to energy, and c is the velocity of light (3×10^{10} cm/sec).

The ergs of energy may be expressed as calories if we remember that 1 cal = 4.184×10^7 ergs.

1. In nuclear fusion, 0.0246 g gives 5.29×10^{11} cal = 0.529 Tcal.

2. In the Bethe cycle, 0.02892 g gives 6.22×10^{11} cal = 0.622 Tcal.

3. In the nuclear fission, 0.205 g gives 4.41×10^{12} cal = 4.41 Tcal.

RATE OF NUCLEAR DISINTEGRATION

Radioactive substances vary in their activity—that is, in the rate at which they disintegrate. Each radioactive isotope disintegrates at a rate that is unaffected by temperature, pressure, or external conditions. As discussed on pp. 232–234, this rate of disintegration

$$-\frac{dN}{dt}$$

is a first-order process with a rate solely proportional to the number of atoms (N) of the isotope present at any given instant.

$$-\frac{dN}{dt} = kN \tag{26-1}$$

The proportionality constant (the rate constant, k) is different for each isotope, and has the units of reciprocal time (sec^{-1}, min^{-1}, yr^{-1}, etc.). In Chapter 15, we show that the relationship between the number of atoms (N_1) at the outset ($t = 0$), and the number of atoms (N_2) remaining at time t is given by

$$\log \frac{N_1}{N_2} = \frac{kt}{2.30} \tag{26-2}$$

N may be expressed in any convenient units because whatever is used cancels out in the ratio N_1/N_2. A common unit is "counts per minute" (cpm) because, through Equation 26-1, the observed cpm are a direct measure of the number of radioactive atoms present.

Equation 26-2 often is written in the form

$$\log N_2 = -\frac{k}{2.30} t + \log N_1 \tag{26-3}$$

because a plot of log (cpm)$_2$ versus t will yield a straight line whose slope ($-k/2.30$) will enable the evaluation of the rate constant k. This procedure is illustrated on p. 234.

It is further shown in Chapter 15 that the needed rate constant is traditionally provided in terms of the half-life ($t_{\frac{1}{2}}$), the time needed for one-half of the isotope to disintegrate. The needed relationship between k and $t_{\frac{1}{2}}$ is given by

$$k = \frac{0.693}{t_{\frac{1}{2}}} \tag{26-4}$$

TABLE 26-1
Half-Lives of Selected Radioactive Elements

Isotope	Half-life ($t_{\frac{1}{2}}$)	Particle emitted	Isotope	Half-life ($t_{\frac{1}{2}}$)	Particle emitted
$^{14}_{6}C$	5.73×10^3 yr	β	$^{226}_{88}Ra$	1.60×10^3 yr	α
$^{35}_{16}S$	88.0 days	β	$^{227}_{89}Ac$	21.6 yr	α
$^{128}_{53}I$	25.08 min	β	$^{232}_{90}Th$	1.41×10^{10} yr	α
$^{211}_{82}Pb$	36.1 min	β	$^{238}_{92}U$	4.51×10^9 yr	α
$^{210}_{84}Po$	138.4 days	α	$^{237}_{93}Np$	2.14×10^6 yr	β
$^{211}_{85}At$	7.21 hr	α	$^{239}_{94}Pu$	2.44×10^4 yr	α

The half-lives of a few selected radioactive isotopes are given in Table 26-1, along with the particles that are emitted in each case.

PROBLEM:
You have 0.200 g of $^{210}_{84}$Po. How much of it will remain 21.0 days from now? Write the nuclear equation for the reaction.

SOLUTION:
Table 26-1 gives the half-life (138.4 days) and shows that $^{210}_{84}$Po emits α particles. The nuclear reaction therefore is

$$^{210}_{84}\text{Po} \rightarrow {}^{206}_{82}\text{Pb} + {}^{4}_{2}\text{He}$$

Because the number of atoms is proportional to the number of grams, we can use Equation 26-2 to obtain

$$\log \frac{N_1}{N_2} = \log \frac{0.200}{N_2} = \frac{kt}{2.30} = \frac{0.693t}{2.30t_{\frac{1}{2}}}$$

$$= \frac{(0.693)(21.0 \text{ days})}{(2.30)(138.4 \text{ days})} = 0.0457$$

Taking the antilog of both sides, we get

$$\frac{0.200}{N_2} = 1.11$$

$$N_2 = \frac{0.200}{1.11} = 0.180 \text{ g remaining}$$

PROBLEM:
How much time must pass for 0.90 of a sample of $^{227}_{89}$Ac to disintegrate?

SOLUTION:
We take the half-life (21.6 yr) from Table 26-1 to find the rate constant:

$$k = \frac{0.693}{t_{\frac{1}{2}}} = \frac{0.693}{21.6 \text{ yr}} = 3.21 \times 10^{-2} \text{ yr}^{-1}$$

Then we use this value of k in the rearranged form of Equation 26-2 to find t:

$$t = \frac{2.30}{k} \log \frac{N_1}{N_2} = \frac{2.30}{3.21 \times 10^{-2}} \log \frac{1}{0.10} = \frac{2.30}{3.21 \times 10^{-2}} \log 10$$

$$= \frac{(2.30)(1.00)}{(3.21 \times 10^{-2} \text{ yr}^{-1})} = 71.7 \text{ yr}$$

PROBLEM:
What is the half-life of an isotope if a sample of it gives 10,000 cpm (counts per minute), and 3.50 hrs later it gives 8335 cpm?

SOLUTION:

Because the cpm are proportional to the number of atoms of the isotope present, we may write

$$\log \frac{10{,}000}{8335} = \frac{k \times 3.50}{2.30}$$

$$k = \frac{2.30}{3.50} \log \frac{10{,}000}{8335} = \frac{(2.30)(0.0791)}{3.50 \text{ hr}}$$

$$= 0.0520 \text{ hr}^{-1}$$

$$t_{\frac{1}{2}} = \frac{0.693}{k} = \frac{0.693}{0.0520 \text{ hr}^{-1}} = 13.3 \text{ hrs}$$

This is actually a *poor* way to determine the half-life of an isotope, because any given measurement of cpm is subject to such a wide natural variation. It is much better to take a series of measurements, make a least-squares fit of log (cpm) versus t according to Equation 26-3, and then determine the half-life from the slope, as on p. 234.

A problem not mentioned in Chapter 15 is one that is very special for radioactive decay when the elapsed time given in the problem is insignificant in comparison with the half-life. Under such circumstances, Equation 26-2 is totally inappropriate, and the proper equation to use is Equation 26-1. In this case, consider $-dN$ to be the number of atoms that disintegrate in a finite period of time dt, which is negligible compared to $t_{\frac{1}{2}}$; consider also that N remains constant during this same period of time. The following problem shows this application of Equation 26-1.

PROBLEM:

How much of a 1.00 g sample of $^{238}_{92}U$ will disintegrate in a period of 10 years?

SOLUTION:

Ten years is an infinitesimal period of time compared to the half-life of 4.51×10^9 yrs. If we tried using Equation 26-2, we would have

$$\log \frac{N_1}{N_2} = \frac{kt}{2.30} = \frac{0.693t}{2.30 t_{\frac{1}{2}}} = \frac{(0.693)(10.0 \text{ yrs})}{(2.30)(4.51 \times 10^9 \text{ yrs})}$$

$$= 0.000000000668$$

It is impractical to evaluate N_2 in this way because the log is far smaller than that shown in any normal log table, or that can be handled by a hand calculator. Instead, we use Equation 26-1:

$$-dN = k \cdot N \cdot dt$$

where $-dN$ = the number of atoms (or grams) that disintegrate; N = the number of atoms (or grams) present (an amount that stays virtually constant in 10 years);

and $dt = 10.0$ yrs (an infinitesimal period of time in this problem). Consequently,

$$-dN = \left(\frac{0.693}{4.51 \times 10^9 \text{ yrs}}\right)(1.00 \text{ g})(10.0 \text{ yrs})$$

$$= 1.54 \times 10^{-9} \text{ g of } {}^{238}_{92}\text{U that disintegrates}$$

PROBLEMS A

1. Write nuclear equations to show the disintegration of the first six radioactive isotopes listed in Table 26-1.

2. The following isotopes are artificially radioactive. They emit the particles shown in parentheses. Write nuclear equations to show the disintegration of these isotopes.
   (a) ${}^{27}_{14}\text{Si}\ ({}^{0}_{1}\text{e})$
   (b) ${}^{28}_{13}\text{Al}\ (-{}^{0}_{1}\text{e})$
   (c) ${}^{30}_{15}\text{P}\ ({}^{0}_{1}\text{e})$
   (d) ${}^{24}_{11}\text{Na}\ (-{}^{0}_{1}\text{e})$
   (e) ${}^{17}_{9}\text{F}\ ({}^{0}_{1}\text{e})$

3. Bombardment reactions are often summarized in a terse form, such as ${}^{9}_{4}\text{Be}(\alpha,\text{n})$. This means that the target (${}^{9}_{4}\text{Be}$) is bombarded by α particles (${}^{4}_{2}\text{He}$), and that neutrons (${}^{1}_{0}\text{n}$) are produced. By the rules for balancing nuclear equations, we know that ${}^{12}_{6}\text{C}$ also is produced. Give the complete balanced nuclear equation for each of the following transmutation bombardments (p stands for proton, and d stands for deuteron in this notation).
   (a) ${}^{14}_{7}\text{N}(\text{n},\text{p})$
   (b) ${}^{26}_{12}\text{Mg}(\text{N},\alpha)$
   (c) ${}^{59}_{27}\text{Co}(\text{d},\text{p})$
   (d) ${}^{14}_{7}\text{N}(\alpha,\text{p})$
   (e) ${}^{63}_{29}\text{Cu}(\text{p},\text{n})$

4. (a) Calculate the energy liberated per mole of lithium when the following reaction takes place:

$$ {}^{7}_{3}\text{Li} + {}^{1}_{1}\text{H} \rightarrow 2\ {}^{4}_{2}\text{He} $$

   The actual mole weights involved are ${}^{7}_{3}\text{Li} = 7.01818$ g; ${}^{1}_{1}\text{H} = 1.00813$ g; ${}^{4}_{2}\text{He} = 4.00386$ g.
   (b) How many tons of carbon would have to be burned to give as much energy as 100 g of ${}^{7}_{3}\text{Li}$ consumed by this reaction? (Carbon gives 94.1 kcal/mole when burned to CO_2.)

5. The halogen astatine can be obtained only by artificial methods through bombardment. It has been found useful for the treatment of certain types of cancer of the thyroid gland, for it migrates to this gland just as iodine does. If a sample containing 0.1 mg of ${}^{211}_{85}\text{At}$ is given to a person at 9 A.M. one morning, how much will remain in the body at 9 A.M. the following morning?

6. A purified sample of a radioactive compound is found to have 1365 cpm at 10 A.M., but only 832 cpm at 1 P.M. the same day. What is the half-life of this sample?

7. If in the explosion of one atom bomb 50.0 g of ${}^{239}_{94}\text{Pu}$ were scattered about in the atmosphere before it had a chance to undergo nuclear fission, how much

would be left in the atmosphere after 1000 yrs? (Assume that none of it settles out.)

8. If you seal a 100 g sample of very pure uranium in a container for safekeeping, and if in the distant future a scientist dissolves it and finds it to contain 0.800 g $^{206}_{82}Pb$, how many years have elapsed since you sealed the sample? (Lead is the stable endproduct of the radioactive decay of uranium).

9. Willard Libby and colleagues have found that a very small amount of the CO_2 in the air is radioactive, as a result of continuous bombardment of nitrogen in the upper atmosphere by neutrons of cosmic origin. This radioactive $^{14}_{6}C$ is uniformly distributed among all forms of carbon that are in equilibrium with the atmosphere, and it is found to show 15.3 cpm/gram of carbon. All plants use CO_2 for growth, and while living are in equilibrium with the atmosphere. When plants die, they no longer remain in equilibrium with the air, and the carbon slowly loses its activity. Cyprus wood from the ancient Egyptian tomb of Sneferu at Meydum has an activity of 6.88 cpm/g of carbon. Estimate the age of this wood (and presumably the age of the tomb).

10. A 0.500 g sample of $^{226}_{88}Ra$ is sealed into a very thin-walled tube, so that the α particles emitted from the radium and its decay products can penetrate and be collected in an evacuated volume of 25.0 ml. What helium pressure at 20.0°C will build up in 100 yrs? (Five α particles are produced as radium disintegrates to a stable isotope of lead.)

11. Solve Problems 1, 2, 3, and 4 on p 247.

PROBLEMS B

12. Write nuclear equations to show the disintegration of the last six radioactive isotopes listed in Table 26-1.

13. The following isotopes are artificially radioactive. They emit the particles shown in parentheses. Write nuclear equations to show the disintegration of these isotopes.
    (a) $^{23}_{10}Ne$ ($_{-1}^{0}e$)
    (b) $^{41}_{18}Ar$ ($_{-1}^{0}e$)
    (c) $^{38}_{19}K$ ($^{0}_{1}e$)
    (d) $^{84}_{37}Rb$ ($_{-1}^{0}e$)
    (e) $^{13}_{7}N$ ($^{0}_{1}e$)

14. Bombardment reactions often are summarized in a terse form, as explained in Problem 3. Give the complete balanced nuclear equation for each of the following transmutation bombardments.
    (a) $^{9}_{4}Be(p,d)$
    (b) $^{16}_{8}O(d,\alpha)$
    (c) $^{27}_{13}Al(n,\alpha)$
    (d) $^{40}_{18}Ar(n,\gamma)$
    (e) $^{25}_{12}Mg(\alpha,p)$

15. (a) Calculate the energy liberated per mole of He when the following reaction takes place:

$$^{3}_{1}H + ^{2}_{1}H \rightarrow ^{4}_{2}He + ^{1}_{0}n$$

The actual mole weights involved are $^3_1H = 3.01710$ g; $^2_1H = 2.01470$ g; $^4_2He = 4.00386$ g; $^1_0n = 1.00897$ g.

(b) How many tons of carbon would have to be burned to give as much energy as 1 lb of hydrogen consumed by this reaction? (Carbon gives 94.1 kcal/ mole when burned to CO_2.)

16. Radioactive sulfur has been used as a tracer in the study of the action of sulfur as an insecticide for red spider (a citrus-fruit pest). If 10.0 mg of $^{35}_{16}S$ are absorbed by an orange, how much will remain after 6.00 months in storage?

17. A pure sample of a radioactive compound is found to have 1555 cpm at 3:30 P.M., and 960 cpm at 4:00 P.M. the same day. What is the half-life of this isotope?

18. If you seal a 100 g sample of very pure $^{232}_{90}Th$ in a container for safekeeping, and if in the distant future a scientist dissolves it and finds it to contain 0.600 g of $^{208}_{82}Pb$, how many years have elapsed since you sealed the sample? (Lead is the stable endproduct of the radioactive decay of thorium.)

19. Willard Libby and colleagues have shown that a very small amount of the CO_2 in the air is radioactive, as explained in Problem 9. Acacia wood from the ancient Egyptian tomb of Zoser and Sakkara has an activity of 7.62 cpm/gram of carbon. Estimate the age of this wood (and presumably the age of the tomb).

20. If a sample of rock contains 2.00 mg of helium for each 100 mg of $^{232}_{90}Th$, calculate the minimal age of this thorium deposit. (Six α particles are produced as thorium disintegrates to a stable isotope of lead. Assume that none of the helium escaped from the rock.)

21. How many grams of $^{232}_{90}Th$ will disintegrate in 1.00 yr if, at the outset, you have a 1.00 g sample?

22. The positrons produced in the Bethe cycle (the production of solar energy) react with electrons to produce γ radiation. This "annihilation" reaction is

$$_{-1}^{0}e + {}_{1}^{0}e \rightarrow \gamma$$

(a) Assuming that the mass of both $_{-1}^{0}e$ and $_{1}^{0}e$ is $\frac{1}{1845}$ that of a proton, calculate the energy produced in the annihilation of a single electron by a single positron. The gram-atomic weight of $_1^1H$ is 1.00813 g.

(b) If a single γ ray is produced by this annihilation, calculate its wavelength from the fundamental equation $E = h\nu$, where

E = energy of the quantum of γ radiation, in ergs;
h = Planck's constant = 6.626×10^{-27} erg $\times$ sec;
ν = frequency of the radiation, in sec^{-1}.

[The fundamental relation between wavelength (λ), velocity (c), and frequency (ν) is $\lambda = c/\nu$.]

23. Two nongaseous isotopes, A and B, are involved in the same radioactive series, which (over the past eons) has established radioactive equilibrium. The half-life of A is 10^6 yrs, and that of B is only 1 yr. What would be the relative

amounts of A and B in a given ore sample that has lain undisturbed and unprocessed since its creation? Show your reasoning.

24. A 100 ml sample of a 0.0100 M $Ba(NO_3)_2$ solution is mixed with 100 ml of a 0.0100 M Na_2SO_4 solution that has been labeled with $^{35}_{16}S$ and that has 25,000 cpm/ml of solution. After precipitation, the supernatant liquid is filtered off, and it is found that 1.00 ml of solution now has only 26.0 cpm (corrected for background). From these data, calculate the solubility product for $BaSO_4$.

25. It is a common experience when working with radioactive materials that a sample under investigation contains two or more radioisotopes. Assume that you have a sample with just two radioisotopes, one with a half-life about 10 times that of the other. The amounts of the two in the sample are too small to separate chemically.
    (a) How could you study the sample so as to determine the half-lives of the two isotopes?
    (b) Apply your method to the accompanying data, which were taken from such a sample. The cpm values have been corrected for background. The times in minutes are followed by cpm in parentheses: 0 (3160), 2 (2512), 4 (2085), 6 (1512), 8 (1147), 10 (834), 12 (603), 14 (519), 17 (457), 20 (407), 25 (388), 30 (346), 40 (327), 50 (320), 60 (308), 70 (288), 80 (282), 90 (269), 105 (269), 120 (240), 135 (228), 150 (229), 165 (214).

26. Solve Problems 15, 16, 17, and 18 on p 250.

Reactions: Prediction and Synthesis

Two important questions asked in the study of chemistry are (1) ''Will a reaction take place when you mix reactants A and B?'' and (2) ''How do you prepare a given compound, X?''

The first question concerns *prediction*, and the second concerns *synthesis*. They are related. In the first you are given the reactants (the lefthand side of the equation) and are asked to predict the products. In the second you are given a product (the righthand side of the equation) and are asked to suggest suitable reactants. For synthesis you must use a reaction that is predicted to ''go''; the products of a reaction that is predicted to ''go'' actually correspond to a chemical synthesis. This chapter gives an organized approach to answering these questions, including writing the chemical equations for the reactions involved.

PREDICTION

Fundamentally, a reaction ''goes'' if it is accompanied by a decrease in free energy (ΔG is negative), as discussed in Chapter 17. The experimental conditions that lead to the fulfillment of this principle are summarized in the following statement.

A reaction will "go" for any one of the following reasons.

1. For electron-transfer reactions, if the electrode potential (E) for the reducing half-reaction is more negative than that for the oxidizing half-reaction.

2. For non-electron-transfer reactions, if
   a. an insoluble product is formed ($K_e > 1$ because $K_{sp} < 1$), or
   b. a weak (or non-) electrolyte is formed ($K_e > 1$ because $K_i < 1$ or $K_{inst} < 1$), or
   c. a gas is formed in an open vessel (loss of gas product continually shifts equilibrium to right).

Guidelines

Some simple, general guidelines for using these principles are listed here.

1. *Electrode potentials* for half-reactions are most easily obtained from a table of standard electrode potentials (Table 17-1, p 275); they may be modified for concentrations appreciably different from unit activity (p 276). A list of features characteristic of electron-transfer reactions is given on p 300.

2. *General solubility rules* for water as a solvent must be stated in two parts, one emphasizing the negative ions of the salts, and the other emphasizing the positive ions, as follows.
   a. All nitrates, acetates, perchlorates, halides (Cl^-, Br^-, and I^-), and sulfates *are soluble,* with the following important exceptions:
      (1) the halides of Ag^+, Hg_2^{2+}, and Pb^{2+} are insoluble;
      (2) the sulfates of Ba^{2+} and Pb^{2+} are insoluble.
   b. All alkali metal (Li^+, Na^+, K^+, Rb^+, and Cs^+) and ammonium (NH_4^+) salts and bases *are soluble*. $Ba(OH)_2$, a fairly important base, also is soluble.
   c. In general, all other common types of inorganic compounds are insoluble.

3. *Strong electrolytes* (those that ionize in water) include with minor exceptions, the following compounds:
   a. all soluble salts (this does not include acids and bases);
   b. all strong acids (there are eight common ones: HCl, HBr, HI, HNO_3, $HClO_4$, H_2SO_4, $H(NH_2)SO_3$, and H_3PO_4—the latter is included only because it loses its first H^+ readily);
   c. all strong bases (these are the alkali metal and Ba^{2+} hydroxides).

4. *Weak electrolytes* include all those substances that are not classed as strong electrolytes. A particularly important weak electrolyte is water.

5. *Common gases* at room temperature are H_2, O_2, N_2, CO_2, CO, NO, N_2O, NO_2, Cl_2, NH_3, HCl, HCN, and H_2S. Other substances (particularly H_2O) may be driven off as gases at elevated temperatures.

Balancing Ionic Equations

A properly balanced chemical equation shows all the information we have just discussed. Soluble salts and strong electrolytes in aqueous solution are always written in ionic form—for example, $Na^+ + Cl^-$ (not NaCl) or H^+ (or H_3O^+) + ClO_4^- (not $HClO_4$). Insoluble salts and weak electrolytes are not written in ionic form, even though a minor fraction of each ion may actually exist in solution. For example, we indicate the formula for calcium carbonate as $CaCO_3$ (not Ca^{2+} + CO_3^{2-}). Similarly, for slightly dissociated acetic acid, we use the molecular formula $HC_2H_3O_2$ instead of $H^+ + C_2H_3O_2^-$.

Insoluble products usually are indicated by $\downarrow$, and evolved gases by $\uparrow$. The following equations illustrate correct procedures:

$$BaCO_3 + 2HC_2H_3O_2 \rightarrow Ba^{2+} + 2C_2H_3O_2^- + H_2O + CO_2\uparrow$$

$$BaCO_3 + 2H^+ + SO_4^{2-} \rightarrow BaSO_4\downarrow + H_2O + CO_2\uparrow$$

$$BaCO_3 + 2H^+ + 2Cl^- \rightarrow Ba^{2+} + 2Cl^- + H_2O + CO_2\uparrow$$

Whenever there are chemical species in solution that are *identical* on both sides of the equation (such as Cl^- in the last example), this species does not participate in any way in the reaction, and it may be omitted from the final balanced equation:

$$BaCO_3 + 2H^+ \rightarrow Ba^{2+} + H_2O + CO_2\uparrow$$

Such equations often emphasize the generality of a reaction. This last one makes it obvious that *any* strong acid will dissolve $BaCO_3$ to produce CO_2 and H_2O.

There are times when the conditions of an experiment determine whether a reaction goes. In the reaction

$$CdS + 2H^+ \rightleftarrows Cd^{2+} + H_2S\uparrow$$

CdS dissolves in an excess of strong acid because the excess H^+ shifts the equilibrium to the right, and because H_2S (a weak acid) is evolved to the atmosphere. However, if we should saturate an aqueous solution of Cd^{2+} (with no added acid) with H_2S, we should actually obtain a precipitate of CdS. The reduced acidity shifts the equilibrium to the left, as does maintaining a constant high pressure of H_2S.

Remember that a weak electrolyte need not be a molecule; it also may be a weakly dissociated ion. For example, $Ca_3(PO_4)_2$ dissolves in an excess of strong acid, because the weak acid ion HPO_4^{2-} (as well as some $H_2PO_4^-$) is formed by the reaction

$$Ca_3(PO_4)_2 + 2H^+ \rightarrow 3Ca^{2+} + 2HPO_4^{2-}$$

In general, you could expect the insoluble salt of a weak acid to dissolve in an excess of strong acid for the reasons just given. (Extremely insoluble substances, such as some of the metal sulfides, are exceptions because H_2S is not a weak enough acid to take the S^{2-} away from the metal ion.)

The last two examples illustrate how precipitation or solution of an insoluble compound may be controlled by controlling the pH (see also pp 378–383).

Summary of Prediction and Equation-Balancing Procedures

1. First, decide whether an electron-transfer reaction is possible, using approximate half-reaction potentials (p 301) and/or the characteristics of electron-transfer reactions (p 300). If it is possible, follow the equation-balancing procedure outlined on pp 295–299.

2. If the reaction is non-electron-transfer, then treat it as a metathesis reaction.

   a. Write a preliminary balanced molecular equation in which the reaction partners have been interchanged:

   $$A_2B + 2CD \rightarrow 2AD + BC_2$$

   b. Study each compound in the molecular equation and decide whether it is an insoluble solid, a weak (or non-) electrolyte, or a gas. Rewrite the equation as appropriate. If A_2B is a strong electrolyte, CD a weak electrolyte, AD a gas, and BC_2 an insoluble solid, then

   $$2A^+ + B^{2-} + 2CD \rightleftarrows 2AD\uparrow + BC_2\downarrow$$

   Eliminate all entities that are identical in form on both sides to get the final balanced ionic equation.

   c. The reaction will go if any *one* of the products is a gas, a solid, or a weak electrolyte. If there is competition (insoluble solids or weak electrolytes on *both* sides), you must consider (1) whether one of the reactants is used in excess or high concentration to favor a shift to the right, and (2) the relative insolubilities of solids or weaknesses of electrolytes.

SYNTHESIS

To prepare compounds, you must use reactions that (by the principles just outlined) are predicted to "go." Not all reactions that take place, however, are suitable for synthesis, and attention must be paid to purity of product, cost of reactants, and ease of preparation. We comment on these factors in the answers to the A problems, and in the following discussion of the preparation of a compound X. If your laboratory work includes qualitative analysis, you might well consider each evolved gas and each precipitate to be a chemical preparation, even though the prime purpose of the project before you is separation or identification.

The general methods of synthesis outlined below are merely a collection of simple, common sense, practical techniques presented in an organized manner. Some of the methods will already be familiar to you from previous experience, and you will probably gain experience in additional methods as you progress through your present chemistry course.

X Is a Gas

For equipment, connect together a generator, a drying agent (which usually is needed), and a reservoir for collection, as shown schematically in Figure 27-1. If possible, add one reagent to the other in a controlled manner through a funnel whose tip lies below the liquid level of the reaction mixture, as in Figure 27-2a. Warm if necessary. If the reaction involves heating a solid or a mixture of solids, then the generator shown in Figure 27-2b can be used. Discard the first volume of gas generated, as it sweeps the air out of the generator and connecting tubes. If possible, use relatively nonvolatile reagents.

If water is produced, or if water solutions are used in the generator, then an appropriate drying agent (desiccant) should be used. Some common desiccants are listed in Table 27-1, along with their principal limitations. Appropriate containers for the different types of desiccants are shown in Figure 27-3. Always pay attention to the *capacity* of the drying apparatus and chemicals, to avoid plugging up the desiccant tube as water accumulates, or else some part of the generator system will blow apart.

The most satisfactory way to collect a gas is by upward displacement over mercury, if mercury is available and if the gas is unreactive toward it. If a gas is collected over water, don't bother to dry it first. Of course, collection over

FIGURE 27-1
Schematic arrangement of apparatus needed for preparing a pure gas.

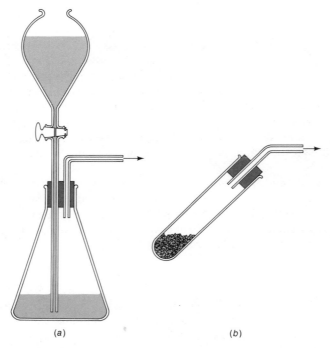

FIGURE 27-2
Simple gas generators.

TABLE 27-1
Desiccants

Formula	Physical state	Type of container in Figure 27-3	Cannot be used to dry	Removes H_2O by forming
$Mg(ClO_4)_2$	Solid	a		$Mg(ClO_4)_2 \cdot 6\,H_2O$
P_4O_{10}	Solid	a	Basic gases	H_3PO_4
$CaCl_2^*$	Solid	a		$CaCl_2 \cdot 6\,H_2O$
CaO^*	Solid	a	Acidic gases	$Ca(OH)_2$
NaOH	Solid	a	Acidic gases	NaOH solution
SiO_2 (silica gel)*	Solid	a		Adsorbed water
H_2SO_4 (concentrated)	Liquid	b	Basic gases	Dilute H_2SO_4
Dry Ice $(CO_2)^*$	Solid	c	Gases with b.p. $< -78°C$	Solid H_2O
Liquid N_2	Liquid	c	Gases with b.p. $< -195°C$	Solid H_2O

* Relatively cheap, safe, and simple to use.

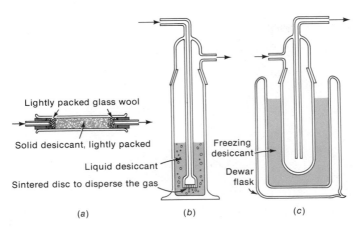

Lightly packed glass wool

Solid desiccant, lightly packed

Liquid desiccant

Sintered disc to disperse the gas

Freezing desiccant

Dewar flask

(a)   (b)   (c)

FIGURE 27-3
Simple methods for drying gases.

water is impossible if the gas is very soluble in water. If the gas is nontoxic and abundant, it may be collected by the upward displacement of air. For whatever upward displacement method you employ, the collection bottle in Figure 27-4a may be used, filled at the outset with mercury, water, or air. The bottle in Figure 27-4b may be used for collecting gases more dense than air by downward displacement. Either of these bottles can be capped with a glass plate and is simple to use in qualitative experiments. The reservoir in Figure 27-4c illustrates one way in which a gas may be collected and then used as desired at a later time.

Do not try to prepare gases such as CO_2 and SO_2 by burning C or S in an excess of air (or even pure O_2); the product will be contaminated by the huge excess of N_2, excess O_2, and other components in the air. A better method is to add an excess of acid to a carbonate or a sulfite contained in the generator of Figure 27-2a.

X Is a Salt

1. *X is insoluble*. Choose two soluble reactants that by exchange of partners will give a precipitate of X; the other product must be soluble. No care is needed in measuring quantities, because X can be filtered from the excesses in solution, washed with distilled water, and dried.

2. *X is soluble*.
   a. Exactly neutralize the appropriate acid with the appropriate base if both are soluble, and evaporate the water from the resulting salt. An indicator or a pH meter may be used to determine the endpoint (see pp 365–366).

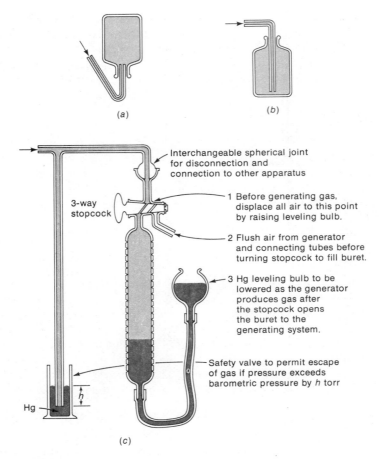

(a)   (b)

Interchangeable spherical joint
for disconnection and
connection to other apparatus

3-way
stopcock

1 Before generating gas,
displace all air to this point
by raising leveling bulb.

2 Flush air from generator
and connecting tubes before
turning stopcock to fill buret.

3 Hg leveling bulb to be
lowered as the generator
produces gas after
the stopcock opens
the buret to the
generating system.

Safety valve to permit escape
of gas if pressure exceeds
barometric pressure by h torr

Hg

(c)

FIGURE 27-4
Simple methods for collecting gases.

b. Add insufficient acid to an insoluble base (or vice versa, if appropriate) for complete reaction with the base, filter off and discard the excess base, and evaporate the water from the filtrate to get X. In this context, a base is defined as an oxide, a hydroxide, or a carbonate. The acid is completely used up, and therefore does not contaminate the filtrate.

c. Add insufficient acid to a metal whose half-reaction (Table 17-1) lies above the H_2–H^+ half-reaction, filter off and discard the excess metal, and evaporate the water from the filtrate to get X. The evolved H_2 gas should be kept away from flames.

d. Add an excess of HNO_3 to a metal (almost any metal except Al, Pt, Sn, or Au) to produce the soluble metal nitrate. To this acidic solution, add an excess of appropriate base to precipitate the metal hydroxide. (If the metal forms a soluble NH_3 complex, do *not* use NH_3; if it is amphoteric and forms a soluble OH^- complex, do *not* use NaOH. Table 25-1 shows lists of ammine and hydroxo complexes.) Treat this hydroxide as in method b to get the desired salt.

X Is an Acid

Many of the common inorganic acids can be prepared easily by heating the appropriate salt with a concentrated, nonvolatile acid—such as H_2SO_4 or H_3PO_4. In the following general scheme, the most commonly available or practical salt has been selected as a starting material, but Na^+ or almost any other metal ion could be used as well.

$$\left.\begin{array}{l} CaF_2 \\ NaCl \\ NaBr \\ NaI \\ NaNO_3 \\ FeS \\ NH_4CN \\ CaCO_3 \\ Na_2SO_3 \end{array}\right\} + \begin{array}{c} \text{conc'd } H_2SO_4 \\ \text{or} \\ \text{conc'd } H_3PO_4 \end{array} \xrightarrow{\Delta} \left.\begin{array}{l} HF \\ HCl \\ HBr \\ HI \\ HNO_3 \\ H_2S \\ HCN \\ CO_2 \\ SO_2 \end{array}\right\} \uparrow + \begin{array}{c} \text{metal } SO_4^{2-}, \\ HSO_4^- \\ \text{or} \\ \text{metal } PO_4^{3-}, \\ HPO_4^{2-}, \\ H_2PO_4^- \end{array}$$

A few specific comments on this preparative method are in order.

1. All the acids prepared are gases at room temperature except HF and HNO_3.

2. Only HCl, HBr, HI, and HNO_3 require *vigorous* heating. The others require little or none, depending on the salt used.

3. H_2S and HCN are *HIGHLY TOXIC* gases, and must be produced with proper ventilation. With the exception of CO_2, all the gases are extremely irritating, corrosive, and obnoxious.

4. An all-glass apparatus must be used for HNO_3 because it is corrosive to rubber, cork, and other materials.

5. Hydrogen fluoride etches glass:

$$4HF + SiO_2 \rightarrow SiF_4\uparrow + 2H_2O$$

Therefore all equipment for generation and collection of HF must be made of polyethylene, paraffin-coated glass, metallic lead, or something equally unreactive to HF.

6. H_3PO_4 is used in preparing HBr and HI in order to avoid the strong oxidizing action of hot concentrated H_2SO_4, which will also yield such products as the halogens, SO_2, and H_2S.

7. In order to get H_2CO_3 and H_2SO_3, the gases CO_2 and SO_2 must be passed into H_2O.

Many organic acids are insoluble in water, and so can be prepared from their soluble salts by simply adding a dilute solution of a strong inorganic acid to their aqueous solutions. Hot concentrated H_2SO_4 and H_3PO_4 should never be used with organic acids. All you would probably get is a mess of carbon. Two inorganic acids, H_3BO_3 and H_4SiO_4 precipitate on acidification of solutions of their soluble salts.

X Is a Hydroxide

As noted by the solubility rules, most hydroxides are only slightly soluble and would be prepared as in method 1 for a salt. NaOH, KOH, and NH_3 are commonly available, but the choice of base must take into account the possibility that soluble OH^- or NH_3 complexes may form (see Table 25-1).

NH_3 is easily prepared, as a gas, by warming any NH_4^+ salt with an excess of NaOH, using a generator of the type shown in Figure 27-2a.

X Is a Miscellaneous Compound

1. *Metal oxides*. With the exception of the alkali metals, prepare by vigorously heating the metal hydroxide to eliminate water. Do *not* try to burn the metal in air.

2. *Sulfides of alkaline earth metals, Al, and Cr*. Prepare by heating the pulverized metals with an excess of sulfur and with protection from the air. They cannot be precipitated by H_2S from solution; they hydrolyze when they come in contact with water.

3. *Acid salts,* such as $NaHCO_3$, KH_2PO_4, or K_2HPO_4. Calculate the pH of the solution you wish to make (see p 362). Choose an indicator (Table 23-2) whose K_i is close to the calculated $[H^+]$ in this solution (see p 366). Titrate the acid to the color change, then stop and evaporate off the water. An alternative method for $NaHCO_3$, $NaHSO_3$, and NaHS is simply to saturate an NaOH solution with the appropriate gas (CO_2, SO_2, or H_2S); the resulting solution will contain the desired salt.

4. *Metasalts and pyrosalts*. Many solid acid salts serve as starting materials for the preparation of *meta*salts and *pyro*salts because they lose water when heated vigorously. For example, dihydrogen salts will yield *meta*salts, such as sodium *meta*arsenate,

$$NaH_2AsO_4 \xrightarrow{\Delta} NaAsO_3 + H_2O\uparrow$$

and monohydrogen salts will yield *pyro* salts, such as sodium *pyro*phosphate,

$$2NaHSO_4 \xrightarrow{\Delta} Na_2S_2O_7 + H_2O\uparrow$$

The formulas of *meta* and *pyro* acids are related to the *ortho* acids that you learned on p 107 (the exception is HNO_3, which is *meta* nitric acid), as follows:

$$\text{ortho acid} - H_2O \rightarrow meta \text{ acid}$$

$$2 \text{ ortho acid} - H_2O \rightarrow pyro \text{ acid}$$

These statements show the relationship between the formulas; the acids *cannot* be made this way.

5. *Metal hydrides*. These compounds usually can be prepared by heating the metal in an excess of H_2 and cooling it in H_2. The temperature of treatment varies from about 150.0°C for some alkali metals to 700.0°C for others. Extreme care must be used when heating with H_2: the system should be absolutely leakproof against air; flames should not be used for heating or be in the vicinity; and the H_2 flow should be safely vented. Further, some hydrides are spontaneously inflammable in air; they should all be handled with extreme care, and never be thoughtlessly added to water, with which they react to form H_2. The alkali and alkaline earth metals form straightforward hydrides (MH and MH_2) composed of M^+ (or M^{2+}) and H^-. The hydrides of most other metals are nonstoichiometric (such as $CeH_{2.69}$ or $TaH_{0.76}$), with the H atoms occupying interstitial positions in the metal's crystal lattice.

FRACTIONAL CRYSTALLIZATION

The preceding methods emphasize procedures resulting in a single product that can be removed as an insoluble compound or a gas, or that will be present by itself in water solution. When such a procedure cannot be used (or found), it may be necessary to separate the desired compound from a mixture of two soluble substances by a process known as fractional crystallization. This method is based on the differences in solubility that the two substances possess at high and low temperatures. The following problem illustrates the method.

PROBLEM:

The solubility of KNO_3 is 13.3 g/100 ml water at 0.0°C, and 247 g/100 ml at 100.0°C. The solubility of NaCl is 35.7 g/100 ml water at 0.0°C, and 39.1 g/100 ml at 100.0°C. What weights of the pure substances can be obtained from a mixture that contains 500 g each of KNO_3 and NaCl, using but one cycle of fractional crystallization (that is, using one high-temperature and one low-temperature crystallization).

SOLUTION:
The mixture is treated with just enough water to dissolve *all* of the more soluble component (KNO_3) at 100.0°C. Part of the NaCl also is dissolved, but a large fraction of it remains as a residue of pure NaCl, which is separated by filtration at 100.0°C. The filtrate is diluted with enough water to dissolve *all* of the remaining NaCl at 0.0°C, and then it is cooled to 0.0°C where a precipitate of pure KNO_3 is obtained and filtered off at 0.0°C.

Step 1. Treat the mixture with enough water to dissolve 500 g of KNO_3 at 100.0°C.

$$\text{Vol of } H_2O \text{ required} = (500 \text{ g } KNO_3)\left(\frac{100 \text{ ml } H_2O}{247 \text{ g } KNO_3}\right) = 202 \text{ ml } H_2O$$

$$\text{Wt of NaCl dissolved in 202 ml} = (202 \text{ ml } H_2O)\left(\frac{39.1 \text{ g NaCl}}{100 \text{ ml } H_2O}\right) = 79.0 \text{ g NaCl}$$

The undissolved NaCl (500 g − 79.0 g = 421 g pure NaCl) is filtered off.

Step 2. Dilute the filtrate to the volume required to dissolve 79.0 g NaCl at 0.0°C.

$$\text{Vol of } H_2O \text{ required} = (79.0 \text{ g NaCl})\left(\frac{100 \text{ ml } H_2O}{35.7 \text{ g NaCl}}\right) = 221 \text{ ml } H_2O$$

$$221 \text{ ml} - 202 \text{ ml} = 19.0 \text{ ml water added to filtrate}$$

$$\text{Wt of } KNO_3 \text{ dissolved in 221 ml} = (221 \text{ ml } H_2O)\left(\frac{13.3 \text{ g } KNO_3}{100 \text{ ml } H_2O}\right) = 29.4 \text{ g } KNO_3$$

The undissolved KNO_3 (500 g − 29.4 g = 470.6 g pure KNO_3) is filtered off.

PROBLEM:
What weights of pure salts can be recovered by one more cycle of fractional crystallization?

SOLUTION:
You have a solution that contains 29.4 g KNO_3 and 79.0 g NaCl. The simplest thing to do is evaporate off all the water, then repeat the cycle of the previous problem.

Step 3. Treat the solid mixture with enough water to dissolve 29.4 g KNO_3 at 100.0°C.

$$\text{Vol of } H_2O \text{ required} = (29.4 \text{ g } KNO_3)\left(\frac{100 \text{ ml } H_2O}{247 \text{ g } KNO_3}\right) = 11.9 \text{ ml } H_2O$$

$$\text{Wt of NaCl dissolved in 11.9 ml} = (11.9 \text{ ml } H_2O)\left(\frac{39.1 \text{ g NaCl}}{100 \text{ ml } H_2O}\right) = 4.7 \text{ g NaCl}$$

The undissolved NaCl (79.0 g − 4.7 g = 74.3 g pure NaCl) is filtered off.

Step 4. Dilute the filtrate to the volume required to dissolve 4.7 g NaCl at 0.0°C.

$$\text{Vol of } H_2O \text{ required} = (4.7 \text{ g NaCl}) \left(\frac{100 \text{ ml } H_2O}{35.7 \text{ g NaCl}} \right) = 13.2 \text{ ml } H_2O$$

$$13.2 \text{ ml} - 11.9 \text{ ml} = 1.3 \text{ ml water added to the filtrate}$$

$$\text{Wt of } KNO_3 \text{ dissolved in } 13.2 \text{ ml} = (13.2 \text{ ml } H_2O) \left(\frac{13.3 \text{ g } KNO_3}{100 \text{ ml } H_2O} \right) = 1.8 \text{ g } KNO_3$$

The undissolved KNO_3 (29.4 g $-$ 1.8 g $=$ 27.6 g pure KNO_3) is filtered off. At the end of two cycles of fractional crystallization, you have

$$421.0 + 74.3 = 495.3 \text{ g pure NaCl}$$

$$470.6 + 27.6 = 498.2 \text{ g pure } KNO_3$$

In practice, you would have to use volumes of water slightly greater than the calculated amounts to avoid inevitable contamination of one salt by the other; the yields of pure salts therefore would not be quite as high. Great care must be taken to keep the solutions at 0.0°C and 100.0°C during the filtrations.

PROBLEMS A

1. None of the reactions listed below takes place. Explain why.
    (a) $2KNO_3 + ZnBr_2 \rightarrow 2KBr + Zn(NO_3)_2$
    (b) $NaCl + H_2O \rightarrow NaOH + HCl$
    (c) $Br_2 + 2NaCl \rightarrow 2NaBr + Cl_2$
    (d) $CaCO_3 + 2NaCl \rightarrow CaCl_2 + Na_2CO_3$
    (e) $2Au + 2H_3PO_4 \rightarrow 2AuPO_4 + 3H_2$
    (f) $KMnO_4 + 5Fe(NO_3)_3 + 8HNO_3$
    $$\rightarrow Mn(NO_3)_2 + 5Fe(NO_3)_4 + KNO_3 + 4H_2O$$
    (g) $H_2S + MgCl_2 \rightarrow MgS + 2HCl$

2. The following reactions go to the right. Rewrite each in ionic form, and tell why it goes to the right. An excess of the second reactant is used.
    (a) $Mg(OH)_2 + 2NH_4Cl \rightarrow MgCl_2 + 2NH_3 + 2H_2O$
    (b) $AgCl + 2NH_3 \rightarrow Ag(NH_3)_2Cl$
    (c) $Ag_2S + 4KCN \rightarrow K_2S + 2KAg(CN)_2$
    (d) $AgCl + KI \rightarrow AgI + KCl$
    (e) $SrSO_4 + Na_2CO_3 \rightarrow SrCO_3 + Na_2SO_4$
    (f) $FeSO_4 + H_2S + 2NaOH \rightarrow FeS + Na_2SO_4 + 2H_2O$
    (g) $NH_4Cl + NaOH \rightarrow NH_3 + H_2O + NaCl$
    (h) $KBr + H_3PO_4(\text{conc.}) \rightarrow HBr + KH_2PO_4$
    (i) $3CuS + 8HNO_3 \rightarrow 3Cu(NO_3)_2 + 3S + 2NO + 4H_2O$
    (j) $Ag(NH_3)_2Cl + KI \rightarrow AgI + KCl + 2NH_3$

3. Give balanced ionic equations for the following reactions in aqueous solution. If a reaction does not occur, write NR. Indicate precipitates by ↓ and gases by ↑.

(a) $Pb(NO_3)_2 + NaCl \rightarrow$	(n) $KMnO_4 + K_2SO_4 \rightarrow$
(b) $CuSO_4 + NaCl \rightarrow$	(o) $HC_2H_3O_2 + HCl \rightarrow$
(c) $AgNO_3 + H_2S \rightarrow$	(p) $AgC_2H_3O_2 + HCl \rightarrow$
(d) $Cd + H_3PO_4 \rightarrow$	(q) $Fe + Cu(NO_3)_2 \rightarrow$
(e) $BaCO_3 + HNO_3 \rightarrow$	(r) $CuO + NaOH \rightarrow$
(f) $Au + HNO_3 \rightarrow$	(s) $KMnO_4 + H_3AsO_3 \rightarrow$
(g) $Ba_3(PO_4)_2 + HCl \rightarrow$	(t) $CuSO_4 + HCl \rightarrow$
(h) $AlCl_3 + NH_3 \rightarrow$	(u) $Ba(NO_3)_2 + H_2SO_4 \rightarrow$
(i) $Fe(OH)_3 + HCl \rightarrow$	(v) $ZnS + H_2SO_4 \rightarrow$
(j) $FeS + NaCl \rightarrow$	(w) $ZnSO_4 + Na_2S \rightarrow$
(k) $Ca_3(PO_4)_2 + NaNO_3 \rightarrow$	(x) $Cl_2 + KBr \rightarrow$
(l) $CuO + H_2SO_4 \rightarrow$	(y) $KClO_3 + KI + HCl \rightarrow$
(m) $NH_3 + HCl \rightarrow$	(z) $Br_2 + KCl \rightarrow$

4. Write balanced ionic equations for reactions that would be suitable for the practical *laboratory* preparation of each of the following compounds in pure form and good yield. You may have to supplement some equations with a few words of explanation. For gases, name a suitable desiccant if one is needed, and state what kind of a generator is used.

(a) $PbSO_4$	(f) $Cu(OH)_2$	(k) CO_2
(b) $KC_2H_3O_2$	(g) $NaHSO_4$	(l) Br_2
(c) CdS	(h) Al_2S_3	(m) Ag
(d) $KHCO_3$	(i) Fe_2O_3	(n) $KAsO_2$
(e) NH_3	(j) Hg_2Cl_2	(o) KH

5. Devise methods for the laboratory preparation and isolation of each of the following compounds in pure form and in good yield. Use pure metals as starting materials.

(a) $Al(NO_3)_3$	(c) $Ni(OH)_2$	(e) ZnS
(b) $Cu_3(PO_4)_2$	(d) $HgCl_2$	(f) $Fe_2(SO_4)_3$

6. Describe the stepwise procedure (stating volumes and temperatures) that you would use for the separation by two cycles of fractional crystallization for each of the following solid mixtures. Calculate the number of grams of pure salts obtained in each case. Solubility data are given in parentheses after each compound; the first figure is for 0.0°C and the second for 100.0°C, both in grams/100 ml of water: $Ce_2(SO_4)_3$ (10.1, 2.3); NH_4Cl (29.7, 75.8); KCl (27.9, 56.7); $KClO_3$ (3.8, 58.1); $K_2Cr_2O_7$ (4.9, 102); KNO_3 (13.3, 247); $NaCl$ (35.7, 39.1); NH_4ClO_4 (10.7, 47.3).

(a) 1 mole each of KCl and NH_4ClO_4
(b) 100 g each of $K_2Cr_2O_7$ and $NaCl$
(c) 50 g each of KNO_3 and $Ce_2(SO_4)_3$
(d) 50 g $KClO_3$ and 100 g NH_4Cl

PROBLEMS B

7. None of the reactions listed below takes place. Explain why.
   (a) $Na_2SO_4 + CuCl_2 \rightarrow CuSO_4 + 2NaCl$

(b) $Cl_2 + 2KF \rightarrow F_2 + 2KCl$
(c) $H_2S + 2NaCl \rightarrow Na_2S + 2HCl$
(d) $2H_2O + K_2SO_4 \rightarrow 2KOH + H_2SO_4$
(e) $H_2SO_4 + 2Ag \rightarrow Ag_2SO_4 + H_2$
(f) $Na_2Cr_2O_7 + 6Mg(NO_3)_2 + 14HNO_3 \rightarrow 2NaNO_3 + 6Mg(NO_3)_3 + 7H_2O +$
$2Cr(NO_3)_3$
(g) $Cu + ZnSO_4 \rightarrow CuSO_4 + Zn$

8. The following reactions go to the right. Rewrite each in ionic form, and tell why it goes to the right. An excess of the second reactant is used.
(a) $Fe(OH)_2 + H_2S \rightarrow FeS + 2H_2O$
(b) $Al(OH)_3 + NaOH \rightarrow NaAlO_2 + 2H_2O$
(c) $Al(OH)_3 + 3HCl \rightarrow AlCl_3 + 3H_2O$
(d) $BiCl_3 + 3NH_3 + 3H_2O \rightarrow Bi(OH)_3 + 3NH_4Cl$
(e) $K_2Cr_2O_7 + 2Pb(NO_3)_2 + H_2O \rightarrow 2PbCrO_4 + 2KNO_3 + 2HNO_3$
(f) $FeCl_3 + 6KSCN \rightarrow K_3Fe(SCN)_6 + 3KCl$
(g) $H_2S + Br_2 \rightarrow S + 2HBr$
(h) $PbSO_4 + H_2S \rightarrow PbS + H_2SO_4$
(i) $3HgS + 2HNO_3 + 6HCl \rightarrow 3HgCl_2 + 2NO + 3S + 4H_2O$
(j) $SnS_2 + 2NH_4HS \rightarrow (NH_4)_2SnS_3 + H_2S$

9. Give balanced ionic equations for the following reactions in aqueous solution. If a reaction does not occur, write NR. Indicate precipitates by ↓ and gases by ↑.
(a) $NaOH + HNO_3 \rightarrow$
(b) $Cd + AgNO_3 \rightarrow$
(c) $Cu + HNO_3 \rightarrow$
(d) $Cu + HCl \rightarrow$
(e) $CuSO_4 + NaCl \rightarrow$
(f) $CaCl_2 + Na_3PO_4 \rightarrow$
(g) $Ca_3(PO_4)_2 + HCl \rightarrow$
(h) $Na_2CO_3 + KCl \rightarrow$
(i) $Na_2CO_3 + BaCl_2 \rightarrow$
(j) $H_2S + FeSO_4 \rightarrow$
(k) $KMnO_4 + HCl \rightarrow$
(l) $FeS + H_2SO_4 \rightarrow$
(m) $CaCl_2 + Na_2C_2O_4 \rightarrow$
(n) $CaC_2O_4 + HCl \rightarrow$
(o) $KNO_3 + NaCl \rightarrow$
(p) $AgNO_3 + NaCl \rightarrow$
(q) $Na_2Cr_2O_7 + SnSO_4 + H_2SO_4 \rightarrow$
(r) $PbO_2 + HI \rightarrow$
(s) $AgNO_3 + KBr \rightarrow$
(t) $AgCl + NH_3 \rightarrow$
(u) $Na_2CrO_4 + CuSO_4 \rightarrow$
(v) $CuSO_4 + NaOH \rightarrow$
(w) $CuSO_4 + H_3PO_4 \rightarrow$
(x) $Zn + H_2SO_4 \rightarrow$
(y) $ZnO + H_2SO_4 \rightarrow$
(z) $CuCrO_4 + HNO_3 \rightarrow$

10. Write balanced ionic equations for reactions that would be suitable for the practical *laboratory* preparation of each of the following compounds in pure form and good yield. You may have to supplement some equations with a few words of explanation. For gases, name a suitable desiccant if one is needed, and state what kind of generator is used.
(a) SO_2
(b) $AgCl$
(c) MgO
(d) BaS
(e) $KHSO_4$
(f) $Cd(OH)_2$
(g) N_2O
(h) $NaHSO_3$
(i) Bi_2S_3
(j) $NaC_2H_3O_2$
(k) $BaSO_4$
(l) Cu
(m) Cl_2
(n) $K_4P_2O_7$
(o) BaH_2

11. Devise methods for the laboratory preparation and isolation of each of the following compounds in pure form and in good yield. Use pure metals as starting materials.

   (a) $Pb(C_2H_3O_2)_2$ (c) Ag_2SO_4 (e) $SnCl_2$
   (b) $MgSO_4$ (d) CdS   (f) Al_2O_3

12. Describe the stepwise procedure (stating volumes and temperatures) that you would use for the separation by two cycles of fractional crystallization for each of the following solid mixtures. Calculate the number of grams of pure salts obtained in each case. Solubility data are given in parentheses after each compound; the first figure is for 0.0°C and the second for 100.0°C, both in grams/100 ml of water: $Ce_2(SO_4)_3$ (10.1, 2.3); $NaBrO_3$ (27.5, 90.9); NH_4ClO_4 (10.7, 47.3); $Na_4P_2O_7$ (3.2, 40.3); NH_4Cl (29.7, 75.8); KCl (27.9, 56.7); $KClO_3$ (3.8, 58.1); KNO_3 (13.3, 247).

   (a) 100 g each of $KClO_3$ and $Ce_2(SO_4)_3$
   (b) 100 g of KCl and 50 g of $NaBrO_3$
   (c) 50 g each of $Na_4P_2O_7$ and NH_4Cl
   (d) 1 mole each of KNO_3 and NH_4ClO_4

Answers to Problems of the A Groups

CHAPTER 2

1. (a) 214.08
   (b) −8,950,798,157
   (c) −93.24
   (d) 1.9690054×10^9

   (e) $1.6636351 \times 10^{-12}$
   (f) 1.6576570×10^5
   (g) $-4.5677782 \times 10^{-5}$
   (h) -1.2586666×10^3

2. (a) 1.1452130×10^{-3}
   (b) 2.7624309×10^4
   (c) 2.3432236×10^{-7}

   (d) 2.4431957×10^{15}
   (e) $1.6603021 \times 10^{-24}$

3. (a) $5.3 \times 10^{-5} \times 8.7 \times 10^{-13} = 4.611 \times 10^{-17}$
   (b) $5.34 \times 10^{14} \times 8.7 \times 10^6 = 4.6458 \times 10^{21}$
   (c) $(5.34 \times 10^{15}) \div (8.7 \times 10^{-14}) = 6.1379310 \times 10^{28}$
   (d) 1.7615847×10^{-6}
   (e) 8.1400511×10^3
   (f) 4.487×10^{-5}

4. (a) 2.78784×10^7
   (b) 72.663608
   (c) 2.78784×10^{-23}
   (d) 2.2978250×10^{-5}
   (e) 1.7221621×10^{81}
   (f) 2.9570461
   (g) 3.6088261×10^4
   (h) $2.4379992 \times 10^{-18}$
   (i) 43.299903

   (j) 1.1270092×10^2
   (k) 5.3191433×10^9
   (l) 1.0166475×10^3
   (m) 1.2148059×10^3
   (n) 1.8752568×10^{-2}
   (o) 4.0352504×10^{23}
   (p) 2.7719053×10^{-2}
   (q) 0.33620266

5. (a) 1.6757783
   (b) 2.5646606
   (c) −2.28399666
   (d) −6.05898576
   (e) 3.19645254
   (f) 9.55954756
   (g) It is not possible to take the log of a negative number.
   (h) 1.34025042
   (i) 8.59137259
   (j) −7.25589532
   (k) −12.28079619
   (l) 5.47646355
   (m) 64.02218160
   (n) It is not possible to take the log of a negative number.
   (o) 5.5613470×10^7
   (p) 2.0000000

   (q) 4.3336117×10^{-7}
   (r) 9.7485494×10^{-9}
   (s) 44.444702
   (t) 3.1763620×10^{-5}
   (u) 48.242481
   (v) 2.0000056
   (w) 0.12878512
   (x) $1.4204502 \times 10^{-42}$
   (y) 1.0067627
   (z) 9.9999509×10^{-2}
   (aa) 5.5613470×10^7
   (bb) 2.0000000
   (cc) 4.3336117×10^{-7}
   (dd) 9.7485494×10^{-9}
   (ee) 1.0168104×10^2
   (ff) 3.1763620×10^{-5}
   (gg) 1.0001462

6. (a) 5.25
   (b) 31.0
   (c) 9.0
   (d) 113
   (e) 5.56008327
   (f) 2.5600000
   (g) 23.59550562
   (h) 0.47222222
   (i) 2.2282812×10^6
   (j) 52.749700
   (k) 2.0642773×10^{-2}
   (l) 4.4794059

   (m) $P = 71.860368$
   (n) $\Delta G = +3.1808155 \times 10^3$
   (o) $M = 12$
   (p) $x = 0.78359368$ and -0.28359368
   (q) $x = -2$ and $+1$
   (r) $x = -2.5$ and $+1$
   (s) $n = 3$ and $-2/3$
   (t) $m = 2$
   (u) $y = -9$ and 1
   (v) $x = 0.65962022$ and -0.19127705
   (w) $x = 3.8332103$ and 0.50363171

(x)

t	Factor	t	Factor
20	0.98740	26	0.98362
21	0.98677	27	0.98299
22	0.98614	28	0.98236
23	0.98551	29	0.98173
24	0.98488	30	0.98110
25	0.98425		

(y)

t	P	t	P
0	4.696	30	32.009
5	6.655	35	42.503
10	9.315	40	55.929
15	12.888	45	72.964
20	17.635	50	94.407
25	23.878		

(z) t	D	t	D
0	0.99983960	8	0.99984802
1	0.99989868	9	0.99978035
2	0.99994008	10	0.99969882
3	0.99996440	11	0.99960377
4	0.99997219	12	0.99949554
5	0.99996398	13	0.99937445
6	0.99994025	14	0.99924078
7	0.99990145	15	0.99909482

CHAPTER 3

1. 2.6784×10^6 sec
2. $F = 8.642 \times 10^4$ sec
3. 20 sec

4. 720 mph
5. 8.342 lbs

6. (a) $F = \left(\dfrac{1 \text{ lb}}{16 \text{ oz}}\right)\left(\dfrac{1 \text{ ton}}{2000 \text{ lb}}\right) = 3.125 \times 10^{-5}$ tons/oz

(b) $F = \left(\dfrac{1 \text{ ft}}{12 \text{ in}}\right)^3 \left(\dfrac{1 \text{ yd}}{3 \text{ ft}}\right)^3 = 2.143 \times 10^{-5}$ yd³/in³

(c) $F = \left(\dfrac{1 \text{ mi}}{5280 \text{ ft}}\right)\left(60 \dfrac{\text{sec}}{\text{min}}\right)\left(60 \dfrac{\text{min}}{\text{hr}}\right) = 0.6818 \dfrac{\text{mi/hr}}{\text{ft/sec}}$

(d) $F = \left(2000 \dfrac{\text{lb}}{\text{ton}}\right)\left(\dfrac{1 \text{ yd}}{3 \text{ ft}}\right)^2 \left(\dfrac{1 \text{ ft}}{12 \text{ in}}\right)^2 = 1.543 \dfrac{\text{lb/in}^2}{\text{ton/yd}^2}$

(e) $F = \left(\dfrac{1 \text{ \$}}{100 \text{ ¢}}\right)\left(2000 \dfrac{\text{lb}}{\text{ton}}\right) = 20 \dfrac{\text{\$/ton}}{\text{¢/lb}}$

(f) $F = \left(\dfrac{1 \text{ min}}{60 \text{ sec}}\right)\left(\dfrac{1 \text{ hr}}{60 \text{ min}}\right)\left(\dfrac{1 \text{ day}}{24 \text{ hr}}\right)\left(\dfrac{1 \text{ wk}}{7 \text{ days}}\right) = 1.653 \times 10^{-6} \dfrac{\text{weeks}}{\text{sec}}$

(g) $F = \left(60 \dfrac{\text{sec}}{\text{min}}\right)\left(12 \dfrac{\text{in}}{\text{ft}}\right)^3 \left(\dfrac{1 \text{ gal}}{231 \text{ in}^3}\right)\left(4 \dfrac{\text{qt}}{\text{gal}}\right) = 1.795 \times 10^3 \dfrac{\text{qt/min}}{\text{ft}^3/\text{sec}}$

(h) $F = \left(5280 \dfrac{\text{ft}}{\text{mi}}\right)\left(\dfrac{1 \text{ fathom}}{6 \text{ ft}}\right) = 880 \dfrac{\text{fathoms}}{\text{mi}}$

(i) $F = \left(3 \dfrac{\text{ft}}{\text{yd}}\right)\left(12 \dfrac{\text{in}}{\text{ft}}\right)\left(1000 \dfrac{\text{mil}}{\text{in}}\right) = 3.60 \times 10^4 \dfrac{\text{mil}}{\text{yd}}$

CHAPTER 4

1. $V = 36$ cm³
   $A = 72$ cm²

2. $V = 95.76$ cm³
   $A = 197.8$ cm²

3. $4A$ ¢
4. 44.1 lb

5. 76.43 mi

6. 23.65 cm

7. (a) 6700 Å and 3700 Å
   (b) 670 nm and 370 nm

8. (a) 1.537395×10^{-8} cm
   (b) 2.28503×10^{-8} cm
   (c) 7.0783×10^{-9} cm
   (d) 2.0862×10^{-9} cm

9. 69.8°F

10. −45.6°C
    227.6 K

11. −37.97°F
    234.33 K

12. −459.76°F

13. −40°C

14. 1.582×10^{-5} light-years

15. 3.333×10^3 m²

16. 0.100 ml

17. $A = 8.87$ cm²
    $D = 1.684$ cm

18. (a) 2.99×10^{-23} ml/molecule
    (b) 1.925 Å

19. (a) 35.09° M (b) −59.20° M

20. 4.405×10^5 tons

21. 0.025 g

22. 200 ml

23. 0.44 mm

24. 4.0×10^{-4} mm

25. (a) 6 cm²
    (b) 60 cm²
    (c) 6×10^6 cm²; 0.135 football fields

26. $\dfrac{g}{cm \times sec}$

27. erg sec

28. Force = mass × acceleration = $\dfrac{g \times cm}{sec^2}$ = dynes

   Pressure = $\dfrac{force}{area}$ = $\dfrac{g \times cm}{sec^2 \times cm^2}$ = $\dfrac{g}{cm \times sec^2}$

   Volume = cm³

   Pressure × volume = $\left(\dfrac{g}{cm \times sec^2}\right)$ (cm³) = $\left(\dfrac{g \times cm}{sec^2}\right)$ (cm)

   = dynes × cm = ergs = work

CHAPTER 5

1. (a) 3
   (b) 3
   (c) 4
   (d) 5
   (e) 3 or 4 (Is final 0 significant?)

   (f) 3
   (g) 2
   (h) 2
   (i) 1
   (j) 1

2. (a) 15.60
   (b) 721.17
   (c) 1308.4
   (d) 0.099

   (e) 1.60×10^3
   (f) 4.01×10^{17}
   (g) 3.400
   (h) 2.472×10^{-42}

3. 0.16 ppm

4. (a) 0.35%, 3.5 ppt
    (b) 0.35%, 3.5 ppt
    (c) 0.039%, 0.39 ppt
    (d) 0.59%, 5.9 ppt

    (e) 0.002%, 0.02 ppt
    (f) 0.87%, 8.7 ppt
    (g) 0.35%, 3.5 ppt

5. Av dev = 5.0, and s = 5.9
    90% confidence interval for a single value = 2643 ± 11.2
    90% confidence interval for the mean = 2643 ± 4.0

6. Av dev = 0.0039
    95% confidence interval for a single value = 5.020 ± 0.011

7. Av dev = 0.10%
    70% confidence limits for a single value = ±0.15%

8. Av dev = 3.7 ppt
    99% confidence limits for the mean = ±4.2 ppt

9. Av dev = 2.9 ppt
    80% confidence limits for the mean = ±2.4 ppt

10. d = 0.997 g/ml
    γ = 72.0 dynes/cm

11. 4 grams or more

CHAPTER 6

1.

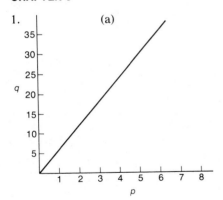

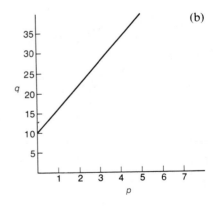

432

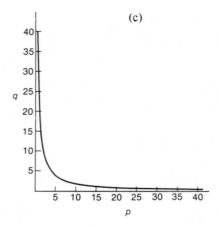

(c)

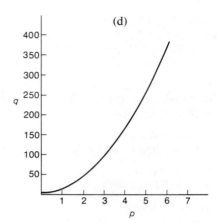

(d)

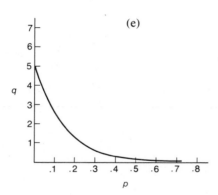

(e)

2.

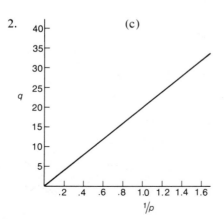

(c)

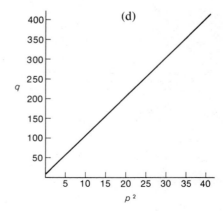

(d)

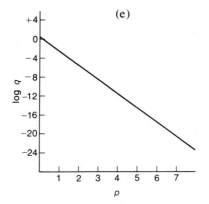

(e)

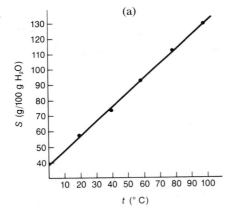

3.

(a)

(b) $s = 0.95t + 37.0$

(c) (i) $s = 0.9420t + 37.52$

(ii) $r = 0.99960$

95% confidence interval of slope $= 0.942 \pm 0.0491$

95% confidence interval of y intercept $= 37.52 \pm 3.26$

4.

(a)

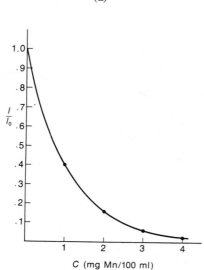

C (mg Mn/100 ml)

(b)

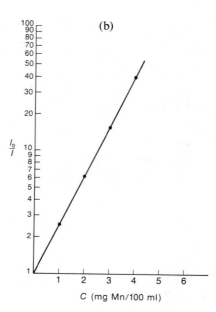

C (mg Mn/100 ml)

(c) $\log \left(\dfrac{I_0}{I} \right) = 0.40C$

(d) (i) $\log \left(\dfrac{I_0}{I} \right) = 0.4028C$

(ii) $r = 0.99846$
95% confidence interval of slope $= 0.4028 \pm 0.0681$

5. (a)

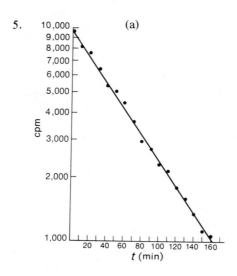

t (min)

(b) $k = 6.12 \times 10^{-3}$

$K = 3.986$

(c) (i) $\log N = -6.174 \times 10^{-3}t + 3.9926$

(ii) $r = 0.99864$

95% confidence interval of slope $= -0.00617 \pm 0.000177$

95% confidence interval of y intercept $= 3.9926 \pm 0.0167$

(d) K is the logarithm of the cpm at $t = 0$.

6.

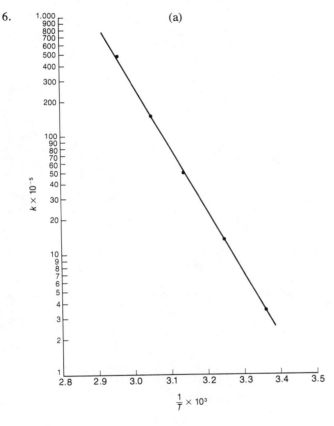

(a)

$k \times 10^{-5}$

$\frac{1}{T} \times 10^3$

(b) $\log k = -5.27 \times 10^3 \times \dfrac{1}{T} + 23.28$

(c) (i) $\log k = -5.392 \times 10^3 \times \dfrac{1}{T} + 23.624$

(ii) $r = 0.99987$

95% confidence interval of slope $= -5392 \pm 161$

95% confidence interval of y intercept $= 23.624 \pm 0.507$

(d) $\Delta H_a = 2.464 \times 10^4$ cal

(e) Q corresponds to the logarithm of the rate constant at infinitely high temperature.

7.

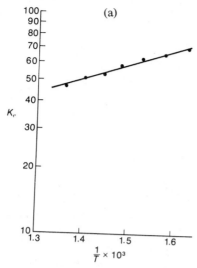

(a)

(b) $\log K_e = 665 \times \dfrac{1}{T} + 0.77$

(c) (i) $\log K_e = 668.5 \times \dfrac{1}{T} + 0.7644$

　　(ii) $r = 0.99821$
　　　95% confidence interval of the slope $= 668.5 \pm 46.1$
　　　95% confidence interval of the y intercept $= 0.7644 \pm 0.0688$

(d) $\Delta H = -3055$ cal

(e) Z corresponds to the logarithm of the equilibrium constant at infinitely high temperature.

CHAPTER 7

1. 33.5 ml

2. 1540 g

3. 7.8 ml

4. Gold

5. 34.0 g

6. 1.66×10^{-4} g/ml

7. 44.3 ¢/ml

8. (a) 497 g
　 (b) 1988 g

9. 0.90 pints

10. 0.15 cm

11. 0.802 g/ml

12. 3.169 g/ml

13. (a) 3.157 g/ml
　　 (b) Second decimal place

14. 9.98 g/ml

15. 25.4036 g

16. 56.3671 g

17. 16.3428 g

18. 1.0041 ml

19. The correction for buoyancy is 0.1 mg, which is less than the 0.2 mg sensitivity of the balance.

20. 24.8989 g

21. 3.14×10^4 lb

CHAPTER 8

1. Calcium hydroxide

2. Silver phosphate

3. Silver thiocyanate

4. Magnesium phthalate

5. Ammonium sulfate

6. Zinc sulfide

7. Cadmium cyanide

8. Barium iodate

9. Cupric sulfite, copper(II) sulfite

10. Cuprous iodide, copper(I) iodide

11. Ferric nitrate, iron(III) nitrate

12. Ferrous oxalate, iron(II) oxalate

13. Mercurous chloride, mercury(I) chloride

14. Manganous carbonate, manganese(II) carbonate

15. Manganic hydroxide, manganese(III) hydroxide

16. Nickelous hypochlorite, nickel(II) hypochlorite

17. Chromic arsenate, chromium(III) arsenate

18. Stannic bromide, tin(IV) bromide

19. Chromous fluoride, chromium(II) fluoride

20. Plumbous permanganate, lead(II) permanganate

21. Sodium silicate

22. Bismuthic oxide, bismuth(V) oxide

23. Aluminum perchlorate

24. Mercuric acetate, mercury(II) acetate

25. Cesium chlorate

26. Strontium hypoiodite

27. Rubidium arsenite

28. Beryllium nitride

29. Calcium bicarbonate, calcium monohydrogen carbonate

30. Antimonous nitrate, antimony(III) nitrate

31. Phosphorus trichloride

32. Bismuthous cyanate, bismuth(III) cyanate

33. Aluminum thiosulfate

34. $Al(BrO_3)_3$

35. Hg_3PO_4

36. Bi_2O_3

37. $Sr(HCO_3)_2$

38. AuI

39. $Cr(IO_3)_3$

40. $Mn(OH)_2$

41. Li_3As

42. $As_2(SO_4)_3$

43. $SnCl_4$

44. $Ni(IO_4)_2$

45. Cl_2O_7

46. $Ag_2C_2O_4$

47. $Cr_3(BO_3)_2$

48. Sb_2S_3

49. $Al(C_2H_3O_2)_3$

50. CaC_2O_4

51. $NaClO_2$

52. $Sn(N_3)_2$

53. $Hg(CN)_2$

54. $(NH_4)_2SO_3$

55. $Co(MnO_4)_2$

56. $PbCO_3$

57. Zn_3P_2

58. Cu_2SiO_4

59. $Ba(IO)_2$

60. (a) ZnSe
    (b) Fr_3PO_4
    (c) $CoSeO_3$
    (d) SeO_2
    (e) Fr_2SeO_4
    (f) SeF_6
    (g) FrH

61. (a) $Be(OH)_2 + 2HSCN \rightarrow Be(SCN)_2 + 2H_2O$
    (b) $5HIO_4 + Sb(OH)_5 \rightarrow Sb(IO_4)_5 + 5H_2O$
    (c) $Hg_2(OH)_2 + 2HC_2H_3O_2 \rightarrow Hg_2(C_2H_3O_2)_2 + 2H_2O$
    (d) $2H_3AsO_3 + 3Cr(OH)_2 \rightarrow Cr_3(AsO_3)_2 + 6H_2O$
    (e) $3Ce(OH)_4 + 4H_3BO_3 \rightarrow Ce_3(BO_3)_4 + 12H_2O$
    (f) $2HN_3 + Fe(OH)_2 \rightarrow Fe(N_3)_2 + 2H_2O$
    (g) $LiOH + HBrO \rightarrow LiBrO + H_2O$

CHAPTER 9

1. Only major resonance forms are considered in providing these approximate answers. The electron-pair geometries on which the sketches are based are referred to by numbers in parentheses.

| (1) | (2) | (3) | (4) | (5) |

Shape	Bond angles	Bond distance (A)	Bond energy (kcal/bond)
(a) O=C=O	180°	1.22	
(b) Tetrahedral (2)	109°28'	1.83	104.3
(c) △-Pyramid (2)	<109°28'	1.70	65.1
(d) F–Xe–F	180°		
(e) Angular (1)	<120°	~1.60	
(f) Angular (2)	<109°28'	~2.32	
(g) O=N=O⁺	180°	1.15	
(h) Angular (1)	<120°	N–F = 1.34	60.6
		N=O = 1.15	
(i) Seesaw (3)	4 angles, 90°	2.16	59.3
	1 angle, <120°		
	1 angle, <180°		
(j) Octahedral (4)	90°	1.81	151.0
(k) △-Coplanar (1)	~120°	N–Cl = 1.69	48.2
		N–O = ~1.25	
(l) Dodecahedral			
(m) Square coplanar (4)	90°	1.78	74.6
(n) Octahedral (4)	90°		
(o) O–C≡N⁻	180°	C–O = 1.43	81.2
		C≡N = 1.15	
(p) Tetrahedral (2)	~109°28'	S–S = 2.08	50.9
		S–O = 1.70	65.1
(q) Pentagonal bypyramid (5)			
(r) △-Coplanar about	O–N–O = ~120°	N–N = 1.40	38.4
each N (1)	O–N–N = ~120°	N–O = 1.36	41.6
(s) △-Coplanar (1)	F–Si–F = <120°	Si–F = 1.81	151.0
		Si=O = 1.62	
(t) Square-base pyramid (4)	90°	1.78	74.6
(u) T-shape (3)	Cl–I–Cl = <90°	2.32	52.8
(v) △-Pyramid (2)	<109°28'	2.25	47.3
(w) Tetrahedral (2)	O–S–O = <109°28'	S–O = 1.70	65.1
	Cl–S–Cl = >109°28'	S–Cl = 2.03	60.2
(x) △-Coplanar about each C;	C–C–C = 120°	C–C = 1.44	
all atoms in same plane	H–C–C = 120°	C–H = 1.07	93.7
		C–H = 1.07	93.7
		C=C = 1.33	

(y)   C=C   Coplanar   ~120°   C–Cl = 1.76   76.3

(z)   109°28'   1.83   104.3

Tetrahedral about each Si

CHAPTER 10

1. (a) 63.6% N, 36.4% O
   (b) 46.7% N, 53.3% O
   (c) 30.4% N, 69.6% O
   (d) 32.4% Na, 22.6% S, 45.0% O
   (e) 29.1% Na, 40.6% S, 30.3% O
   (f) 14.3% Na, 10.0% S, 69.5% O, 6.2% H
   (g) 18.5% Na, 25.9% S, 51.6% O, 4.0% H
   (h) 43.5% Ca, 26.1% N, 30.4% O
   (i) 29.2% N, 12.5% C, 50.0% O, 8.3% H
   (j) 47.4% U, 5.6% N, 44.6% O, 2.4% H
   (k) 56.1% C, 8.2% N, 18.7% O, 7.6% H, 9.4% S

2. (a) 44.0
   (b) 30.0
   (c) 46.0
   (d) 142.1
   (e) 158.2
   (f) 322.1
   (g) 248.2
   (h) 92.1
   (i) 96.0
   (j) 502.1
   (k) 342.1
   All have units of g/mole

3. (a) 10.3
   (b) 15.1
   (c) 9.87
   (d) 3.19
   (e) 2.87
   (f) 1.41
   (g) 1.83
   (h) 4.93
   (i) 4.73
   (j) 0.904
   (k) 1.33
   All have units of moles

4. (a) 8.36×10^{23} molecules
   (b) 4.99×10^{22} molecules
   (c) 3.54×10^{16} molecules
   (d) 3.12×10^{22} molecules
   (e) 1.51×10^{23} molecules

5. 334 g/mole

6. (a) FeO
   (b) Fe_2O_3
   (c) Fe_3O_4
   (d) K_2CrO_4
   (e) $K_2Cr_2O_7$
   (f) CH
   (g) CH_4
   (h) $Mg_2P_2O_7$
   (i) $Ag_4V_2O_7$
   (j) $Sn_2Fe(CN)_6$

7. (a) $CuSO_4 \cdot 5H_2O$
   (b) $Hg(NO_3)_2 \cdot \frac{1}{2}H_2O$
   (c) $Pb(C_2H_3O_2)_2 \cdot 3H_2O$
   (d) $CoCl_2 \cdot 6H_2O$
   (e) $CaSO_4 \cdot 2H_2O$

8. (a) Cu_2S
   (b) Mg_3N_2
   (c) LiH
   (d) $AlCl_3$
   (e) La_2O_3

9. 140.2 g/mole

10. (a) 20.18570
    (b) 32.07249

11. (a) 9, 18, 27, 36
    (b) MCl_3

12. 58.8 or some multiple of this

13. C_6H_6OS

CHAPTER 11

1. 22.0 ml

2. 619 cm

3. 725.6 torr

4. 13.9 lb/in²

5. 240 ml

6. 841 ml

7. 475 torr

8. 107 torr

9. 3.3 atm

10. 3.53×10^{10} molecules

11. 203 ml

12. (a) 402 ml
    (b) 389 ml

13. 498 ml

14. 0.279 g

15. 832 ml

16. 134 g/ml

17. 2.73 g/ml

18. (a) 1.96 g/liter
    (b) 1.73 g/liter

19. 68.2 g/mole

20. 277 ml

21. 34.2 g/mole

22. 0.350 atm

23. (a) 0.075 atm N_2, 0.250 atm N_2O, 0.175 atm CO_2
    (b) resultant pressure is 0.325 atm
        partial pressures are 0.075 atm N_2 and 0.250 atm N_2O

24. 9.54 g/mole

25. 87.8 g/mole

26. $C_4H_8O_2$

27. 36.8 lbs/in²

28. 48.8 sec

29. 1.66 ml/min

30. 11.6 g

31. (a) Balloon would rise
    (b) It would lift 2.9 g

32. 22.1% H_2 and 77.9% O_2

33. 30.6 kg

34. (a) 7.94×10^3 liters/day
    (b) 10.2 kg/day

35. (a) 315 liters/day
    (b) 618 g/day

36. C_2H_4NO

CHAPTER 12

1. The answers to these problems are given as the numerical coefficients needed to balance the equations, and in the order in which the substances appear in the unbalanced equations.
    (a) 2, 2, 1
    (b) 2, 2, 4, 1
    (c) 2, 2, 2, 1
    (d) 3, 4, 1, 4
    (e) 1, 3, 2, 3
    (f) 1, 4, 3, 4

(g) 1, 1, 1   (i) 1, 2, 1, 1   (k) 1, 3, 3, 1, 3
(h) 1, 4, 2, 1, 1   (j) 1, 2, 2, 1, 1   (l) 1, 4, 1, 3, 2

2. (a) (i) 396 g   10. (a) 1.35×10^3 liters
      (ii) 1.17 kg   (b) 6.45×10^3 liters
   (b) 16.3 tons
   (c) 5.35 g NH_4Cl, 2.81 g CaO   11. (a) 7.05 g
   (d) 1.53 kg   (b) 0.357 moles
   (e) 7.81 liters HCl, 5.06 kg $CaCO_3$ (c) 29.7 ml
   (f) (i) 1.03 kg   (d) 5.61 g
       (ii) 561 ml
   (g) (i) 4.45 kg   12. (a) 1.81 g
       (ii) 3.07 liters   (b) 0.0760 moles
   (h) 157 g   (c) 7.60 ml
   (i) $255.34 for NH_4Cl
       $594.03 for $NaNO_3$ 13. 2.41×10^{-4} ml
   (j) 86.1%
   (k) 76.0%   14. 266 g

                                             15. 1.75×10^4 ft³
3. 8.75 g
                                             16. 2.50×10^7 liters
4. (a) 7.62 g
   (b) 0.117 moles   17. 1.94×10^7 ft³

5. 1.50 liters   18. 8.33×10^5 ft³

6. 1×10^6 ft³   19. 14.58%

7. 3.90 liters   20. 0.3969 g

8. (a) 17.4 g   21. (a) 11.46%
   (b) 0.801 mole   (b) $C_6H_6N_2O_3$

9. 241 ml

CHAPTER 13

1. Dissolve each of the following weights in distilled water, then dilute to the volumes
   requested in the problems.
   (a) 132 g   (d) 4.04×10^3 g
   (b) 31.6 g   (e) 20.6 g
   (c) 4.71 g

2. Dilute each of the following volumes with distilled water to the volumes requested in
   the problems.
   (a) 0.500 ml   (d) 0.263 ml
   (b) 56.0 ml   (e) 8.13 ml
   (c) 68.4 ml

3. (a) 8.38 M, 9.60 m   (c) 17.8 M, 194 m
   (b) 11.4 M, 18.6 m   (d) 3.41 M, 4.28 m

(e) 0.870 M, 0.954 *m*
(f) 0.834 M, 0.862 *m*

(g) 0.535 M, 0.534 *m*
(h) 0.805 M, 0.824 *m*

4. (a) 1.59 g
   (b) 0.703 g
   (c) 0.182 g
   (d) 0.173 g

(e) 0.203 g
(f) 486 g
(g) 74.3 g

5. 23.0%

15. 4.930%

6. (a) 90.1%
   (b) 144 *m*

16. 50.37 ml

17. 44.0%

7. 4.2 M

18. 71.37%

8. 767 ml

19. (a) 0.86 g
    (b) 0.35 g
    (c) 25 ml

9. (a) 23.5%
   (b) 1.01 *m*

10. 0.1688 M

20. 51.3 ml

11. 0.4812 M

21. 54.0 ml

12. 0.2151 M

22. (a) 162 g Ag, 333 ml HNO_3
    (b) 13.4 liters NO

13. 36.22 ml

23. 32.77%

14. 0.06563 M

CHAPTER 14

1. (a) 0.193 cal/g °C
   (b) 0.0948 cal/g °C
   (c) 0.0446 cal/g °C

(d) 0.0260 cal/g °C
(e) 0.0299 cal/g °C

2. 56.7°C

6. 93.7 g/mole

3. 27.1°C

7. 0.319 cal/g °C

4. 35.1°C

8. 41.7 cal/°C

5. 0.0°C

9. (a) 0.818 cal/g °C
   (b) 0.917 cal/g °C
   (c) $Pb^{2+}_{(aq)} + 2I^-_{(aq)} \rightarrow PbI_{2(s)}$
       $\Delta H = -15.5$ kcal/mole

(d) $KBrO_{3(s)} + aq \rightarrow K^+_{(aq)} + BrO^-_{3(aq)}$
    $\Delta H = +9.83$ kcal/mole
(e) $C_7H_7I_{(s)} \rightarrow C_7H_7I_{(l)}$
    $\Delta H_f = +4.08$ kcal/mole

10. (a) X_2O_3

(b) 51.00 g/mole

11. (a) YCl_3 and YCl_5

(b) 121.4 g/mole

12. (a) $\Delta H° = -4.62$ kcal $= -19.33$ kj; $\Delta E° = -4.03$ kcal
    (b) $\Delta H° = -216.24$ kcal $= -904.75$ kj; $\Delta E° = -216.83$ kcal

(c) $\Delta H° = -36.43$ kcal $= -152.42$ kj; $\Delta E° = -37.02$ kcal
(d) $\Delta H° = -69.10$ kcal $= -289.11$ kj; $\Delta E° = -68.81$ kcal
(e) $\Delta H° = -33.09$ kcal $= -138.45$ kj; $\Delta E° = -31.91$ kcal
(f) $\Delta H° = -38.98$ kcal $= -163.09$ kj; $\Delta E° = -39.57$ kcal

13. (a) $NH_{3(g)} + aq \rightarrow NH_{3(aq)}$
    $\Delta H° = -8.28$ kcal
    (b) $2Fe_{(s)} + 3O_{2(g)} \rightarrow Fe_2O_{3(s)}$
    $\Delta H° = -196.5$ kcal
    (c) $C_6H_{6(l)} + \frac{15}{2}O_{2(g)} \rightarrow 6CO_{2(g)} + 3H_2O_{(l)}$
    $\Delta H° = -780.98$ kcal
    (d) $Cd_{(s)} + 2H^+_{(aq)} \rightarrow Cd^{2+}_{(aq)} + H_{2(g)}$
    $\Delta H° = -17.30$ kcal
    (e) $2H^+_{(aq)} + S^{2-}_{(aq)} \rightarrow H_2S_{(g)}$
    $\Delta H° = -14.82$ kcal

14. $\Delta H°_T = -80$ cal/mole
    $\Delta S°_T = -0.27$ cal/mole °C

15. $\Delta H°_v = +10.12$ kcal/mole
    $\Delta S°_v = +34.0$ cal/mole °C

16. (a) 123.12 kcal/bond   (c) 84.29 kcal/bond
    (b) 110.57 kcal/bond   (d) 63.12 kcal/bond

17. (a) $C_{12}H_{22}O_{11(s)} + 12\ O_{2(g)} \rightarrow 12CO_{2(g)} + 11\ H_2O_{(l)}$
    $\Delta H° = -1345$ kcal/mole
    (b) $\Delta H°_f = -535.12$ kcal/mole
    (c) 20.45 kcal/teaspoonful
        1785.47 kcal/lb

CHAPTER 15

1. 48.8 min

2. 0.20 cpm

3. 99.57 mg

4. 2.076×10^3 yr

5. (a) first order with respect to BOOP
   (b) $k = 3.54 \times 10^{-4}$ M^{-1} min^{-1}

6. (a) Second-order reaction
   (b) $k = 1.44 \times 10^{-4}$ torr^{-1} min^{-1}

7. (a) Second-order reaction
   (b) $k = 9.12 \times 10^{-5}$ torr^{-1} min^{-1}

8. (a) First-order reaction
   (b) $k = -0.0187$ min^{-1}

9. (a) Approx 900 cal/mole for each
   (b) 2.44×10^{19} molecules/ml for each

10. (a) 5.99×10^{-15}
    (b) 3.30×10^{-10}
    (c) 7.74×10^{-8}
    (d) 2.05×10^{-6}
    (e) 1.43×10^{-3}

11. 11 times faster

12. (a) Second-order with respect to N (b) Zero-order with respect to M

13. (a) 31.2 kcal
   (b) 1.10×10^{-11}
   (c) 85.4 kcal

(d)

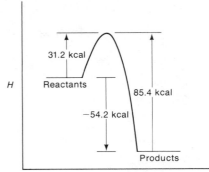

Time course of reaction

14. (a) 26.9 kcal
   (b) 5.57×10^{-12}
   (c) 13.4 kcal

(d)

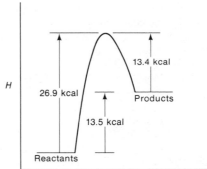

Time course of reaction

CHAPTER 16

1.

	A		B		C	
R	0	0	0	L	0
R	+	R	+	L	−
R	0	R	0	R	0
0	0	0	0	0	0
R	0	0	0	L	0

2. 1.27 atm

3. 1.10×10^{-13} atm

4. (a) 2.26×10^{-4} M and 4.19×10^{-4} M   (c) 3.98×10^{-5} and 8.93×10^{-4}
   (b) 2.51×10^4 and 1.12×10^3

5. (a) $4.2 \times 10^{-7}\%$ (b) 0.28%

6. approximately 8.0%

7. (a) CO and H_2O are each 22.9%
       CO_2 and H_2 are each 27.1%
   (b) CO_2 and H_2 are each 14.5%
       CO is 2.2%
       H_2O is 68.8%

8. (a) 10.5%
   (b) 18.3%
   (c) 0.704 M^{-1}

9. -3.14 kcal

10. -44.9 kcal

11. (a) $+39.1$ kcal
    (b) -31.4 kcal

12. 8.05×10^{-6} atm^{-1}

CHAPTER 17

1. Arrange beakers and salt bridge as in Figure 17-1. Use solutions and electrodes as follows.
   (a) Mg electrode in Mg^{2+} solution in left beaker; Ag electrode in Ag^+ solution in right beaker.
   (b) Cu electrode in Cu^{2+} solution in left beaker. With pool of Hg in bottom of right beaker containing Hg_2^{2+} solution, use Pt electrode as in right beaker of Figure 17-2.
   (c) In both beakers use Pt electrode as in right beaker of Figure 17-2. In left beaker put solution containing Sn^{2+} and Sn^{4+}; in right beaker put solution containing $S_2O_8^{2-}$ and SO_4^{2-}.
   (d) In both beakers use Pt electrode as in right beaker of Figure 17-2. In left beaker put solution containing Sn^{2+} and Sn^{4+}; in right beaker put solution containing Br^-, and enough Br_2 to have a slight excess of liquid Br_2 lying on the bottom.
   (e) Fe electrode in Fe^{2+} solution in left beaker. Pt electrode in solution of Fe^{2+} and Fe^{3+} as in right side of Figure 17-2.

2. (a) $Mg/Mg^{2+}//Ag^+/Ag$ (d) $Pt/Sn^{2+},Sn^{4+}//Br^-,Br_2/Pt$
   (b) $Cu/Cu^{2+}//Hg_2^{2+}/Hg,Pt$ (e) $Fe/Fe^{2+}//Fe^{2+},Fe^{3+}/Pt$
   (c) $Pt/Sn^{2+},Sn^{4+}//S_2O_8^{2-},SO_4^{2-}/Pt$

3. (*i*) (a) 3.16 v   (*ii*) (a) 3.16 v
        (b) 0.45 v   (b) 0.51 v
        (c) 1.86 v   (c) 1.92 v
        (d) 0.92 v   (d) 1.04 v
        (e) 1.21 v   (e) 1.33 v

4. (a) 8.66×10^{106} (d) 1.36×10^{31}
   (b) 1.69×10^{15} (e) 8.86×10^{40}
   (c) 8.79×10^{62}

5. (a) -146 kcal   (d) -42.4 kcal
   (b) -20.8 kcal   (e) -55.8 kcal
   (c) -85.8 kcal

6. At negative electrode:
   $2e^- + Fe(OH)_2 \rightleftarrows 2OH^- + Fe$

At positive electrode:

$e^- + Ni(OH)_3 \rightleftarrows OH^- + Ni(OH)_2$

7. (a) 6.48 kcal   (d) 10.08 kcal
   (b) 6.48 kcal   (e) 3.04 kcal
   (c) 16.28 kcal

8. (a) $\Delta G^\circ = -249.07$ kcal; $K_e = 7.68 \times 10^{182}$; spontaneous
   (b) $\Delta G^\circ = -335.82$ kcal; $K_e = 3.84 \times 10^{246}$; spontaneous
   (c) $\Delta G^\circ = 16.66$ kcal; $K_e = 5.85 \times 10^{-13}$; not spontaneous
   (d) $\Delta G^\circ = 25.06$ kcal; $K_e = 3.97 \times 10^{-19}$; not spontaneous

9. (a) $\Delta S^\circ = -43.91$ cal/°C   (c) $\Delta S^\circ = +34.74$ cal/°C
   (b) $\Delta S^\circ = -38.60$ cal/°C   (d) $\Delta S^\circ = +10.58$ cal/°C

10. (a) $\Delta H^\circ = -262.16$ kcal   (c) $\Delta H^\circ = +27.01$ kcal
    (b) $\Delta H^\circ = -347.32$ kcal   (d) $\Delta H^\circ = +28.21$ kcal

11. (a) 4.45×10^{-13} (c) 1.68×10^{37}
    (b) 1.77×10^{83}

12. $+30.0$ cal/mole °C

CHAPTER 18

1. The oxidation numbers are given for the atoms in the order in which they occur in the given molecule or ion.
   (a) $+4, -2$ (f) $+1, -1$
   (b) $+1, +5, -2$ (g) $+3, -2$
   (c) $+2, -1$ (h) $+4, -1$
   (d) $+3, +6, -2$ (i) $+3, -2$
   (e) $+2, +5, -2$ (j) $+6, -2$

2. (a) $HNO_2 + H_2O \rightleftarrows 3H^+ + NO_3^- + 2e^-$
   (b) $H_3AsO_3 + H_2O \rightleftarrows H_3AsO_4 + 2H^+ + 2e^-$
   (c) $Al \rightleftarrows Al^{3+} + 3e^-$
   (d) $Ni \rightleftarrows Ni^{2+} + 2e^-$
   (e) $Hg_2^{2+} \rightleftarrows 2Hg^{2+} + 2e^-$
   (f) $H_2O_2 \rightleftarrows O_2 + 2H^+ + 2e^-$
   (g) $2I^- \rightleftarrows I_2 + 2e^-$

3. (a) $2e^- + PbO_2 + 4H^+ \rightleftarrows Pb^{2+} + 2H_2O$
   (b) $e^- + NO_3^- + 2H^+ \rightleftarrows NO_2 + H_2O$ (conc'd acid)
       $3e^- + NO_3^- + 4H^+ \rightleftarrows NO + 2H_2O$ (dilute acid)
   (c) $e^- + Co^{3+} \rightleftarrows Co^{2+}$
   (d) $8e^- + ClO_4^- + 8H^+ \rightleftarrows Cl^- + 4H_2O$
   (e) $2e^- + BrO^- + 2H^+ \rightleftarrows Br^- + H_2O$
   (f) $e^- + Ag^+ \rightleftarrows Ag$
   (g) $2e^- + F_2 \rightleftarrows 2F^-$
   (h) $2e^- + Sn^{2+} \rightleftarrows Sn$

4. (a) $Zn + Cu^{2+} \rightarrow Zn^{2+} + Cu$

   (b) $Zn + 2H^+ \rightarrow Zn^{2+} + H_2$

   (c) $Cr_2O_7^{2-} + 6I^- + 14H^+ \rightarrow 2Cr^{3+} + 3I_2 + 7H_2O$

   (d) $2MnO_4^- + 10Cl^- + 16H^+ \rightarrow 2Mn^{2+} + 5Cl_2 + 8H_2O$

   (e) $ClO_3^- + 6Br^- + 6H^+ \rightarrow Cl^- + 3Br_2 + 3H_2O$

   (f) $2MnO_4^- + 5H_2O_2 + 6H^+ \rightarrow 2Mn^{2+} + 5O_2 + 8H_2O$

   (g) $MnO_2 + 4H^+ + 2Cl^- \rightarrow Mn^{2+} + Cl_2 + 2H_2O$

   (h) $3Ag + 4H^+ + NO_3^- \rightarrow 3Ag^+ + NO + 2H_2O$

   (i) $PbO_2 + Sn^{2+} + 4H^+ \rightarrow Pb^{2+} + Sn^{4+} + 2H_2O$

   (j) $2MnO_4^- + 5H_2C_2O_4 + 6H^+ \rightarrow 2Mn^{2+} + 10CO_2 + 8H_2O$

   (k) $H_2O_2 + HNO_2 \rightarrow H^+ + NO_3^- + H_2O$

   (l) $Fe + Cu^{2+} \rightarrow Fe^{2+} + Cu$

   (m) $2Fe^{3+} + 2I^- \rightarrow 2Fe^{2+} + I_2$

   (n) $ClO_4^- + 4H_3AsO_3 \rightarrow Cl^- + 4H_3AsO_4$

   (o) $H_2S + ClO^- \rightarrow S + Cl^- + H_2O$

5. (a) $Zn + CuSO_4 \rightarrow ZnSO_4 + Cu$

   (b) $Zn + H_2SO_4 \rightarrow ZnSO_4 + H_2$

   (c) $K_2Cr_2O_7 + 6KI + 7H_2SO_4 \rightarrow Cr_2(SO_4)_3 + 3I_2 + 4K_2SO_4 + 7H_2O$

   (d) $2KMnO_4 + 10KCl + 8H_2SO_4 \rightarrow 2MnSO_4 + 5Cl_2 + 6K_2SO_4 + 8H_2O$

   (e) $KClO_3 + 6KBr + 3H_2SO_4 \rightarrow KCl + 3Br_2 + 3K_2SO_4 + 3H_2O$

   (f) $2KMnO_4 + 5H_2O_2 + 3H_2SO_4 \rightarrow 2MnSO_4 + 5O_2 + K_2SO_4 + 8H_2O$

   (g) $MnO_2 + 2KCl + 2H_2SO_4 \rightarrow MnSO_4 + Cl_2 + K_2SO_4 + 2H_2O$

   (h) $3Ag + 4HNO_3 \rightarrow 3AgNO_3 + NO + 2H_2O$

   (i) $PbO_2 + SnSO_4 + 2H_2SO_4 \rightarrow PbSO_4 + Sn(SO_4)_2 + 2H_2O$

   (j) $2KMnO_4 + 5H_2C_2O_4 + 3H_2SO_4 \rightarrow 2MnSO_4 + 10CO_2 + K_2SO_4 + 8H_2O$

   (k) $H_2O_2 + HNO_2 \rightarrow HNO_3 + H_2O$

   (l) $Fe + CuSO_4 \rightarrow FeSO_4 + Cu$

   (m) $2Fe_2(SO_4)_3 + 2KI \rightarrow 2FeSO_4 + I_2 + K_2SO_4$

   (n) $KClO_4 + 4H_3AsO_3 \rightarrow KCl + 4H_3AsO_4$

   (o) $H_2S + KClO \rightarrow S + KCl + H_2O$

6. (a) $CH_2O + Ag_2O + OH^- \rightarrow 2Ag + HCO_2^- + H_2O$

   (b) $C_2H_2 + 2MnO_4^- + 6H^+ \rightarrow 2CO_2 + 2Mn^{2+} + 4H_2O$

   (c) $2C_2H_3OCl + 3Cr_2O_7^{2-} + 24H^+ \rightarrow 4CO_2 + Cl_2 + 6Cr^{3+} + 15H_2O$

   (d) $6Ag^+ + AsH_3 + 3H_2O \rightarrow 6Ag + H_3AsO_3 + 6H^+$

   (e) $2OH^- + CN^- + 2Fe(CN)_6^{3-} \rightarrow CNO^- + 2Fe(CN)_6^{4-} + H_2O$

   (f) $3C_2H_4O + 2NO_3^- + 2H^+ \rightarrow 2NO + 3C_2H_4O_2 + H_2O$

7. (a) $4Zn + 10H^+ + NO_3^- \rightarrow 4Zn^{2+} + NH_4^+ + 3H_2O$

   (b) $2S_2O_3^{2-} + I_2 \rightarrow S_4O_6^{2-} + 2I^-$

   (c) $IO_3^- + 5I^- + 6H^+ \rightarrow 3I_2 + 3H_2O$

   (d) $Cu + 2H_2SO_4 \rightarrow CuSO_4 + SO_2 + 2H_2O$

   (e) $Zn + 2Fe^{3+} \rightarrow 2Fe^{2+} + Zn^{2+}$

   (f) $8I^- + 5H_2SO_4 \rightarrow 4I_2 + H_2S + 4SO_4^{2-} + 4H_2O$

   (g) $3ClO^- \rightarrow 2Cl^- + ClO_3^-$

8. See page 456 for answers to this problem.

CHAPTER 19

1.

	Negative Electrode			Positive Electrode	
	Ni	0.25 mole		Ni^{2+}	0.25 mole
	H_2	5.6 liters		I_2	0.25 mole
	H_2	5.6 liters		O_2	2.80 liters
	Au	0.167 mole		O_2	2.80 liters
	H_2	5.6 liters		Cu^{2+}	0.25 mole

2. (a) 1.60 g
   (b) 7.10 g
   (c) 2.43 g
   (d) 5.89 g
   (e) 23.92 g

3. 2.98 amp

4. 3 hrs 12 min 58 sec

5. 0.0411 M

6. 2.30 M

7. 33.0 hrs

8. 1.128×10^5 amp hrs

9. 265 amp hrs

CHAPTER 20

1. Dissolve the following weights in distilled water and dilute to the mark in a suitable volumetric flask.
   (a) 2.43 g   (c) 234 g
   (b) 15.8 g

2. Dilute each of the following volumes with distilled water to the volumes asked for in the problems.
   (a) 21.5 ml   (c) 96.0 ml
   (b) 1.74 ml

3. 8.33 liters

4. (a) 26.67 ml
   (b) 224.4 mg
   (c) 68.0 mg
   (d) 130.8 mg
   (e) 3.08 ml
   (f) 40.00 ml

5. 53.12 ml

6. 89.99%

7. 0.1545 M

8. (a) 0.07540 N
   (b) 0.07540 M

9. KOH: 0.1454 N and 0.1454 M
   $KMnO_4$: 0.08484 N and 0.01697 M

10. 0.1565 N

11. 0.02940 M $Ba(OH)_2$
    0.06486 M HNO_3

12. (a) 0.3103 N
    (b) 0.62 ml

13. 0.2963 N

14. 194 ml

15. (a) 115.4 g
    (b) 500 ml
    (c) 37.2 liters

16. 46.00%

17. Answers to these problems are given on p 443.

CHAPTER 21

1. (a) 5.12°C
   (b) 0.36°C
   (c) 0.928
   (d) 92.8 torr
   (e) 82.68°C

2. 98.5 g/mole

3. −0.604°C

4. 84.65°C

5. 169 g/mole

6. S_8

7. $C_8H_{10}O_2$

8. $C_{10}H_6N_2O_4$

9. 48.6

10. (a) $m = 1.21$, mole fraction = 0.0213
    (b) $m = 2.92$, mole fraction = 0.149
    (c) $m = 1.30$, mole fraction = 0.0564
    (d) $m = 0.333$, mole fraction = 0.00596
    (e) $m = 4.39$, mole fraction = 0.255

11. 13.9 qt

12. 45.45 g in one beaker,
    204.55 g in the other

13. 6.2 torr

14. The apparent mole weight of 121.8 in water indicates a very slight dissociation into benzoate and H^+ ions. The apparent molecular weight of 244 in benzene indicates a polymerization into a compound with the formula $(C_7H_6O_2)_2$.

15. 41.1 atm

16. 3.51×10^4 g/mole

17. 51 units

18. 20.9 torr

19. 2.15%

20. 79.0%

21. −0.0194°C

22. 78.5%

23. 91.6%

CHAPTER 22

1. (a) 4
   (b) 6
   (c) 8
   (d) −1
   (e) 1.92

   (f) 1.05
   (g) 4.43
   (h) 7.19
   (i) −0.54
   (j) 0.30

2. (a) 10
   (b) 8
   (c) 6
   (d) 15
   (e) 12.40
   (f) 12.90
   (g) 9.67
   (h) 6.41
   (i) 14.81
   (j) 13.86

3. (a) 2.45×10^{-4} M
   (b) 3.02×10^{-8} M
   (c) 3.72×10^{-14} M
   (d) 0.170 M
   (e) 3.55×10^{-7} M
   (f) 1.10×10^{-9} M
   (g) 1.00 M
   (h) 1.58×10^{-3} M
   (i) 3.98 M
   (j) 1.58×10^{-15} M

4. (a) 10
   (b) 8
   (c) 6
   (d) 15
   (e) 12.08
   (f) 12.95
   (g) 9.57
   (h) 6.81
   (i) 14.54
   (j) 13.70

5. 12.85

6. 0.96

7. (a) 1.00
   (b) 1.18
   (c) 1.48
   (d) 2.28
   (e) 3.00
   (f) 4.00
   (g) 7.00
   (h) 10.00
   (i) 11.00
   (j) 11.68

CHAPTER 23

1. (a) 6.30×10^{-10}
   (b) 1.73×10^{-5}
   (c) 1.74×10^{-5}
   (d) 1.75×10^{-5}
   (e) 1.70×10^{-4}

2. (a) 1.73×10^{-5}
   (b) 6.31×10^{-10}
   (c) 1.77×10^{-5}
   (d) 2.21×10^{-4}
   (e) 4.49×10^{-4}

3. (a) 1.98
   (b) 2.86
   (c) 5.61
   (d) 2.68
   (e) 1.71
   (f) 11.71
   (g) 12.20
   (h) 9.24
   (i) 10.47
   (j) 11.94

4. (a) 0.0459%
   (b) 1.68%
   (c) 2.16%
   (d) 0.14%
   (e) 0.12%

5. (a) 2.77
   (b) 4.62

6. 113 g

7. 4.94

8. 8.44

9. 294 g

10. 9.14

11. 4.76

12. 9.06

13. (a) 5.74×10^{-10}
    (b) 1.57×10^{-5}
    (c) 2.19×10^{-11}
    (d) 5.75×10^{-11}
    (e) 5.76×10^{-10}

14. (a) 9.16
    (b) 4.53
    (c) 8.09
    (d) 7.00
    (e) 5.68
    (f) 7.00
    (g) 8.73
    (h) 7.00 The [H$^+$] from H_2O is more
        important than that from the HCl.

15. (a) 7.00
    (b) 5.36
    (c) 11.10
    (d) 8.82

16. (a) Bromthymol blue
    (b) Methyl red
    (c) Thymolphthalein
    (d) Phenolphthalein

17. 5.16

18. Any solution in which the molar ratio of NH_4^+ to NH_3 is 17.4

19. (a) 9.24
    (b) 9.33
    (c) 9.15
    (d) 14.00
    (e) 0.00
    (f) Small amounts of strong acid and
        base, when added to water
        (pH = 7), normally cause a tre-
        mendous change in pH, as in (d)
        and (e). In a buffer solution, small
        amounts of acid and base cause
        only a very small change in pH, as
        in (b) and (c).

20. Construct a titration curve with ml of NaOH as abscissa and pH as ordinate, using
    the following values.

ml NaOH added	pH
0.0	2.88
5.0	4.16
10.0	4.58
24.0	6.14
24.9	7.15
25.0	8.73
25.1	10.30
26.0	11.29
30.0	11.96

21. Construct a titration curve with ml of NaOH as abscissa and pH as ordinate, using
    the following values.

ml NaOH added	pH
0.0	1.00
5.0	1.17
10.0	1.37
24.0	2.69
24.9	3.70
25.0	7.00
25.1	10.30
26.0	11.29
30.0	11.96

CHAPTER 24

1. (a) 8.31×10^{-17}
   (b) 1.07×10^{-10}
   (c) 3.97×10^{-15}
   (d) 2.47×10^{-9}
   (e) 1.43×10^{-16}
   (f) 2.51×10^{-27}
   (g) 5.50×10^{-10}
   (h) 1.90×10^{-93}

2. $MnCO_3$, $SrCrO_4$, $Zn(OH)_2$, BaC_2O_4 will dissolve

3. (a) no ppt
   (b) no ppt
   (c) ppt
   (d) no ppt
   (e) no ppt
   (f) no ppt
   (g) ppt

4. (a) 1.26×10^{-26} M
   (b) 4.73×10^{-3} M
   (c) 4.79×10^{-5} M
   (d) 2.47×10^{-17} M
   (e) 1.17×10^{-3} M
   (f) 4.29×10^{-6} M

5. (a) 6.13×10^{-3} g
   (b) 2.93×10^{-6} g
   (c) 2.93×10^{-5} g
   (d) slightly more than 6.13×10^{-3} g

6. (a) 0.135 g
   (b) 4.63×10^{-2} g
   (c) 1.86×10^{-3} g

7. (a) 6.00×10^{-35} M
   (b) 6.00×10^{-42} M
   (c) 1.25×10^{-17} M
   (d) 4.64×10^{-31} M
   (e) 1.60×10^{-46} M

8. (a) 3.17×10^{-9} M
   (b) 3.73×10^{-10} M
   (c) 1.44×10^{-7} M
   (d) 7.94×10^{-14} M
   (e) 1.19×10^{-10} M

9. (a) Pb^{2+} precipitates first
   (b) 3.04×10^{-6} M

10. 32.3 mg

11. 4.67×10^{-7} M

12. $7.35 \times 10^{-5}\%$

13. 1.50×10^{-11} M

14. 0.138 M

15. 16.1 g

16. 0.0443 mole

17. 7.00

18. (a) -0.59
    (b) 1.00×10^{-4} M

19. -0.79

20. (a) -0.84
    (b) -4.37
    (c) 0.11
    (d) 7.70

21. AgI is converted to Ag_2S because the $[Ag^+]$ available from AgI and the $[S^{2-}]$ possible in 5 M HCl exceed the K_{sp} for Ag_2S.

22. 131 liters

23. (a) 2.00×10^{-16} M   (c) 5.33×10^{-5} M
    (b) 2.67×10^{-7} M

24. 1.01×10^{-13} M

25. (a) $[H^+]^2[CO_3^{2-}] = 7.11 \times 10^{-19}$
    (b) 5.29
    (c) 2.29×10^{-6} M
    (d) The method isn't practical because there is not enough difference between the K_{sp} values for the various carbonates.

26. $Mn(OH)_2$ will precipitate first.

27. (a) 4.56×10^{-11} M   (b) 3.19×10^{-6} M

CHAPTER 25

1. (a) hexaaquo magnesium(II)
    (b) hexahydroxo antimonate(III)
    (c) dithiosulfato silver(I)
    (d) dioxalato nickelate(II)
    (e) hexaammine cobalt(III)
    (f) dinitrodiammine nickel(II)
    (g) tetraiodo mercurate(II)

2. 4.77×10^{-16} M

3. (a) 0.100 M
    (b) 1.80 M
    (c) 3.09×10^{-10} M
    (d) 0.100 M
    (e) 1.79×10^{-12} M
    (f) 5.60×10^{-3} M

4. 3.33 g

5. 0.653 g

6. 0.836 M

7. 12.5 M

8. 1.65×10^{-4}

9. 3.55 M

10. 9.51×10^{-8} M

11. Initial concentration must be 0.0201 M
    Final concentration must be 9.81×10^{-5} M

12. 5.03

13. (a) 4.17   (b) 1.62

14. (a) $[OH^-]$ = 1.00×10^{-5} M
    $[Pb^{2+}]$ = 5.00×10^{-6} M
    $[Pb(OH)_3^-]$ = 4.00×10^{-7} M
    (b) $[Pb^{2+}]$ = 5.00×10^{-16} M
    $[Pb(OH)_3^-]$ = 0.0400 M

(c) $[Sn^{2+}]$ = 3.19×10^{-26} M
    $[Sn(OH)_3^-]$ = 0.0556 M
(d) 5.20 liters

CHAPTER 26

1. (a) $^{14}_{6}C \rightarrow {}^{14}_{7}N + {}^{0}_{-1}e$
   (b) $^{35}_{16}S \rightarrow {}^{35}_{17}Cl + {}^{0}_{-1}e$
   (c) $^{128}_{53}I \rightarrow {}^{128}_{54}Xe + {}^{0}_{-1}e$
   (d) $^{211}_{82}Pb \rightarrow {}^{211}_{83}Bi + {}^{0}_{-1}e$
   (e) $^{210}_{84}Po \rightarrow {}^{206}_{82}Pb + {}^{4}_{2}He$
   (f) $^{211}_{85}At \rightarrow {}^{207}_{83}Bi + {}^{4}_{2}He$

2. (a) $^{27}_{14}Si \rightarrow {}^{27}_{13}Al + {}^{0}_{1}e$
   (b) $^{28}_{13}Al \rightarrow {}^{28}_{14}Si + {}^{0}_{-1}e$
   (c) $^{30}_{15}P \rightarrow {}^{30}_{14}Si + {}^{0}_{1}e$
   (d) $^{24}_{11}Na \rightarrow {}^{24}_{12}Mg + {}^{0}_{-1}e$
   (e) $^{17}_{9}F \rightarrow {}^{17}_{8}O + {}^{0}_{1}e$

3. (a) $^{14}_{7}N + {}^{1}_{0}n \rightarrow {}^{1}_{1}H + {}^{14}_{6}C$
   (b) $^{26}_{12}Mg + {}^{1}_{0}n \rightarrow {}^{4}_{2}He + {}^{23}_{10}Ne$
   (c) $^{59}_{27}Co + {}^{2}_{1}H \rightarrow {}^{1}_{1}H + {}^{60}_{27}Co$
   (d) $^{14}_{7}N + {}^{4}_{2}He \rightarrow {}^{1}_{1}H + {}^{17}_{8}O$
   (e) $^{63}_{29}Cu + {}^{1}_{1}H \rightarrow {}^{1}_{0}n + {}^{63}_{30}Zn$

11. See p 444 for the answers to these problems.

4. (a) 4.00×10^8 kcal
   (b) 802 tons

5. 9.9×10^{-3} mg

6. 4.20 hrs

7. 48.6 g

8. 6.04×10^7 yrs

9. 6.60×10^3 yrs

10. 343 torr

CHAPTER 27

1. In (a), (b), (d), and (g), the reactions do not take place because none of the products is insoluble, unionized, or goes off as a gas. In (c) and (e), the cell potentials are negative, and the reaction would take place in the reverse direction. In (f), the reactants are all oxidizing agents, and there is no reducing agent to be oxidized; the apparent oxidation of Fe^{3+} is false, because the higher oxidation state of 4+ does not exist.

2. Many of the statements made below are true because an excess of reagent is used; for example, NH_4^+ in (a). This excess shifts the equilibrium to the right.
   (a) $Mg(OH)_2 + 2NH_4^+ \rightarrow Mg^{2+} + 2NH_3 + 2H_2O$
   OH^- removed more completely by excess NH_4^+ than by Mg^{2+}.
   (b) $AgCl + 2NH_3 \rightarrow Ag(NH_3)_2^+ + Cl^-$
   Ag^+ more completely removed by slightly ionized $Ag(NH_3)_2^+$ than by insoluble $AgCl$, especially in excess NH_3.
   (c) $Ag_2S + 4CN^- \rightarrow 2Ag(CN)_2^- + S^{2-}$
   With excess CN^-, Ag^+ more completely removed by slightly ionized $Ag(CN)_2^-$ than by insoluble Ag_2S.
   (d) $AgCl + I^- \rightarrow AgI + Cl^-$
   AgI is more insoluble than $AgCl$; excess I^- helps.
   (e) $SrSO_4 + CO_3^{2-} \rightarrow SrCO_3 + SO_4^{2-}$
   $SrCO_3$ is more insoluble than $SrSO_4$; excess CO_3^{2-} helps.

(f) $Fe^{2+} + H_2S + 2OH^- \rightarrow FeS + 2H_2O$

One of the products (FeS) is insoluble, and the other (H_2O) is only slightly ionized.

(g) $NH_4^+ + OH^- \rightarrow NH_3 + H_2O$

One of the products (NH_3) is a gas, the other (H_2O) is slightly ionized.

(h) $Br^- + H_3PO_4$ (conc'd) $\rightarrow HBr + H_2PO_4^-$

One of the products (HBr) is a gas that evolves easily on warming.

(i) $3CuS + 8H^+ + 2NO_3^- \rightarrow 3Cu^{2+} + 3S + 2NO + 4H_2O$

An electron-transfer reaction for which the cell potential is positive. Also, S is a solid and NO is a gas.

(j) $Ag(NH_3)_2^+ + I^- \rightarrow AgI + 2NH_3$

AgI is more insoluble than $Ag(NH_3)_2^+$ is unionized, and excess I^- helps shift the equilibrium.

3. (a) $Pb^{2+} + 2Cl^- \rightarrow PbCl_2\downarrow$

(b) NR

(c) $2Ag^+ + H_2S \rightarrow Ag_2S\downarrow + 2H^+$

(d) $3Cd + 2H^+ + 2H_2PO_4^- \rightarrow Cd_3(PO_4)_2\downarrow + 3H_2\uparrow$

(e) $BaCO_3 + 2H^+ \rightarrow Ba^{2+} + H_2O + CO_2\uparrow$

(f) NR

(g) $Ba_3(PO_4)_2 + 4H^+ \rightarrow 3Ba^{2+} + 2H_2PO_4^-$

(h) $Al^{3+} + 3NH_3 + 3H_2O \rightarrow Al(OH)_3\downarrow + 3NH_4^+$

(i) $Fe(OH)_3 + 3H^+ \rightarrow Fe^{3+} + 3H_2O$

(j) NR

(k) NR

(l) $CuO + 2H^+ \rightarrow Cu^{2+} + H_2O$

(m) $NH_3 + H^+ \rightarrow NH_4^+$

(n) NR

(o) NR

(p) $Ag^+ + C_2H_3O_2^- + H^+ + Cl^- \rightarrow AgCl\downarrow + HC_2H_3O_2$

(q) $Fe + Cu^{2+} \rightarrow Fe^{2+} + Cu$

(r) NR

(s) $2MnO_4^- + 5H_3AsO_3 + 6H^+ \rightarrow 2Mn^{2+} + 5H_3AsO_4 + 3H_2O$

(t) NR

(u) $Ba^{2+} + SO_4^{2-} \rightarrow BaSO_4\downarrow$

(v) $ZnS + 2H^+ \rightarrow Zn^{2+} + H_2S\uparrow$

(w) $Zn^{2+} + S^{2-} \rightarrow ZnS\downarrow$

(x) $Cl_2 + 2Br^- \rightarrow 2Cl^- + Br_2$

(y) $ClO_3^- + 6I^- + 6H^+ \rightarrow Cl^- + 3I_2 + 3H_2O$

(z) NR

4. (a) Mix solutions of $Pb(NO_3)_2$ and Na_2SO_4 and filter off precipitate.

$$Pb^{2+} + SO_4^{2-} \rightarrow PbSO_4\downarrow$$

(b) Neutralize KOH with $HC_2H_3O_2$, then evaporate off water.

$$K^+ + OH^- + HC_2H_3O_2 \rightarrow K^+ + C_2H_3O_2^- + H_2O$$

(c) Pass H_2S gas into a dilute $Cd(NO_3)_2$ solution, and filter off CdS.

$$Cd^{2+} + H_2S \rightarrow CdS\downarrow + 2H^+$$

(d) Saturate a KOH solution with CO_2 gas, then evaporate off water.

$$K^+ + OH^- + CO_2 \rightarrow K^+ + HCO_3^-$$

(e) Heat a mixture of solid NH_4Cl and CaO [or $Ca(OH)_2$] in generator of Figure 27-2b, or add NaOH solution to NH_4Cl in generator of Figure 27-2a, and warm. Dry NH_3 with solid CaO, NaOH, or $CaCl_2$.

$$2NH_4^+ + 2Cl^- + CaO \rightarrow 2NH_3\uparrow + Ca^{2+} + 2Cl^- + H_2O$$

(f) Add NaOH to a $CuSO_4$ solution, and filter off the precipitate. NH_3 cannot be used.

$$Cu^{2+} + 2OH^- \rightarrow Cu(OH)_2\downarrow$$

(g) Divide a given volume of H_2SO_4 in half. Neutralize one-half with NaOH, then add the remaining half of H_2SO_4 to the neutralized solution. Evaporate off the water.

$$2Na^+ + 2OH^- + 2H^+ + SO_4^{2-} \rightarrow 2Na^+ + SO_4^{2-} + 2H_2O$$

$$2Na^+ + SO_4^{2-} + 2H^+ + SO_4^{2-} \rightarrow 2Na^+ + 2HSO_4^-$$

(h) Heat Al metal with excess of S, and distill off excess S. Al_2S_3 will not precipitate from solution with H_2S.

$$2Al + 3S \xrightarrow{\Delta} Al_2S_3$$

(i) Add excess NH_3 to a $FeCl_3$ solution. Filter off the precipitate of $Fe(OH)_3$, then heat it vigorously.

$$Fe^{3+} + 3NH_3 + 3H_2O \rightarrow Fe(OH)_3\downarrow + 3NH_4^+$$

$$2Fe(OH)_3 \xrightarrow{\Delta} Fe_2O_3 + 3H_2O\uparrow$$

(j) Add HCl or NaCl solution to solution of $Hg_2(NO_3)_2$, then filter off precipitate.

$$Hg_2^{2+} + 2Cl^- \rightarrow Hg_2Cl_2\downarrow$$

(k) Add H_2SO_4 to $CaCO_3$ using generator of Figure 27-2a. Dry the CO_2 with P_4O_{10}, SiO_2, concentrated H_2SO_4, or dry ice.

$$2H^+ + SO_4^{2-} + CaCO_3 \rightarrow CaSO_4\downarrow + H_2O + CO_2\uparrow$$

(l) Add HBr to MnO_2 or $KMnO_4$. Warm. Condense Br_2 in very cold container. Perform under a hood.

$$MnO_2 + 4H^+ + 2Br^- \rightarrow Mn^{2+} + Br_2\uparrow + 2H_2O$$

(m) Add excess metallic Zn to $AgNO_3$ solution to reduce Ag^+, then add excess dilute HNO_3 to dissolve the excess Zn (Ag will not dissolve). Filter off the Ag.

$$Zn + 2Ag^+ \rightarrow Zn^{2+} + 2Ag$$

$$8H^+ + 2NO_3^- + 3Zn \rightarrow 3Zn^{2+} + 2NO\uparrow + 4H_2O$$

(n) Vigorously heat KH_2AsO_3.

$$KH_2AsO_3 \xrightarrow{\Delta} KAsO_2 + H_2O\uparrow$$

(o) Heat metallic K in a stream of H_2, in a tube well protected from the air.

$$2K + H_2 \xrightarrow{\Delta} 2KH$$

5. (a) Al will not dissolve directly in HNO_3. Therefore, dissolve Al in HCl, neutralize with NH_3, and with excess NH_3 precipitate $Al(OH)_3$. NaOH can't be used. Filter off the $Al(OH)_3$ and then *just* neutralize it with HNO_3 (add HNO_3 until the precipitate *just* dissolves). Evaporate off water.

$$2Al + 6H^+ \rightarrow 2Al^{3+} + 3H_2\uparrow$$
$$Al^{3+} + 3NH_3 + 3H_2O \rightarrow Al(OH)_3\downarrow + 3NH_4^+$$
$$Al(OH)_3 + 3H^+ + 3NO_3^- \rightarrow Al^{3+} + 3NO_3^- + 3H_2O$$

(b) Dissolve Cu in HNO_3, then make basic with NaOH to precipitate $Cu(OH)_2$. NH_3 can't be used. Filter off $Cu(OH)_2$ and then just neutralize with H_3PO_4.

$$3Cu + 8H^+ + 2NO_3^- \rightarrow 3Cu^{2+} + 2NO\uparrow + 4H_2O$$
$$Cu^{2+} + 2OH^- \rightarrow Cu(OH)_2\downarrow$$
$$3Cu(OH)_2 + 2H^+ + 2H_2PO_4^- \rightarrow Cu_3(PO_4)_2 + 6H_2O$$

(c) Dissolve Ni in HCl, then make basic with excess NaOH. NH_3 can't be used.

$$Ni + 2H^+ \rightarrow Ni^{2+} + H_2\uparrow$$
$$Ni^{2+} + 2OH^- \rightarrow Ni(OH)_2\downarrow$$

(d) Dissolve Hg in concentrated HNO_3, then make basic with excess NaOH. NH_3 can't be used. *Just* dissolve the precipitate with HCl, then evaporate off the water.

$$Hg + 4H^+ + 2NO_3^- \rightarrow Hg^{2+} + 2NO_2\uparrow + 2H_2O$$
$$Hg^{2+} + 2OH^- \rightarrow Hg(OH)_2\downarrow$$
$$Hg(OH)_2 + 2H^+ + 2Cl^- \rightarrow Hg^{2+} + 2Cl^- + 2H_2O$$

(e) Dissolve Zn in HCl, then make slightly basic with NH_3 and saturate with H_2S. Filter off ZnS.

$$Zn + 2H^+ \rightarrow Zn^{2+} + H_2\uparrow$$
$$Zn^{2+} + H_2S \rightarrow ZnS\downarrow + 2H^+$$

(f) Dissolve Fe in HNO_3 to get Fe in 3+ state, then add excess NH_3 (or NaOH) to precipitate $Fe(OH)_3$. Filter off $Fe(OH)_3$ and *just* neutralize with H_2SO_4, then evaporate off water.

$$Fe + 6H^+ + 3NO_3^- \rightarrow Fe^{3+} + 3NO_2\uparrow + 3H_2O$$
$$Fe^{3+} + 3NH_3 + 3H_2O \rightarrow Fe(OH)_3\downarrow + 3NH_4^+$$
$$2Fe(OH)_3 + 6H^+ + 3SO_4^{2-} \rightarrow 2Fe^{3+} + 3SO_4^{2-} + 6H_2O$$

6. In each of these problems, it is assumed that the calculated amount of water has been added or removed in going from one step to the next. In many cases (when volumes are small, or when *some* of water must be removed), it is easier to evaporate off *all* the water, then add back the desired amount.
   (a) *Step 1.* Treat the mixture with enough water (W) to dissolve 74.6 g (1 mole) KCl (K) at 0.0°C.

$$\text{Vol W} = (74.6 \text{ g K}) \left(\frac{100 \text{ ml W}}{27.9 \text{ g K}} \right) = 267.4 \text{ ml W}$$

$$\text{Wt NH}_4\text{ClO}_4 \text{ (S) dissolved in 267.4 ml} = (267.4 \text{ ml W}) \left(\frac{10.7 \text{ g S}}{100 \text{ ml W}} \right) = 28.6 \text{ g S}$$

Filter off $117.5 - 28.6 = 88.9$ g pure S at 0.0°C

Step 2. Adjust water volume to that required to dissolve 28.6 g S at 100.0°C.

$$\text{Vol W} = (28.6 \text{ g S}) \left(\frac{100 \text{ ml W}}{47.3 \text{ g S}} \right) = 60.5 \text{ ml W}$$

$$\text{Wt K dissolved in 60.5 ml} = (60.5 \text{ ml W}) \left(\frac{56.7 \text{ g K}}{100 \text{ ml W}} \right) = 34.3 \text{ g K}$$

Filter off $74.6 - 34.3 = 40.3$ g pure K at 100.0°C

Step 3. Adjust water volume to that required to dissolve 34.3 g K at 0.0°C.

$$\text{Vol W} = (34.3 \text{ g K}) \left(\frac{100 \text{ ml W}}{27.9 \text{ g K}} \right) = 123 \text{ ml W}$$

$$\text{Wt S dissolved in 123 ml} = (123 \text{ ml W}) \left(\frac{10.7 \text{ g S}}{100 \text{ ml W}} \right) = 13.2 \text{ g S}$$

Filter off $28.6 - 13.2 = 15.4$ g pure S at 0.0°C

Step 4. Adjust water volume to that required to dissolve 13.2 g S at 100.0°C.

$$\text{Vol W} = (13.2 \text{ g S}) \left(\frac{100 \text{ ml W}}{47.3 \text{ g S}} \right) = 27.9 \text{ ml W}$$

$$\text{Wt K dissolved in 27.9 ml W} = (27.9 \text{ ml W}) \left(\frac{56.7 \text{ g K}}{100 \text{ ml W}} \right) = 15.8 \text{ g K}$$

Filter off $34.3 - 15.8 = 18.5$ g pure K at 100.0°C

After 2 cycles, you have $88.9 + 15.4 = 104.3$ g pure NH_4ClO_4 (S)
and $40.3 + 18.5 = 58.8$ g pure KCl (K).

(b) *Step 1.* Treat the mixture with enough water (W) to dissolve 100 g NaCl (N) at 0.0°C.

$$\text{Vol W} = (100 \text{ g N}) \left(\frac{100 \text{ ml W}}{35.7 \text{ g N}} \right) = 280 \text{ ml W}$$

$$\text{Wt K}_2\text{Cr}_2\text{O}_7 \text{ (K) dissolved in 280 ml} = (280 \text{ ml W}) \left(\frac{4.9 \text{ g K}}{100 \text{ ml W}} \right) = 13.7 \text{ g K}$$

Filter off $100 - 13.7 = 86.3$ g pure K at 0.0°C

Step 2. Adjust the water volume to that needed to dissolve 13.7 g K at 100.0°C.

$$\text{Vol W} = (13.7 \text{ g K}) \left(\frac{100 \text{ ml W}}{102 \text{ g K}} \right) = 13.4 \text{ ml W}$$

Wt N that dissolves in 13.4 ml $= (13.4 \text{ ml W}) \left(\dfrac{39.1 \text{ g N}}{100 \text{ ml W}} \right) = 5.3 \text{ g N}$

Filter off $100 - 5.3 = 94.7$ g pure N at 100.0°C

Step 3. Adjust water volume to that required to dissolve 5.3 g N at 0.0°C.

$$\text{Vol W} = (5.3 \text{ g N}) \left(\dfrac{100 \text{ ml W}}{35.7 \text{ g N}} \right) = 14.9 \text{ ml W}$$

Wt K that dissolves in 14.9 ml $= (14.9 \text{ ml}) \left(\dfrac{4.9 \text{ g K}}{100 \text{ ml W}} \right) = 0.73 \text{ g K}$

Filter off $13.7 - 0.7 = 13.0$ g pure K at 0.0°C

Step 4. Adjust water volume to that required to dissolve 0.73 g K at 100.0°C.

$$\text{Vol W} = (0.73 \text{ g K}) \left(\dfrac{100 \text{ ml W}}{102 \text{ g K}} \right) = 0.72 \text{ ml W}$$

Wt N that dissolves in 0.80 ml $= (0.80 \text{ ml W}) \left(\dfrac{39.1 \text{ g N}}{100 \text{ ml W}} \right) = 0.31 \text{ g N}$

Filter off $5.3 - 0.3 = 5.0$ g pure N at 100.0°C

After 2 cycles, you now have $86.3 + 13 = 99.3$ g pure $K_2Cr_2O_7$
and $94.7 + 5 = 99.7$ g pure NaCl

(c) *Step 1.* Treat the mixture with enough water (W) to dissolve 50 g KNO_3 (K) at 100.0°C.

$$\text{Vol W} = (50 \text{ g K}) \left(\dfrac{100 \text{ ml W}}{247 \text{ g K}} \right) = 20.3 \text{ ml W}$$

Wt $Ce_2(SO_4)_3$ (C) dissolved in 20.3 ml $= (20.3 \text{ ml W}) \left(\dfrac{2.3 \text{ g C}}{100 \text{ ml W}} \right) = 0.50 \text{ g C}$

Filter off $50.0 - 0.5 = 49.5$ g pure $Ce_2(SO_4)_3$ at 100°C

Step 2. Adjust water volume to that needed to dissolve 0.5 g C at 0.0°C.

$$\text{Vol W} = (0.5 \text{ g C}) \left(\dfrac{100 \text{ ml W}}{10.1 \text{ g C}} \right) = 5.0 \text{ ml W}$$

Wt K dissolved in 5 ml $= (5.0 \text{ ml W}) \left(\dfrac{13.3 \text{ g K}}{100 \text{ ml W}} \right) = 0.7 \text{ g K}$

Filter off $50.0 - 0.7 = 49.3$ g pure KNO_3 at 0.0°C

There is no point in carrying this recrystallization any further.
(d) *Step 1.* Treat the mixture with enough water (W) to dissolve 100 g NH_4Cl (N) at 0.0°C.

$$\text{Vol W} = (100 \text{ g N}) \left(\dfrac{100 \text{ ml W}}{29.7 \text{ g N}} \right) = 336.7 \text{ ml W}$$

Wt $KClO_3$ (K) that dissolves in 336.7 ml = (336.7 ml W) $\left(\dfrac{3.8 \text{ g K}}{100 \text{ ml W}} \right)$ = 12.8 g K

Filter off 50.0 − 12.8 = 37.2 g pure K at 0.0°C

Step 2. Adjust water volume to that needed to dissolve 12.8 g K at 100.0°C.

$$\text{Vol W} = (12.8 \text{ g K}) \left(\frac{100 \text{ ml W}}{58.1 \text{ g K}} \right) = 22 \text{ ml W}$$

Wt N that dissolves in 22 ml = (22 ml W) $\left(\dfrac{75.8 \text{ g N}}{100 \text{ ml W}} \right)$ = 16.7 g N

Filter off 100.0 − 16.7 = 83.3 g pure N at 100.0°C

Step 3. Adjust water volume to that needed to dissolve 16.7 g N at 0.0°C.

$$\text{Vol W} = (16.7 \text{ g N}) \left(\frac{100 \text{ ml W}}{29.7 \text{ g N}} \right) = 56.2 \text{ ml W}$$

Wt K that dissolves in 56.2 ml = (56.2 ml W) $\left(\dfrac{3.8 \text{ g K}}{100 \text{ ml W}} \right)$ = 2.1 g K

Filter off 12.8 − 2.1 = 10.7 g pure K at 0.0°C

Step 4. Adjust water volume to that needed to dissolve 2.1 g K at 100.0°C.

$$\text{Vol W} = (2.1 \text{ g K}) \left(\frac{100 \text{ ml W}}{58.1 \text{ g K}} \right) = 3.6 \text{ ml W}$$

Wt N that dissolves in 4 ml = (4.0 ml W) $\left(\dfrac{75.8 \text{ g N}}{100 \text{ ml W}} \right)$ = 3.0 g N

Filter off 16.7 − 3.0 = 13.7 g pure N at 100.0°C

After 2 cycles, you have 83.3 + 13.7 = 97.0 g pure NH_4Cl
and 37.2 + 10.7 = 47.9 g pure $KClO_3$

Four-Place Logarithms

N	0	1	2	3	4	5	6	7	8	9	Proportional parts								
											1	2	3	4	5	6	7	8	9
10	0000	0043	0086	0128	0170	0212	0253	0294	0334	0374	4	8	12	17	21	25	29	33	37
11	0414	0453	0492	0531	0569	0607	0645	0682	0719	0755	4	8	11	15	19	23	26	30	34
12	0792	0828	0864	0899	0934	0969	1004	1038	1072	1106	3	7	10	14	17	21	24	28	31
13	1139	1173	1206	1239	1271	1303	1335	1367	1399	1430	3	6	10	13	16	19	23	26	29
14	1461	1492	1523	1553	1584	1614	1644	1673	1703	1732	3	6	9	12	15	18	21	24	27
15	1761	1790	1818	1847	1875	1903	1931	1959	1987	2014	3	6	8	11	14	17	20	22	25
16	2041	2068	2095	2122	2148	2175	2201	2227	2253	2279	3	5	8	11	13	16	18	21	24
17	2304	2330	2355	2380	2405	2430	2455	2480	2504	2529	2	5	7	10	12	15	17	20	22
18	2533	2577	2601	2625	2648	2672	2695	2718	2742	2765	2	5	7	9	12	14	16	19	21
19	2788	2810	2833	2856	2878	2900	2923	2945	2967	2989	2	4	7	9	11	13	16	18	20
20	3010	3032	3054	3075	3096	3118	3139	3160	3181	3201	2	4	6	8	11	13	15	17	19
21	3222	3243	3263	3284	3304	3324	3345	3365	3385	3404	2	4	6	8	10	12	14	16	18
22	3424	3444	3464	3483	3502	3522	3541	3560	3579	3598	2	4	6	8	10	12	14	15	17
23	3617	3636	3655	3674	3692	3711	3729	3747	3766	3784	2	4	6	7	9	11	13	15	17
24	3802	3820	3838	3856	3874	3892	3909	3927	3945	3962	2	4	5	7	9	11	12	14	16
25	3979	3997	4014	4031	4048	4065	4082	4099	4116	4133	2	3	5	7	9	10	12	14	15
26	4150	4166	4183	4200	4216	4232	4249	4265	4281	4298	2	3	5	7	8	10	11	13	15
27	4314	4330	4346	4362	4378	4393	4409	4425	4440	4456	2	3	5	6	8	9	11	13	14
28	4472	4487	4502	4518	4533	4548	4564	4579	4594	4609	2	3	5	6	8	9	11	12	14
29	4624	4639	4654	4669	4683	4698	4713	4728	4742	4757	1	3	4	6	7	9	10	12	13
30	4771	4786	4800	4814	4829	4843	4857	4871	4886	4900	1	3	4	6	7	9	10	11	13
31	4914	4928	4942	4955	4969	4983	4997	5011	5024	5038	1	3	4	6	7	8	10	11	12
32	5051	5065	5079	5092	5105	5119	5132	5145	5159	5172	1	3	4	5	7	8	9	11	12
33	5185	5198	5211	5224	5237	5250	5263	5276	5289	5302	1	3	4	5	6	8	9	10	12
34	5315	5328	5340	5353	5366	5378	5391	5403	5416	5428	1	3	4	5	6	8	9	10	11
35	5441	5453	5465	5478	5490	5502	5514	5527	5539	5551	1	2	4	5	6	7	9	10	11
36	5563	5575	5587	5599	5611	5623	5635	5647	5658	5670	1	2	4	5	6	7	8	10	11
37	5682	5694	5705	5717	5729	5740	5752	5763	5775	5786	1	2	3	5	6	7	8	9	10
38	5798	5809	5821	5832	5843	5855	5866	5877	5888	5899	1	2	3	5	6	7	8	9	10
39	5911	5922	5933	5944	5955	5966	5977	5988	5999	6010	1	2	3	4	5	7	8	9	10
40	6021	6031	6042	6053	6064	6075	6085	6096	6107	6117	1	2	3	4	5	6	8	9	10
41	6128	6138	6149	6160	6170	6180	6191	6201	6212	6222	1	2	3	4	5	6	7	8	9
42	6232	6243	6253	6263	6274	6284	6294	6304	6314	6325	1	2	3	4	5	6	7	8	9
43	6335	6345	6355	6365	6375	6385	6395	6405	6415	6425	1	2	3	4	5	6	7	8	9
44	6435	6444	6454	6464	6474	6484	6493	6503	6513	6522	1	2	3	4	5	6	7	8	9
45	6532	6542	6551	6561	6571	6580	6590	6599	6609	6618	1	2	3	4	5	6	7	8	9
46	6628	6637	6646	6656	6665	6675	6684	6693	6702	6712	1	2	3	4	5	6	7	7	8
47	6721	6730	6739	6749	6758	6767	6776	6785	6794	6803	1	2	3	4	5	5	6	7	8
48	6812	6821	6830	6839	6848	6857	6866	6875	6884	6893	1	2	3	4	4	5	6	7	8
49	6902	6911	6920	6928	6937	6946	6955	6964	6972	6981	1	2	3	4	4	5	6	7	8
50	6990	6998	7007	7016	7024	7033	7042	7050	7059	7067	1	2	3	3	4	5	6	7	8
51	7076	7084	7093	7101	7110	7118	7126	7135	7143	7152	1	2	3	3	4	5	6	7	8
52	7160	7168	7177	7185	7193	7202	7210	7218	7226	7235	1	2	3	3	4	5	6	7	7
53	7243	7251	7259	7267	7275	7284	7292	7300	7308	7316	1	2	2	3	4	5	6	6	7
54	7324	7332	7340	7348	7356	7364	7372	7380	7388	7396	1	2	2	3	4	5	6	6	7

N	0	1	2	3	4	5	6	7	8	9	Proportional parts 1	2	3	4	5	6	7	8	9
55	7404	7412	7419	7427	7435	7443	7451	7459	7466	7474	1	2	2	3	4	5	5	6	7
56	7482	7490	7497	7505	7513	7520	7528	7536	7543	7551	1	2	2	3	4	5	5	6	7
57	7559	7566	7574	7582	7589	7597	7604	7612	7619	7627	1	2	2	3	4	5	5	6	7
58	7634	7642	7649	7657	7664	7672	7679	7686	7694	7701	1	1	2	3	4	4	5	6	7
59	7709	7716	7723	7731	7738	7745	7752	7760	7767	7774	1	1	2	3	4	4	5	6	7
60	7782	7789	7796	7803	7810	7818	7825	7832	7839	7846	1	1	2	3	4	4	5	6	6
61	7853	7860	7868	7875	7882	7889	7896	7903	7910	7917	1	1	2	3	4	4	5	6	6
62	7924	7931	7938	7945	7952	7959	7966	7973	7980	7987	1	1	2	3	3	4	5	6	6
63	7993	8000	8007	8014	8021	8028	8035	8041	8048	8055	1	1	2	3	3	4	5	5	6
64	8062	8069	8075	8082	8089	8096	8102	8109	8116	8122	1	1	2	3	3	4	5	5	6
65	8129	8136	8142	8149	8156	8162	8169	8176	8182	8189	1	1	2	3	3	4	5	5	6
66	8195	8202	8209	8215	8222	8228	8235	8241	8248	8254	1	1	2	3	3	4	5	5	6
67	8261	8267	8274	8280	8287	8293	8299	8306	8312	8319	1	1	2	3	3	4	5	5	6
68	8325	8331	8338	8344	8351	8357	8363	8370	8376	8382	1	1	2	3	3	4	4	5	6
69	8388	8395	8401	8407	8414	8420	8426	8432	8439	8445	1	1	2	2	3	4	4	5	6
70	8451	8457	8463	8470	8476	8482	8488	8494	8500	8506	1	1	2	2	3	4	4	5	6
71	8513	8519	8525	8531	8537	8543	8549	8555	8561	8567	1	1	2	2	3	4	4	5	5
72	8573	8579	8585	8591	8597	8603	8609	8615	8621	8627	1	1	2	2	3	4	4	5	5
73	8633	8639	8645	8651	8657	8663	8669	8675	8681	8686	1	1	2	2	3	4	4	5	5
74	8692	8698	8704	8710	8716	8722	8727	8733	8739	8745	1	1	2	2	3	4	4	5	5
75	8751	8756	8762	8768	8774	8779	8785	8791	8797	8802	1	1	2	2	3	4	4	5	5
76	8808	8814	8820	8825	8831	8837	8842	8848	8854	8859	1	1	2	2	3	4	4	5	5
77	8865	8871	8876	8882	8887	8893	8899	8904	8910	8915	1	1	2	2	3	3	4	4	5
78	8921	8927	8932	8938	8943	8949	8954	8960	8965	8971	1	1	2	2	3	3	4	4	5
79	8976	8982	8987	8993	8998	9004	9009	9015	9020	9025	1	1	2	2	3	3	4	4	5
80	9031	9036	9042	9047	9053	9058	9063	9069	9074	9079	1	1	2	2	3	3	4	4	5
81	9085	9090	9096	9101	9106	9112	9117	9122	9128	9133	1	1	2	2	3	3	4	4	5
82	9138	9143	9149	9154	9159	9165	9170	9175	9180	9186	1	1	2	2	3	3	4	4	5
83	9191	9196	9201	9206	9212	9217	9222	9227	9232	9238	1	1	2	2	3	3	4	4	5
84	9243	9248	9253	9258	9263	9269	9274	9279	9284	9289	1	1	2	2	3	3	4	4	5
85	9294	9299	9304	9309	9315	9320	9325	9330	9335	9340	1	1	2	2	3	3	4	4	5
86	9345	9350	9355	9360	9365	9370	9375	9380	9385	9390	1	1	2	2	3	3	4	4	5
87	9395	9400	9405	9410	9415	9420	9425	9430	9435	9440	0	1	1	2	2	3	3	4	4
88	9445	9450	9455	9460	9465	9469	9474	9479	9484	9489	0	1	1	2	2	3	3	4	4
89	9494	9499	9504	9509	9513	9518	9523	9528	9533	9538	0	1	1	2	2	3	3	4	4
90	9542	9547	9552	9557	9562	9566	9571	9576	9581	9586	0	1	1	2	2	3	3	4	4
91	9590	9595	9600	9605	9609	9614	9619	9624	9628	9633	0	1	1	2	2	3	3	4	4
92	9638	9643	9647	9652	9657	9661	9666	9671	9675	9680	0	1	1	2	2	3	3	4	4
93	9685	9689	9694	9699	9703	9708	9713	9717	9722	9727	0	1	1	2	2	3	3	4	4
94	9731	9736	9741	9745	9750	9754	9759	9763	9768	9773	0	1	1	2	2	3	3	4	4
95	9777	9782	9786	9791	9795	9800	9805	9809	9814	9818	0	1	1	2	2	3	3	4	4
96	9823	9827	9832	9836	9841	9845	9850	9854	9859	9863	0	1	1	2	2	3	3	4	4
97	9868	9872	9877	9881	9886	9890	9894	9899	9903	9908	0	1	1	2	2	3	3	4	4
98	9912	9917	9921	9926	9930	9934	9939	9943	9948	9952	0	1	1	2	2	3	3	4	4
99	9956	9961	9965	9969	9974	9978	9983	9987	9991	9996	0	1	1	2	2	3	3	3	4

Index

The Système International (SI)

A. The Base Units of SI

Unit	Symbol	Quantity measured
Meter	m	Length
Kilogram	kg	Mass
Second	s	Time
Ampere	A	Electric current
Kelvin	K	Thermodynamic temperature
Mole	mol	Amount of substance
Candela	cd	Luminous intensity

B. The Derived Units of SI

Unit	Symbol	Quantity measured	Relation to other units
Newton	N	Force	$kg \cdot m/s^2$
Pascal	Pa	Pressure	N/m^2
Joule	J	Work, energy, heat	$N \cdot m$
Watt	W	Power, or radiant energy flux	J/s
Volt	V	Electric potential, electromotive force	W/A
Ohm	Ω	Electric resistance	V/A
Siemens	S	Electric conductance	A/V
Coulomb	C	Quantity of electric charge	$A \cdot s$
Farad	F	Electric capacitance	C/V
Weber	Wb	Magnetic flux	$V \cdot s$
Henry	H	Inductance	Wb/A
Tesla	T	Magnetic flux density	Wb/m^2
Lumen	lm	Luminous flux	$cd \cdot sr$
Lux	lx	Illuminance	lm/m^2
Becquerel	Bq	Activity of a radioactive isotope	1 disintegration/s
Hertz	Hz	Frequency of a periodic phenomenon	1 cycle/s
Gray	Gy	Absorbed dose of ionizing radiation	J/kg

NOTE: There are also two dimensionless units: rad for radian (plane angle), and sr for steradian (solid angle).

Useful Constants

Ideal gas constant $R = 62.36$ torr liters/mole K

$= 1.987$ cal/mole K

$= 0.08206$ liter atm/mole K

$= 8.314 \times 10^7$ ergs/mole K

Avogadro's number $N = 6.023 \times 10^{23}$ mole^{-1} (or molecules/mole)

Faraday's constant $F = 96,487$ coulombs/mole of electrons

Planck's constant $h = 6.626 \times 10^{-27}$ erg sec

Velocity of light $c = 2.9979 \times 10^{10}$ cm/sec

Acceleration of gravity $g = 980.7$ cm/sec^2

Common Conversion Factors

1 atmosphere $= 760$ torr

$= 14.7$ lbs/in^2

$= 1.013 \times 10^6$ dynes/cm^2

1 calorie $= 4.184$ joules

1 joule $= 10^7$ ergs

1 angstrom $= 10^{-8}$ cm

1 quart $= 0.946$ liter

1 inch $= 2.54$ cm

1 pound $= 454$ g

$°C = \frac{5}{9}(°F - 32)$

Geometrical Formulas

Area of circle of radius r	$A = \pi r^2$
Area of cylinder of radius r and length l	$A = 2\pi r l + 2\pi r^2$
Area of sphere of radius r	$A = 4\pi r^2$
Area of triangle of base b and height h	$A = \frac{1}{2}hb$
Volume of cylinder of radius r and length l	$V = \pi r^2 l$
Volume of sphere of radius r	$V = \frac{4}{3}\pi r^3$